Radar Remote Sensing of Urban Areas

Remote Sensing and Digital Image Processing

VOLUME 15

For other titles published in this series, go to
http://www.springer.com/series/6477

Radar Remote Sensing of Urban Areas

Uwe Soergel
Editor

Leibniz Universität Hannover
Institute of Photogrammetry and GeoInformation, Germany

Springer

Editor
Uwe Soergel
Leibniz Universität Hannover
Institute of Photogrammetry and GeoInformation
Nienburger Str. 1
30167 Hannover
Germany
soergel@ipi.uni-hannover.de

Cover illustration: Fig. 7 in Chapter 11 in this book

Responsible Series Editor: André Marçal

ISSN 1567-3200
ISBN 978-90-481-3750-3 e-ISBN 978-90-481-3751-0
DOI 10.1007/978-90-481-3751-0
Springer Dordrecht Heidelberg London New York

Library of Congress Control Number: 2010922878

Cover design: deblik, Berlin

Printed on acid-free paper

Springer is part of Springer Science+Business Media (www.springer.com)

Preface

One of the key milestones of radar remote sensing for civil applications was the launch of the European Remote Sensing Satellite 1 (ERS 1) in 1991. The platform carried a variety of sensors; the Synthetic Aperture Radar (SAR) is widely considered to be the most important. This active sensing technique provides all-day and all-weather mapping capability of considerably fine spatial resolution. ERS 1 and its sister system ERS 2 (launch 1995) were primarily designed for ocean applications, but soon the focus of attention turned to onshore mapping. Examples for typical applications are land cover classification also in tropical zones and monitoring of glaciers or urban growth. In parallel, international Space Shuttle Missions dedicated to radar remote sensing were conducted starting already in the 1980s. The most prominent were the SIR-C/X-SAR mission focussing on the investigation of multi-frequency and multi-polarization SAR data and the famous Shuttle Radar Topography Mission (SRTM). Data acquired during the latter enabled to derive a DEM of almost global coverage by means of SAR Interferometry. It is indispensable even today and for many regions the best elevation model available. Differential SAR Interferometry based on time series of imagery of the ERS satellites and their successor Envisat became an important and unique technique for surface deformation monitoring.

The spatial resolution of those devices is in the order of some tens of meters. Image interpretation from such data is usually restricted to radiometric properties, which limits the characterization of urban scenes to rather general categories, for example, the discrimination of suburban areas from city cores. The advent of a new sensor generation changed this situation fundamentally. Systems like TerraSAR-X (Germany) and COSMO-SkyMed (Italy) achieve geometric resolution of about 1 m. In addition, these sophisticated systems are more agile and provide several modes tailored for specific tasks. This offers the opportunity to extend the analysis to individual urban objects and their geometrical set-up, for instance, infrastructure elements like roads and bridges, as well as buildings. In this book, potentials and limits of SAR for urban mapping are described, including SAR Polarimetry and SAR Interferometry. Applications addressed comprise rapid mapping in case of time critical events, road detection, traffic monitoring, fusion, building reconstruction, SAR image simulation, and deformation monitoring.

Audience

This book is intended to provide a comprehensive overview of the state-of-the art of urban mapping and monitoring by modern satellite and airborne SAR sensors. The reader is assumed to have a background in geosciences or engineering and to be familiar with remote sensing concepts. Basics of SAR and an overview of different techniques and applications are given in Chapter 1. All chapters following thereafter focus on certain applications, which are presented in great detail by well known experts in these fields.

In case of natural disaster or political crisis rapid mapping is a key issue (Chapter 2). An approach for automated extraction of roads and entire road networks is presented in Chapter 3. A topic closely related to road extraction is traffic monitoring. In case of SAR, Along-Track Interferometry is a promising technique for this task, which is discussed in Chapter 4. Reflections at surface boundaries may alter the polarization plane of the signal. In Chapter 5, this effect is exploited for object recognition from a set of SAR images of different polarization states at transmit and receive. Often, up-to-date SAR data has to be compared with archived imagery of complementing spectral domains. A method for fusion of SAR and optical images aiming at classification of settlements is described in Chapter 6. The opportunity to determine the object height above ground from SAR Interferometry is of course attractive for building recognition. Approaches designed for mono-aspect and multi-aspect SAR data are proposed in Chapters 7 and 8, respectively. Such methods may benefit from image simulation techniques that are also useful for education. In Chapter 9, a methodology optimized for real-time requirements is presented. Monitoring of surface deformation suffers from temporal signal decorrelation especially in vegetated areas. However, in cities many temporally persistent scattering objects are present, which allow tracking of deformation processes even for periods of several years. This technique is discussed in Chapter 10. Finally, in Chapter 11, design constraints of a modern airborne SAR sensor are discussed for the case of an existing device together with examples of high-quality imagery that state-of-the-art systems can provide.

Uwe Soergel

Contents

Contributors

Fabio Dell'Acqua
Department of Electronics, University of Pavia, Via Ferrata, 1-I-27100 Pavia
fabio.dellacqua@unipv.it

Timo Balz
State Key Laboratory of Information Engineering in Surveying, Mapping and Remote Sensing – Wuhan University, China
balz@lmars.whu.edu.cn

Michele Crosetto
Institute of Geomatics, Av. Canal Olímpic s/n, 08860 Castelldefels (Barcelona), Spain
michele.crosetto@ideg.es

Helmut Essen
FGAN- Research Institute for High Frequency Physics and Radar Techniques, Department Millimeterwave Radar and High Frequency Sensors (MHS), Neuenahrer Str. 20, D-53343 Wachtberg-Werthhoven, Germany
essen@fgan.de

Paolo Gamba
Department of Electronics, University of Pavia. Via Ferrata, 1-I-27100 Pavia
paolo.gamba@unipv.it

Ronny Hänsch
Technische Universität, Berlin Computer Vision and Remote Sensing, Franklinstr, 28/29, 10587 Berlin, Germany
rhaensch@fpk.tu-berlin.de

Karin Hedman
Institute of Astronomical and Physical Geodesy, Technische Universitaet Muenchen, Arcisstrasse 21, 80333 Munich, Germany
karin.hedman@bv.tum.de

Olaf Hellwich
Technische Universität, Berlin Computer Vision and Remote Sensing, Franklinstr. 28/29, 10587 Berlin, Germany
hellwich@cs.tu-berlin.de

Gerardo Herrera
Instituto Geológico y Minero de España (IGME), Rios Rosas 23, 28003 Madrid, Spain
g.herrera@igme.es

Stefan Hinz
Remote Sensing and Computer Vision, University of Karlsruhe, Germany
stefan.hinz@ipf.uni-karlsruhe.de

Franz Kurz
Remote Sensing Technology Institute, German Aerospace Center DLR, Germany

Oriol Monserrat
Institute of Geomatics, Av. Canal Olímpic s/n, 08860 Castelldefels (Barcelona), Spain
oriol.monserrat@ideg.es

Uwe Soergel
Institute of Photogrammetry and GeoInformation, Leibniz Universität Hannover, 30167 Hannover, Germany
soergel@ipi.uni-hannover.de

Uwe Stilla
Institute of Photogrammetry and Cartography, Technische Universitaet Muenchen, Arcisstrasse 21, 80333 Munich, Germany
stilla@bv.tum.de

Steffen Suchandt
Remote Sensing Technology Institute, German Aerospace Center DLR, Germany

Antje Thiele
Fraunhofer-IOSB, Sceneanalysis, 76275 Ettlingen, Germany
Karlsruhe Institute of Technology (KIT), Institute of Photogrammetry and Remote Sensing (IPF), 76128 Karlsruhe, Germany
antje.thiele@kit.edu

Céline Tison
CNES, DCT/SI/AR, 18 avenue Edouard Belin, 31 400 Toulouse, France
celine.tison@cnes.fr

Florence Tupin
Institut TELECOM, TELECOM ParisTech, CNRS LTCI, 46 rue Barrault, 75 013 Paris, France
florence.tupin@telecom-paristech.fr

Jan Dirk Wegner
IPI Institute of Photogrammetry and GeoInformation, Leibniz Universität Hannover, 30167 Hannover, Germany
wegner@ipi.uni-hannover.de

Diana Weihing
Remote Sensing Technology, TU Muenchen, Germany

Chapter 1
Review of Radar Remote Sensing on Urban Areas

Uwe Soergel

1.1 Introduction

Synthetic Aperture Radar (SAR) is an active remote sensing technique capable of providing high-resolution imagery independent from daytime and to great extent unimpaired by weather conditions. However, SAR inevitably requires an oblique scene illumination resulting in undesired occlusion and layover especially in urban areas. As a consequence, SAR is without any doubt not the first choice for providing complete coverage of urban areas. For such purpose, sensors being capable of acquiring high-resolution data in nadir view are better suited like optical cameras or airborne laserscanning devices. Nevertheless, there are at least two kinds of application scenarios concerning city monitoring where the advantages of SAR play a key role: firstly, time critical events and, secondly, the necessity to gather gap-less and regular spaced time series of imagery of a scene of interest.

Considering time critical events (e.g., natural hazard, political crisis), fast data acquisition and processing are of utmost importance. Satellite sensors have the advantage of providing almost global data coverage, but the limitation of being tied to a predefined sequence of orbits, which determine the potential time slots and the aspect of observation (ascending or descending orbit) to gather data of a certain area of interest. On the other hand, airborne sensors are more flexible, but have to be mobilized and transferred to the scene. Both types of SAR sensors have been used in many cases for disaster mitigation and damage assessment in the past, especially during or after flooding (Voigt et al. 2005) and in the aftermath of earthquakes (Takeuchi et al. 2000). One recent example is the Wenchuan Earthquake that hit central China in May 2008. The severe damage of a city caused by landslides triggered by the earthquake was investigated using post-strike images of satellites TerraSAR-X (TSX) and Cosmo-Skymed (Liao et al. 2009).

U. Soergel (✉)
Institute of Photogrammetry and GeoInformation, Leibniz Universität Hannover, Germany
e-mail: soergel@ipi.uni-hannover.de

U. Soergel (ed.), *Radar Remote Sensing of Urban Areas*, Remote Sensing and Digital Image Processing 15, DOI 10.1007/978-90-481-3751-0_1,

Examples for applications that rely on multi-temporal remote sensing images of urban areas are monitoring of surface deformation, land cover classification, and change detection in tropical zones. The most common and economic way to ensure gap-less and regular spaced time series of imagery of a given urban area of interest is the acquisition of repeat-pass data by SAR satellite sensors. Depending on the repeat cycle of the different satellites, the temporal baseline grid for images of approximately the same aspect by the same sensor is, for example, 45 days (ALOS), 35 days (ENVISAT), 24 days (Radarsat 1/2), and 11 days (TSX).

The motivation for this book is to give an overview of different applications and techniques related to remote sensing of urban areas by SAR. The aims of this first chapter are twofold. First, the reader who is not familiar with radar remote sensing is introduced in the fundamentals of conventional SAR and the characteristics of higher-level techniques like SAR Polarimetry and SAR Interferometry. Second, the most important applications with respect to settlement areas and their corresponding state-of-the-art approaches are presented in dedicated sections in preparation of following chapters of the book, which address those issues in more detail.

This chapter is organized as follows. In Section 1.2, the basics of radar remote sensing, the SAR principle, and the appearance of 3d objects in SAR data are discussed. Section 1.3 is dedicated to 2d approaches which rely on image processing, image classification, and object recognition without explicitly modeling the 3d structure of the scene. This includes land cover classification for settlement detection, characterization of urban areas, techniques for segmentation of object primitives, road extraction, SAR Polarimetry, and image fusion. In Section 1.4, the explicit consideration of the 3d structure of the topography is addressed comprising Radargrammetry, stereo techniques, SAR Interferometry, image fusion, and the combination of Interferometry and Polarimetry. The two last sections give an insight into surface deformation monitoring and traffic monitoring.

1.2 Basics

The microwave (MW) domain of the electromagnetic spectrum roughly ranges from wavelength $\lambda = 1\,\text{mm}$ to $1\,\text{m}$, equivalent to signal frequencies $f = 300\,\text{GHz}$ and $300\,\text{MHz}$ ($\lambda \cdot f = c$, with velocity of light c), respectively. In comparison with the visible domain, the wavelength is several orders of magnitude larger. Since the photon energy $E_{ph} = h \cdot f$, with the Planck constant h, is proportional to frequency, microwave signal interacts quite different with matter compared to sunlight. The high energy of the latter leads to material dependent molecular resonance effects (i.e., absorption), which are the main source of colors observed by humans. In this sense, remote sensing in the visible and near infrared spectrum reveals insight into the chemical structure of soil and atmosphere. In contrast, the energy of the MW signal is too low to cause molecular resonance, but still high enough to stimulate resonant rotation of certain dipole molecules (e.g., liquid water) according to the frequency dependent change of the electric field component of the signal. In summery, SAR

Table 1.1 Overview of microwave bands used for remote sensing and a selection of related SAR sensors

Band	P	L	S	C	X	K_u	W
Center frequency (GHz)	0.35	1.3	3.1	5.3	10	35	95
wavelength (cm)	85	23	9.6	5.66	3	0.86	0.32
Examples for SAR space borne and airborne sensors using this band	E-SAR, AIRSAR, RAMSES	ALOS, E-SAR, AIRSAR, RAMSES	RAMSES	ERS 1/2, ENVISAT, Radarsat 1/2, SRTM, E-SAR, AIRSAR, RAMSES	TSX, SRTM, PAMIR, E-SAR, RAMSES	MEMPHIS, RAMSES	MEMPHIS, RAMSES

sensors are rather sensitive to physical properties like surface roughness, morphology, geometry, and permittivity ε. Because liquid water features a considerably high ε value in the MW domain, such sensors are well suited to determine soil moisture.

The MW spectral range subdivides in several bands commonly labeled according to a letter code first used by the US military in World War II. An overview of these bands is given in Table 1.1. The atmospheric loss due to Rayleigh scattering by aerosols or raindrops is proportional to $1/\lambda^4$. Therefore, in practice X-Band is the lower limit for space borne imaging radar in order to ensure all-weather mapping capability. On the other hand, shorter wavelengths have some advantages, too, for example, smaller antenna devices and better angular resolution (Essen 2009, Chapter 11 of this book).

Both, passive and active radar remote sensing sensors exist. Passive radar sensors are called radiometers, providing data useful to estimate the atmospheric vapor content. Active radar sensors can further be subdivided into non-imaging and imaging sensors. Important active non-imaging sensors are radar altimeters and scatterometers. Altimeters profile the globe systematically by repeated pulse run-time measurements along-track towards nadir, which is an important data source to determine the shape of the geoid and its changes. Scatterometer sample the backscatter of large areas on the oceans, from which the radial component of the wind direction is derived, a useful input for weather forecast. In this book, we will focus on high-resolution imaging radar.

1.2.1 Imaging Radar

Limited by diffraction, the aperture angle α of any image-forming sensor is determined by the ratio of its wavelength λ and aperture D. The spatial resolution ∂_α depends on α and the distance r between sensor and scene:

$$\partial_\alpha \propto \alpha \cdot r \approx \frac{\lambda}{D} \cdot r. \tag{1.1}$$

Hence, for given λ and D the angular resolution ∂_α linearly worsens with increasing r. Therefore, imaging radar in nadir view is in practice restricted to low altitude platforms (Klare et al. 2006).

The way to use also high altitude platforms for mapping is to illuminate the scene obliquely. Even though the antenna footprint on ground is still large and covers many objects, it is possible to discriminate the backscatter contributions of individual objects of different distance to the sensor from the runtime of the incoming signal. The term slant range refers to the direction in space along the axis of the beam antenna's 3 dB main lobe that approximately coincides with solid angle α. The slant range resolution ∂_r is not a function of the distance and depends only on the pulse length τ, which is inverse proportional to the pulse signal bandwidth B_r. However, the resolution of the other image coordinate direction perpendicular to the range axis and parallel to the sensor track, called azimuth, is still diffraction limited according to Eq. (1.1). Synthetic Aperture Radar (SAR) overcomes this limitation (Schreier 1993): The scene is illuminated obliquely orthogonal to the carrier path by a sequence of coherent pulses with high spatial overlap of subsequent antenna footprints on ground. High azimuth resolution ∂_a is achieved by signal processing of the entire set of pulses along the flight path which cover a certain point in the scene. In order to focus the image in azimuth direction, the varying distance between sensor and target along the carrier track has to be taken into account. As a consequence, the signal phase has to be delayed according to this distance during focusing. In this manner, all signal contributions originating from a target are integrated to the correct range/azimuth resolution cell. The impulse response $|u(a,r)|$ of an ideal point target located at azimuth/range-coordinates a_0, r_0 to a SAR system can be split into azimuth (u_a) and range (u_r) parts:

$$|u_a(a,r)| = \left| \sqrt{B_a \cdot T_a} \cdot \operatorname{sinc}\left[\pi \cdot \frac{B_a \cdot (a - a_0)}{v}\right] \right|,$$

$$|u_r(a,r)| = \left| \sqrt{B_r \cdot T_r} \cdot \operatorname{sinc}\left[2\pi \cdot \frac{B_r \cdot (r - r_0)}{c}\right] \right|,$$

with bandwidths B_a and B_r, integration times T_a and T_r, and sensor carrier speed v (Moreira 2000; Curlander and McDonough 1991). The magnitude of the impulse response (Fig. 1.1a) follows a 2d sinc function centered at a_0, r_0. Such pattern can often be observed in urban scenes when dominant signal of certain objects covers surrounding clutter of low reflectance for a large number of sidelobes. These undesired sidelobe signals can be suppressed using specific filtering techniques. However, this processing reduces the spatial resolution, which is by convention defined as the extent of the mainlobe 3 dB below its maximum signal power. The standard SAR process (Stripmap mode) yields range and azimuth resolution as:

$$\partial_r \approx \frac{c}{2 \cdot B_r} = \frac{c \cdot \tau}{2}, \quad \partial_{rg} \approx \frac{\partial_r}{\sin(\theta)} \quad \partial_a \approx \frac{v}{B_a} = \frac{D_a}{2}, \tag{1.2}$$

with velocity of light c and antenna size in azimuth direction D_a. The range resolution is constant in slant range, but varies on ground. For a flat scene, the ground

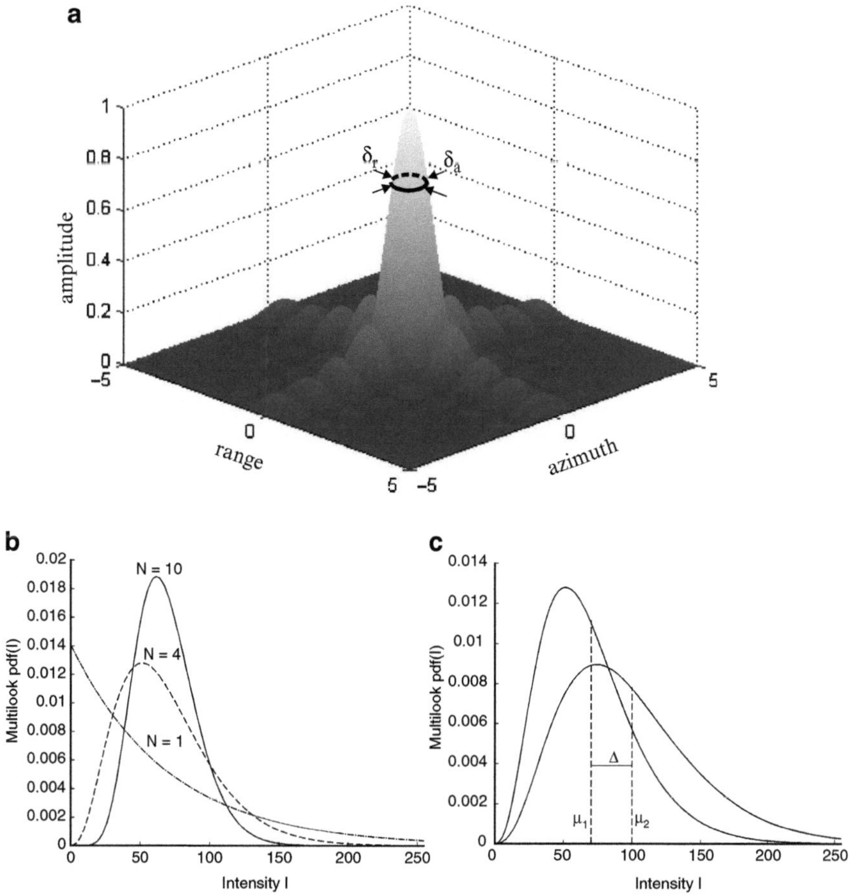

Fig. 1.1 SAR image: (**a**) impulse response, (**b**) spatial, and (**c**) radiometric resolution

range resolution ∂_{rg} depends on the local viewing angle. It is always best in far range. The azimuth resolution can be further enhanced by enlarging the integration time. The antenna is steered in such manner that a small scene of interest is observed for a longer period at the cost of other areas not being covered at all. For instance, the SAR images obtained in TSX Spotlight modes are high-resolution products of this kind. On the contrary, for some applications a large spatial coverage is more important than high spatial resolution. Then, the antenna operates in a so-called ScanSAR mode illuminating the terrain with a series of pulses of different off-nadir angles. In this way, the swath width is enlarged accepting the drawback of a coarser azimuth resolution. In case of TSX, this mode yields a swath width of 100 km compared to 30 km in Stripmap mode and the azimuth resolution is 16 versus 3 m.

Considering the backscatter characteristics of different types of terrain, two classes of targets have to be discriminated. The first one comprises so-called canonical objects (e.g., sphere, dipole, flat plane, dihedral, trihedral) whose *radar cross section* σ (RCS, unit either m^2 or dBm^2) can be determined analytically. Many man-made objects can be modeled as structures of canonical objects. The second class refers to regions of land cover of rather natural type, like agricultural areas and forests. Their appearance is governed by coherent superposition of uncorrelated reflection from a large number of randomly distributed scattering objects located in each resolution cell, which cannot be observed separately. The signal of connected components of homogeneous cover is therefore described by a dimensionless normalized RCS or *backscatter coefficient* σ_0. It is a measure of the average scatterer density.

In order to derive amplitude and phase of the backscatter, the sampled received signal is correlated twice with the transmitted pulse: once directly (in-phase component u_i), the second time after delay of a quarter of a cycle period (quadrature component u_q). Those components are regarded as real and imaginary part of a complex signal u, respectively:

$$u = u_i + j u_q.$$

It is convenient to picture this signal to be a phasor in polar coordinates. The joint probability density function (pdf) of u is modeled to be a complex circular Gaussian process (Goodman 1985) if the contributions of the (many) individual scattering objects are statistically independent of each other. All phasors sum up randomly and the sensor merely measures the final sum phasor. If we move from the Cartesian to the polar coordinate system, we yield magnitude and phase of this phasor. The magnitude of a SAR image is usually expressed in terms of either amplitude (A) or intensity (I) of a pixel:

$$I = u_i^2 + u_q^2, \quad A = \sqrt{u_i^2 + u_q^2}$$

The expectation value of pixel intensity $\bar{I}$ of a homogenous area is proportional to σ_0. For image analysis, it is crucial to consider the image statistics. The amplitude is Rayleigh distributed, while the intensity is exponentially distributed:

$$\bar{I} = E\left[u \cdot u^*\right] \sim \sigma_0, \quad pdf(I) = \frac{1}{\bar{I}} \cdot e^{-\frac{I}{\bar{I}}} \; for \; I \geq 0. \tag{1.3}$$

Phase distribution in both cases is uniform. Hence, the knowledge of the phase value of a certain pixel carries no information about the phase value of any other location within the same image. The benefit of the phase comes as soon as several images of the scene are available: the pixel-by-pixel *difference* of the phase of co-registered images carries information, which is exploited, for example, by SAR Interferometry. The problem with the exponential distribution according to Eq. (1.3) is that the expectation value equals the standard deviation. As a result, connected areas of same natural land cover like grass appear grainy in the image and the larger the average intensity of this region is the more the pixel values fluctuate. This phenomenon is called speckle. Even though speckle is the signal and by no means noise, can it be thought of to be a multiplicative random perturbation S of the underlying deterministic backscatter coefficient of a field covered homogeneously by one crop:

$$\bar{I} \sim \sigma_0 \cdot S. \tag{1.4}$$

For many remote sensing applications, it is important to discriminate adjacent fields of different land cover. Speckle complicates this task. In order to reduce speckle and to enhance the radiometric resolution, multi-looking is often applied. The available bandwidth is divided into several looks (i.e., images of reduced spatial resolution) which are averaged. As a consequence, the standard deviation of the resulting image σ_{ML} drops with the square root of the effective (i.e., independent) number of Looks N. The pdf of the multi-look intensity image is χ^2 distributed:

$$\begin{aligned} \sigma_{ML} &= \frac{\bar{I}}{\sqrt{N}} \\ pdf_{ML}(I, N) &= \frac{I^{(N-1)}}{\left(\frac{\bar{I}}{L_{eff}}\right)^{N} \cdot \Gamma(N)} \cdot e^{\left(-\frac{I \cdot N}{\bar{I}}\right)} \end{aligned} \tag{1.5}$$

In Fig. 1.1b the effect of multi-looking on the distribution of the pixel values is shown for the intensity image processed using the entire bandwidth (the single-look image), a four-look, and a ten-look image of the same area with expectation value 70. According to the central limit theorem for large N we yield a Gaussian distribution ($\mu = 70,\ \sigma_{ML}(N)$). The described model works fine for natural landscape. Nevertheless, in urban areas some of the underlying assumptions are violated, because man-made objects are not distributed randomly but rather regularly and strong scatterers dominate their surroundings. In addition, the small resolution cell of modern sensors leads to a lower number N of scattering objects inside. Many different statistical models for urban scenes have been investigated; Tison et al. (2004), who propose the Fisher distribution, provide an overview.

Similar to multi-looking, speckle reduction can also be achieved by image processing of the single-look image using window-based filtering. A variety of speckle filters have been developed (Lopes et al. 1993). However, also in this case a loss of detail is inevitable. An often-applied performance measure of speckle filtering is the Coefficient of Variation (CoV). It is defined as the ratio of σ and μ of the image. The CoV is also used by some adaptive speckle filter methods to adjust the degree of smoothing according to the local image statistic.

As mentioned above, such speckle filtering or multilook processing enhances the radiometric resolution, ∂_R, which is defined for SAR as the limit for discrimination of two adjacent homogeneous areas whose expectation values are μ and $\mu + \sigma$, respectively (Fig. 1.1c):

$$\delta_R = \frac{\mu + \sigma}{\mu} = 10 \cdot \log_{10}\left(1 + \frac{1 + {}^{1}\!/_{SNR}}{\sqrt{L_{eff}}}\right)$$

1.2.2 Mapping of 3d Objects

If we focus on sensing geometry and neglect other issues for the moment, the mapping process of real world objects to the SAR image can be described most intuitively using a cylindrical coordinate system as sensor model. The coordinates are chosen such that the z-axis coincides with the sensor path and each pulse emitted by the beam antenna in range direction intersects a cone of solid angle α of the cylinder volume (Fig. 1.2).

The set union of subsequent pulses represents all signal contributions of objects located inside a wedge-shaped volume subset of the world. A SAR image can be thought of as projection of the original 3d space (azimuth = z, range, and elevation angle = θ coordinates) onto a 2d image plane (range, azimuth axes) of pixel size $\partial_r \; x \; \partial_a$. This reduction of one dimension is achieved by coherent signal integration in θ direction yielding the complex SAR pixel value. The backscatter contributions of the set of all those objects are summed up, which are located in a certain volume. This volume defined by the area of the resolution cell of size $\partial_r \; x \; \partial_a$ attached to a given $r, \; z$ SAR image coordinate and the segment of a circle of length $r \; x \; \alpha$ along the intersection of the cone and the cylinder barrel. Therefore, the true θ value of an individual object could coincide with any position on this circular segment. In other words, the poor angular resolution ∂_α of a real aperture radar system is still valid for the elevation coordinate. This is the reason for the layover phenomenon: all signal contributions of objects inside the antenna beam sharing the same range and azimuth coordinates are integrated into the same 2d resolution cell of the SAR image although differing in elevation angle. Owing to vertical facades, layover is ubiquitous in urban scenes (Dong et al. 1997). The sketch in Fig. 1.2 visualizes the described mapping process for the example of signal mixture of backscatter from a building and the ground in front of it.

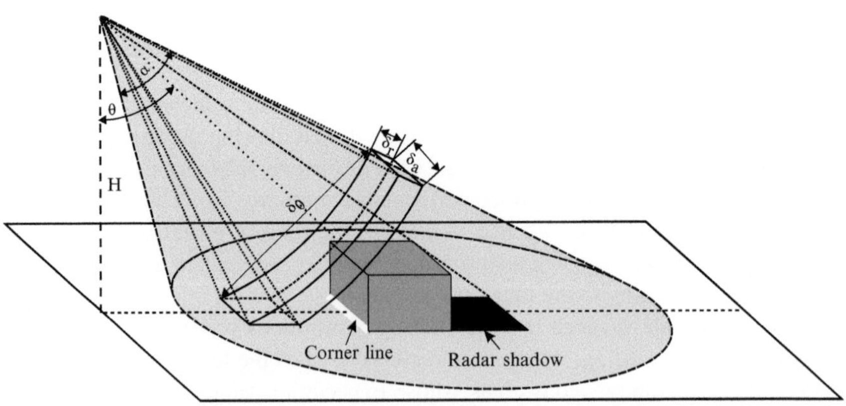

Fig. 1.2 Sketch of SAR principle: 3d volume mapped to a 2d resolution cell and effects of this projection on imaging of buildings

Besides layover, the side-looking illumination leads to occlusion behind buildings. This radar shadow is the most important limitation for road extraction and traffic monitoring by SAR in built-up areas (Soergel et al. 2005). Figure 1.3 depicts

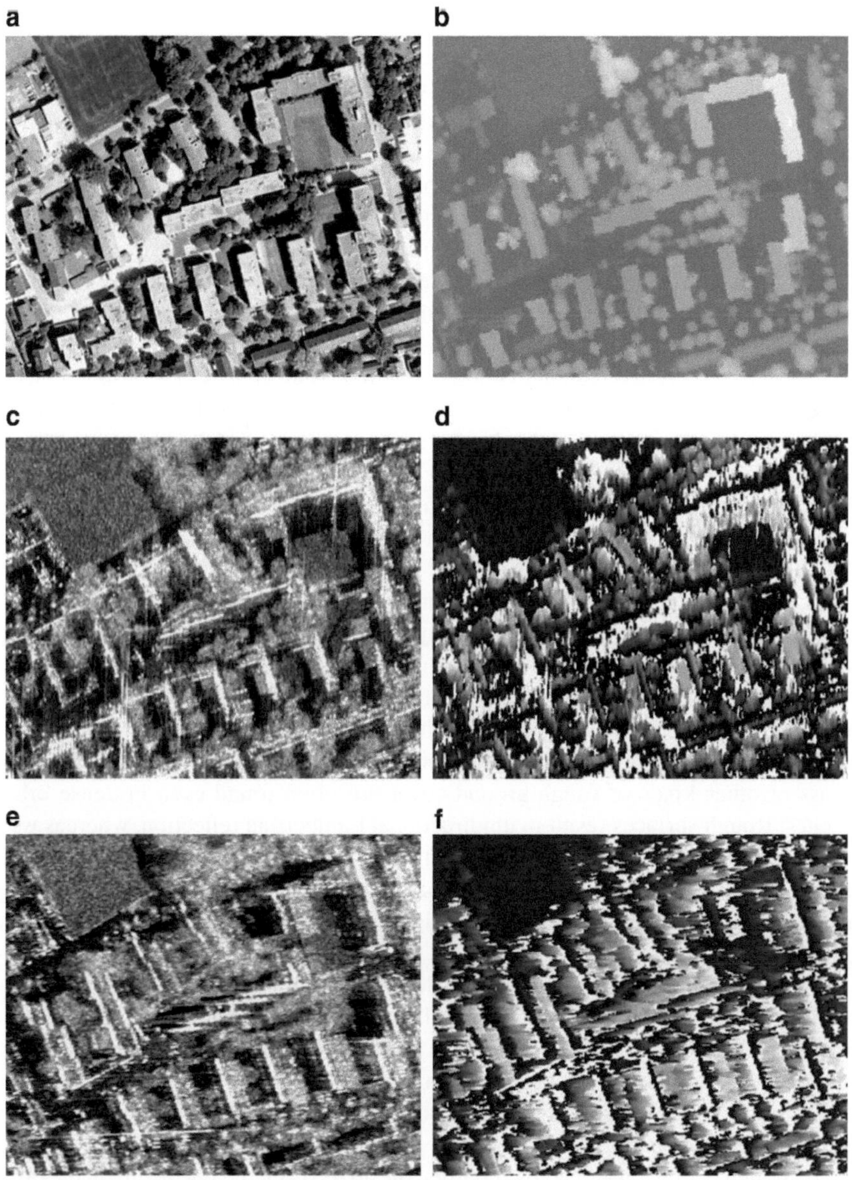

Fig. 1.3 Urban scene: (**a**) orthophoto, (**b**) LIDAR DSM, (**c, d**) amplitude and phase, respectively, of InSAR data taken from North, (**e, f**) as (**c, d**) but illumination from East. The InSAR data have been taken by Intermap, spatial resolution is better than half a meter

two InSAR data sets taken from orthogonal directions along with reference data in form of an orthophoto and a LIDAR DSM. The aspect dependency of the shadow cast on ground is clearly visible in the amplitude images (Fig. 1.3 c, e), for example, at the large building block in the upper right part. Occlusion and layover problems can to some extent be mitigated by the analysis of multi-aspect data (Thiele et al. 2009b, Chapter 8 of this book).

The reflection of planar objects depends on the incidence angle β (the angle between the object plane normal and the viewing angle). Determined by the chosen aspect and illumination angle of the SAR data acquisition, a large portion of the roof planes may cause strong signal due to specular reflection towards the sensor. Especially in the case of roads oriented parallel to the sensor track this effect leads to salient bright lines. Under certain conditions, similar strong signal occurs even for rotated roofs, because of Bragg resonance. If a regular spaced structure (e.g., a lattice fence or tiles of a roof) is observed by a coherent sensor from a viewpoint such that the one-way distance to the individual structure elements is an integer multiple of $\lambda/2$, constructive interference is the consequence.

Due to the preferred rectangular alignment of objects mostly consisting of piecewise planar surface facets, multi-bounce signal propagation is frequently observed. The most prominent effect of this kind often found in cities is double-bounce signal propagation between building walls and ground in front of them. Bright line features, similar to those caused by specular reflection from roof structure elements, appear at the intersection between both planes (i.e., coinciding with part of the building footprint). This line also marks the far end of the layover area. If all objects would behave like mirrors, such feature would be visible only in case of walls oriented in along-track direction. In reality, the effect is most pronounced in this setup, indeed. However, it is still visible for considerable degree of rotation, because neither the façades nor the grounds in front are homogeneously planar. Exterior building walls are often covered by rough coatings and feature subunits of different material and orientation like windows and balconies. Besides smooth asphalt areas grass or other kinds of rough ground cover are often found even in dense urban scenes. Rough surfaces result in unidirectional Lambertian reflection, whereas windows and balconies consisting of planar and rectangular parts may cause aspect dependent strong multi-bounce signal. In addition, also regular façade elements may cause Bragg resonance. Consequently, bright L-shaped structures are often observed in cities.

Gable roof buildings may cause both described bright lines that appear parallel at two borders of the layover area: the first line caused by specular reflection from the roof situated closer to the sensor and the second one resulting from double-bounce reflection located on the opposite layover end. This feature is clearly visible on the left in Fig. 1.3e. Those sets of parallel lines are strong hints to buildings of that kind (Thiele et al. 2009a, b).

Although occlusion and layover burden the analysis on the one hand, on the other hand valuable features for object recognition can be derived from those phenomena, especially in case of building extraction. The sizes of the layover area l in front of a building and the shadow area s behind it depend on the building height h and the local viewing angle θ:

$$l = h \cdot \cot(\theta_l), \quad s = h \cdot \tan(\theta_s). \tag{1.6}$$

In SAR images of spatial resolution better than one meter a large number of bright straight lines and groups of regular spaced point-like building features are visible (Soergel et al. 2006) that are useful for object detection (Michaelsen et al. 2006). Methodologies to exploit the mentioned object features for recognition are explained in the following in more detail.

1.3 2d Approaches

In this section all approaches are summarized which rely on image processing, image classification, and object recognition without explicitly modeling the 3d structure of the scene.

1.3.1 Pre-processing and Segmentation of Primitive Objects

The salt-and-pepper appearance of SAR images burdens image classification and object segmentation. Hence, appropriate pre-processing is a prerequisite for successful information extraction from SAR data. Although land cover classification can be carried out from the original data directly, speckle filtering is often applied previously in order to reduce inner-class variance through the smoothing effect. As a result, in most cases the clusters of the classes in the feature space are more pronounced and easier to be separated. In many approaches land cover classification is an intermediate stage of inference in order to screen the data for regions which seem to be worthwhile to accomplish a focused search for objects of interest based on algorithms of higher complexity.

Typically, three kinds of primitives are of interest in automated image analysis aiming at object detection and recognition: salient isolated points, linear objects, and homogeneous regions. Since SAR data show different distributions than other remote sensing imagery, standard image processing methods cannot be applied without suitable pre-processing. Therefore, special operators have been developed for SAR data that consider the underlying statistical model according to Eq. (1.5).

Many approaches aiming at detection and recognition of man-made objects like roads or buildings rely on an initial segmentation of edge or line primitives.

Touzi et al. (1988) proposed a template-based algorithm to extract edges in SAR amplitude images in four directions (horizontal, vertical, and both diagonals). As explained previously, the standard deviation of a homogenous area in a single-look intensity image equals the expectation value. Thus, speckle can be considered as

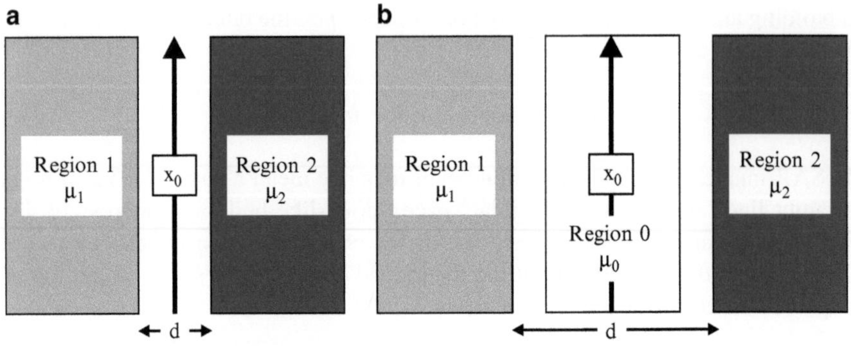

Fig. 1.4 (**a**) Edge detector, (**b**) line detector

a random multiplicative disturbance of the true constant σ_0 attached to this field. Therefore, the operator is based on the ratio of the average pixel values μ_1 and μ_2 of two parallel adjacent rectangular image segments (Fig. 1.4a). The authors show that the pdf of the ratio μ_i to μ_j can be expressed analytically and also that the operator is a *constant false alarm rate (CFAR)* edge detector. One way to determine potential edge pixels is to choose all pixels where the value r_{12} is above a threshold, which can be determined automatically from the user desired false alarm probability:

$$r_{12} = 1 - \min\left(\frac{\mu_1}{\mu_2}, \frac{\mu_2}{\mu_1}\right)$$

This approach was later extended to lines by adding a third stripe structure (Fig. 1.4b) and to assess two edge responses with respect to the middle stripe (Lopes et al. 1993). If the weaker response is above the threshold, the pixel is labeled to lie on a line. Tupin et al. (1998) describe the statistical model of this operator they call D1 and add a second operator D2, which considers also the homogeneity of the pixel values in the segments. Both responses from D1 and D2 are merged to obtain a unique decision whether a pixel is labeled as line.

A drawback of those approaches is high computational load, because the ratios of all possible orientations have to be computed for every pixel. This effort even rises linearly if lines of different width shall be extracted and hence different widths of the centre region have to be tested. Furthermore, the result is an image that still has to be post-processed to find connected components.

Another way to address object extraction is to conduct, first, an adaptive speckle filtering. The resulting initial image is then partitioned into regions of different heterogeneity. Finally, locations of suitable image statistics are determined. The approach of Walessa and Datcu (2000) belongs to this kind of methods. During the speckle reduction in a Markov Random Field framework, potential locations of strong point scatterers and edges are identified and preserved, while regions that are more homogeneous are smoothed. This initial segmentation is of course of high value for subsequent object recognition.

A fundamentally different but popular approach is to change the initial distribution of the data such that image processing methods from the shelf can be applied. One way to achieve this is to take the logarithm of the amplitude or intensity images. Thereby, the multiplicative speckle "disturbance" according to Eq. (1.4) turns into an additive one, which matches the usual concept of image processing of a signal that is corrupted by zero mean additive noise. If one decides to do so, it is reasonable to transfer the data given in digital numbers (DN) right away into the backscatter coefficient σ_0. For this conversion, a sensor and image specific calibration constant K and the local incidence angle have to be considered. Furthermore, σ_0 is usually given in Decibel, a dimensionless quantity ubiquitous in radar remote sensing representing ten times the logarithm to the base of ten of the ratio between the signal power and a reference power value. Sometimes the resulting histogram is clipped to exclude extremely small and large values and then the pixel values are stretched to 256 grey levels (Wessel et al. 2002).

Thereafter, the SAR data are prepared for standard image processing techniques, the most frequently applied are the edge and line detectors proposed by Canny (1986) and Steger (1998), respectively. For example, Thiele et al. (2009b) use the Canny edge operator to find building contours and Hedman et al. (2009) the Steger line detector for road extraction.

One possibility to combine the advantages of approaches tailored for SAR and optical data is to use first an operator best suitable for SAR images, for example, the line detector proposed by Lopes, and than to apply to the resulting image the Steger operator.

After speckle filtering and suitable non-linear logarithmic transformation, region segmentation approaches become feasible, too. For example, region growing (Levine and Shaheen 1981) or watershed segmentation (Vincent and Soille 1991) are often applied to extract homogeneous regions in SAR data. Due to the regular structure of roof and façade elements especially in high-resolution SAR images, salient rows of bright point-like scatterers are frequently observed. Such objects can easily be detected by template-based approaches (bright point embedded in dark surrounding). By subsequent grouping regular spaced rows of point scatterers can be extracted, which are for example useful for building recognition (Michaelsen et al. 2005).

1.3.2 Classification of Single Images

Considering the constraints attached to the sensor principle discussed previously, multi-temporal image analysis is advantageous. This is true for any imaging sensor, but especially for SAR because it provides no spectral information. However, one reason for the analysis of single SAR images (besides cost of data) is the necessity of rapid mapping, for instance, in case of time critical events.

Land cover classification is probably among the most prominent applications of remote sensing. A vast body of literature deals with land cover retrieval using

SAR data. Many different classification methods known from pattern recognition have been applied to this problem like Nearest Neighbour, Minimum Distance, Maximum Likelihood (ML), Bayesian, Markov Random Field (MRF, Tison et al. 2004), Artificial Neural Network (ANN, Tzeng and Chen 1998), Decision Tree (DT, Simard et al. 2000), Support Vector Machine (SVM, Waske and Benediktsson 2007), or object-based approaches (Esch et al. 2005). There is not enough room to discuss this in detail here; the interested reader is referred to the excellent book of Duda et al. (2001) for pattern classification, Lu and Weng (2007), who survey land cover classification methods, and to Smits et al. (1999), who deal with accuracy assessment of land cover classification. In this section, we will focus on the detection of settlements and on approaches to discriminate various kinds of subclasses, for example, villages, sub urban residential areas, industrial areas, and inner city cores.

1.3.2.1 Detection of Settlements

In case of a time critical event, an initial screening is often crucial which results in a coarse but quick partition of the scene into a few classes (e.g., forest, grassland, water, settlement). Areas of no interest are excluded permitting to focus further efforts on regions worthwhile to be investigated in more detail.

Inland water areas usually look dark in SAR images and natural landscape is well characterized by speckle according Eq. (1.5). Urban areas tend to exhibit both higher magnitude values and heterogeneity (Henderson and Mogilski 1987). The large heterogeneity can be explained by the high density of sources of strong reflection leading to many bright pixels or linear objects embedded into dark background. The reason is that man-made objects are often of polyhedral shape (i.e., their boundaries are compound by planar facets). Planar objects appear bright for small incidence angle β or dark in the case of large β because most of the signal is reflected away from the sensor. Therefore, one simple method to identify potential settlement areas in an initial segmentation is to search for connected components of large density of isolated bright pixels, high CoV, or dynamic range.

In dense urban scenes, a method based on isolated bright pixels usually fails when bright pixels appear in close proximity or are even connected. Therefore, approaches that are more sophisticated analyze the local image histogram as approximation of the underlying pdf. Gouinaud and Tupin (1996) developed the ffmax algorithm that detects image regions featuring long-tailed histograms; thresholds are estimated from the image statistics in the vicinity of isolated bright pixels. This algorithm was also applied by He et al. (2006), who run it iteratively with adaptive choice of window size in order to improve the delineation of the urban area. An approach to extract human settlements proposed by Dell'Acqua and Gamba (2009, Chapter 2 of this book) starts with the segmentation of water bodies that are easily detected and excluded from further search. They interpolate the image on a 5 m grid and scale the data to [0,255]; a large difference of the minimum and maximum value in a 5×5 pixel window is considered as hint to a settlement. After morphological closing, a texture analysis is finally carried out to separate settlements from high-rise vegetation. The difficulty to distinguish those two classes was also pointed out by Dekker (2003), who investigated various types of texture measures for ERS data.

The principle drawback of traditional pixel based classification schemes is the neglect of context in the first decision step. It often leads to salt-and-pepper like results instead of desired homogeneous regions. One solution to solve this issue is post-processing, for example, using a sliding window majority vote. There exist also classification methods that consider context from the very beginning. One important class of those approaches are Markov Random Fields (Tison et al. 2004). Usually the classification is conducted in Bayesian manner and the local context is introduced in a Markovian framework by a predefined set of cliques connecting a small number of adjacent pixels. The most probable label set is found iteratively by minimizing an energy function, which is the sum of two contributions. The first one measures how well the estimated labels fit to the data and the second one is a regularization term linked to the cliques steering the desired spatial result. For example, homogeneous regions are enforced by attaching a low cost to equivalent labels within a clique and a high cost for dissimilar labels.

A completely different concept is to begin with a segmentation of regions as pre-processing step and to classify right away those segments instead of the pixels. The most popular approach of his kind is the commercial software eCognition that conducts a multi-scale segmentation and exploits spectral, geometrical, textural, and hierarchical object features for classification. This software has already been applied successfully for the extraction of urban areas in high-resolution airborne SAR data (Esch et al. 2005).

1.3.2.2 Characterization of Settlements

The characterization of settlements may be useful for miscellaneous kinds of purposes. Henderson and Xia (1998) present a comprehensive status report on the applications of SAR for settlement detection, population estimation, assessment of the impact of human activities on the physical environment, mapping and analyzing urban land use patterns, interpretation of socioeconomic characteristics, and change detection. The applicability of SAR for those tasks is of course varying and depends, for instance, on depression and aspect angles, wavelength, polarization, spatial resolution, and radiometric resolution.

Since different urban sub-classes like suburbs, industrial zones, and inner city cores are characterized by diverse sizes, densities, and 3d shapes of objects, such features are also useful to tell them apart. However, it is hard to generalize findings of any kind (e.g., thresholds) from one region to another or even to a different country, due to the large inner-class variety because of diverse historical or cultural reasons that may govern urban structures. Henderson and Xia (1997) report that approaches that worked fine for US cities failed for Germany, where the urban structure is quite different. This is of course a general problem of remote sensing not limited to radar.

The suitable level of detail of the analysis very much depends on the characteristics of the SAR sensor, particularly its spatial resolution. Walessa and Datcu (2000) apply a MRF to an E-SAR image of about 2 m spatial resolution. They carry out several processing steps: de-speckling of the image, segmentation of connected components of similar characteristics, and discrimination of five classes including the urban class. Tison et al. (2004) investigate airborne SAR data of spatial resolution well below half a meter (Intermap Company, AeS-1 sensor). From data of this quality, a finer level of detail is extractable. Therefore, their MRF approach aims at discrimination of three types of roofs (dark, mean, and bright) and three other classes (ground, dark vegetation, and bright vegetation). The classes ground, dark vegetation, and bright roofs can easily be identified, the related diagonal elements of the confusion matrix reach almost 100%. However, those numbers of the remaining classes bright vegetation, dark roof, and mean roof drop to 58–67%. In the discussion of these results, the authors propose to use L-shaped structures as features to discriminate buildings from vegetation.

The problem to distinguish vegetation, especially trees, from buildings is often hard to solve for single images. A multi-temporal analysis (Ban and Wu 2005) is beneficial, because of the variation of important classes of vegetation due to phenological processes, while man-made structures tend to persist for longer periods of time. This issue will be discussed in more detail in the next section.

1.3.3 Classification of Time-Series of Images

The phenological change or farming activities lead to temporal decorrelation of the signal in vegetated regions, whereas the parts of urban areas consisting of buildings and infrastructure stay stable. In order to benefit from this fact, time-series of images taken from the same aspect are required. In case of amplitude imagery, the correlation coefficient is useful to determine the similarity of two images. If complex data are available, the more sensitive magnitude of the complex correlation coefficient can be exploited, which is called coherence (see Section 1.4.2 for more details).

Ban and Wu (2005) investigate a SAR data set of five Radatsat-1 fine beam images (10 m resolution) of different aspect (ascending and descending) and illumination angle. Consequently, the analysis of the complex data is not feasible. Hence, amplitude images are used to discriminate three urban classes (high-density built-up areas, low-density built-up areas, and roads) from six classes of vegetation plus water. The performance of MLC and ANN is compared processing the raw images, de-speckled images, and further texture features. If only single raw images are analyzed, the results are poor (Kappa index of about 0.2), based on the entire image set kappa rises to 0.4, which is still poor. However, the results improve significantly using speckle filtering (kappa about 0.75) and incorporating texture features (up to 0.89).

Another method to benefit from time-series of same aspect data is to stack amplitudes incoherently. In such manner both noise and speckle are mitigated and especially persistent man-made objects appear much clearer in the resulting average image, which is advantageous for segmentation. In contrast to multi-looking the spatial resolution is preserved (assuming that no change occurred).

Strozzi et al. (2000) analyze stacks of 3, 4, and 8 (scene Berne) ERS images suitable for Interferometry of three scenes. The temporal variability of the image amplitude is highest for water, due to wind-induced waves at some dates, moderate for agricultural fields (vegetation growth, farming activities), and very small for forests and urban areas. With respect to long-term coherence (after more than 35 days, that is, more than one ERS repeat cycle) only the urban class shows values larger than 0.3. The authors partition the scene into the four classes water, urban area, forest, and sparse vegetation applying three different approaches: Threshold Scheme (manual chosen thresholds), MLC, and Fuzzy Clustering Segmentation. The results are comparable; overall accuracy is about 75%. This result seems not to be overwhelming especially for the urban class, but the authors point out that the reference data did not reflect any vegetation zones (parks, gardens etc.) inside the urban area. If the reference would be more detailed and realistic, the performance could be improved.

Bruzzone et al. (2004) investigate the eight ERS images over Berne, too. They use an ANN approach to discriminate settlement areas from the three other classes water, fields, and forest based on a set of eight ERS complex SAR images spanning 1 year. The best results (kappa 87%) are obtained exploiting both the temporal variation of the amplitude and the temporal coherence.

1.3.4 Road Extraction

The extraction of roads from remote sensing images is one of the most important applications of cartography. First approaches aiming at automation of this tedious manual task have been proposed already in the seventies (Bajcsy and Tavakoli 1976). The most obvious data sources for road extraction are aerial images taken in nadir view (Baumgartner et al. 1999). Nevertheless, also SAR data were used quite early (Hellwich and Mayer 1996).

Extraction of road networks is usually accomplished in a hierarchical manner, starting with a segmentation of primitive objects, for example straight lines, which are later connected to a network during a higher level of reasoning.

1.3.4.1 Recognition of Roads and of Road Networks

At this stage of the SAR data processing pixels are labeled to be part of an edge or line to some degree of probability. The next step is to segment connected components above a threshold hopefully coinciding with straight or curved object contours. Gaps are bridged and components violating a predefined shape model are rejected.

After such post-processing of initial segmentation results, higher levels of inference start. Only those primitives that actually belong to roads are filtered and connected consistently in a road network.

Wessel et al. (2002) adapt an approach developed for road network recognition in rural areas from aerial images (Baumgartner et al. 1999) to SAR data. A first step is to classify forest and urban areas, which are excluded from further processing. Then, a weighted graph is constructed from the potential dark road segments that have been extracted with the Steger operator; the weight reflects the goodness of the road segment hypothesis in a fuzzy logic manner. The road segments built the edges of the graph and their endpoints the knots. This initial graph contains, in general, gaps because not all road parts are found. Therefore, each gap is also evaluated based on its collinearity, the absolute and the relative gap length. For the network generation, various seed points have to be selected; segments with relatively high weights are chosen. Then, each pair of seed points is connected by calculating the optimal path through the graph. Finally, it is possible to fill remaining gaps by a network analysis, which hypothesize missing road segments in case of large detours. The authors evaluate the approach for two rural test areas based on airborne E-SAR X-band and fully polarimetric L-band data of about 2 m spatial resolution. The completeness of automatically extracted roads compared to manual segmentation varies from 50% to 67%, mainly secondary rounds are hard to find. The correctness is about 50%. Most of the false alarms are other dark linear structures like shadows at the borders of forests and hedges. In later work (Wessel 2004), the approach is extended considering context objects (e.g., rows of trees, cars) and an explicit model of highways.

The approach described in the previous paragraph was further developed by Hedman et al. (2005), who evaluate the quality of road hypotheses more comprehensively. In further developed versions of this approach, the analysis is accomplished using Bayesian Networks, which is explained in more detail in Chapter 3 of this book (Hedman and Stilla 2009).

Dell'Acqua and Gamba (2001) propose a methodology to extract roads in urban areas. First, three basic urban classes (vegetation, roads, and built-up areas) are distinguished using a Fuzzy C Means approach. The urban area is analyzed applying three different algorithms: the connectivity weighted Hough transform, the rotation Hough transform, and the shortest path extraction using dynamic programming. While the first two methods show good results for straight roads, the third approach is capable of detecting curved roads, too. The test data consists of AIRSAR imagery of about 10 m resolution showing parts of Los Angeles featuring the typical regular structure of US cities and wide roads between blocks. Both completeness and correctness of the results are about 80%. In later work of the group of authors, the different segmentation results are combined in order to remove errors and to fill in gaps (Dell'Acqua et al. 2003, 2009; Lisini et al. 2006)

The approach of Tupin et al. (1998) is one of the most comprehensive and elaborated ones. After extraction of potential line segments using a ratio operator (described previously) a MRF is set-up for grouping and incorporation of contextual a priori knowledge. A graph is built from the detected segments and the road identification process aims at the extraction of the optimal graph labeling. As usual for MRF approaches, the clique potentials carry the considered context knowledge chosen here as: (a) roads are long, (b) roads have low curvature, and (c) intersections are rare. The optimal label set is found iteratively by a special version of simulated annealing. In a final post-processing step, the road contours are fit to the data using snakes. The approach is applied to ERS and SIR-C/X-SAR amplitude data of 25 and 10 m resolution, respectively. Despite many initial false road candidates and significant gaps in-between segments, it is possible to extract the main parts of the urban road network.

1.3.4.2 Benefit of Multi-aspect SAR Images for Road Network Extraction

For a given SAR image a significant part of the entire road area of a scene might be either occluded by shadow or covered by layover from adjacent buildings or trees (Soergel et al. 2005). Hence, in dense urban scenes roads oriented in along-track sometimes cannot be seen at all. The dark areas observed in-between building rows are caused by radar shadow from the building row situated closer to the sensor, while the road itself is entirely hidden by layover of the opposite building row. This situation can be improved adding SAR data taken from other aspects. The optimal aspect directions depend on the properties of the scene at hand. In case of a checkerboard pattern type of city structure for example, two orthogonal views along the road directions would be optimal. In this way, problematic areas can be filled in with complementing data from the orthogonal view. In terms of mitigation of occlusion and layover issues, an anti-parallel aspect configuration would be the worst case (Tupin et al. 2002), because occlusion and layover areas would just be exchanged. However, this still offers the opportunity of improving results, due to redundant extraction of the roads visible in both images.

Hedman et al. (2005) analyze two rural areas covered by airborne SAR data of spatial resolution below 1 m taken from orthogonal aspects. They compare the performance of results for individual images and for a fused set of primitives. The fusion is carried out applying the logical OR operator (i.e., take all); the assessment of segments is increased in case of overlap, because the results mutually confirm. In the most recent version the fusion approach is carried out in a Bayesian network (Hedman and Stilla 2009, Chapter 3 of this book). The results improve especially in terms of completeness.

F. Tupin extends her MRF road extraction approach described above to multi-aspect data considering orthogonal and anti-parallel configurations (Tupin 2000; Tupin et al. 2002). Fusion is realized in two different ways. The first method consists of fusion on the level of road networks that have been extracted independently in the images, whereas in the second case fusion takes place at an earlier stage of the approach: the two sets of potential road segments are unified before the MRF is set-up. The second method showed slightly better results. One main problem is the registration of the images, because of the aspect dependent different layover shifts of buildings.

Lisini et al. (2006) present a road extraction method comprising fusion of classification results and structural information in form of segmented lines. Probability values are assigned to both kinds of features that are then fed into a MRF. Two classification approaches are investigated: a Markovian (Tison et al. 2004) and an ANN approach (Gamba and Dell'Acqua 2003). For line extraction, the Tupin operator is used. In order to cope with different road widths, the same line extractor is applied to images at multiple scales. These results are fused later. The approach was tested for airborne SAR data of resolution better than 1 m. The ANN approach seems to perform better with respect to correctness, whereas the Markovian method shows better completeness results.

1.3.5 Detection of Individual Buildings

For building extraction, 3d approaches are usually applied, which are discussed in the Section 1.4 in more detail. However, the segmentation of building primitives and as a consequence the detection of building hypotheses is generally conducted in 2d, that is, in the image space. Probably the best indicators of building locations are bright lines caused by double-bounce between wall and ground (Wegner et al. 2009a; Thiele et al. 2007). For large buildings, these features are already visible in ERS type of SAR data. Those lines indicate the walls visible from the sensor point of view. The building footprint stretches from such line to some extent into the image in the direction of larger range values, depending on the individual building. At the sensor far building side the shadow area is found. Some authors (Bolter 2000; Soergel et al. 2003a) consider this boundary explicitly in order to get more stable building hypotheses by mutual support from several independent features. Finally, a quadrangular building footprint hypothesis can be transferred to a subsequent 3d analysis for building reconstruction.

Tison et al. (2004) and Michaelsen et al. (2006) use rectangular angles built from two orthogonal bright lines as building features. This reduces the large number of bright lines usually found in urban SAR data to those with high probability to coincide with building locations. On the other hand, buildings that cause weaker response leading to the presence of only one of the two bright lines in the image might be lost.

The smaller resolution cells of modern sensors reveal far more individual strong point scatterers that are averaged out by the darker background in data of coarser resolution. Hence, in SAR images of resolution of 1 m and better linear chains of such regular spaced such scatterers appear saliently.

1.3.6 SAR Polarimetry

One means of extracting further information from SAR data of a given point in time is to exploit the complex nature of the signal and the agility of modern SAR sensors that enables to provide data of arbitrary polarimetric states (i.e., by definition the plane in which the electric field component of the electromagnetic wave oscillates).

1.3.6.1 Basics

A comprehensive overview of SAR Polarimetry (PolSAR) principles and applications can be found in Boerner et al. (1998). In radar remote sensing horizontal and vertical polarized signals are usually used. By systematically switching polarization states on transmit and receive the scattering matrix **S** is obtained that transforms the incident (transmit) field vector (subscript i) to the reflected (receive) field vector (r):

$$\begin{bmatrix} E_H^r \\ E_V^r \end{bmatrix} = \overbrace{\frac{e^{jkr}}{\sqrt{4\pi} r} \begin{bmatrix} S_{HH} & S_{HV} \\ S_{VH} & S_{VV} \end{bmatrix}}^{[S]} \begin{bmatrix} E_H^i \\ E_V^i \end{bmatrix}$$

Unfortunately, the order of the indices is non-standardized. Most authors denote the transmit polarization by the right index and the polarization on receive by the left index.

The scattering matrix carries useful information because reflection at object surfaces may change the polarization orientation according to certain constraints of the field components valid at material boundaries. There is no room to treat these issues in detail here; instead we will briefly outline the basic principles of the idealized case of reflection at perfectly conducting metal planes (Fig. 1.5). In such case no transmission occurs, because neither electric nor magnetic fields can exist inside metal. In addition, only the normal E-field component exists on the boundary, because a tangential component would immediately break down due to induced current. Consider specular reflection at a metal plane with incidence angle 0° (Fig. 1.5a): the E-field is always tangential no matter which polarization the incident wave has. Hence, at the boundary the E-field phase flips 180° in order to provide vanishing tangential field there, that means, for instance, matrix components S_{HH} and S_{VV}

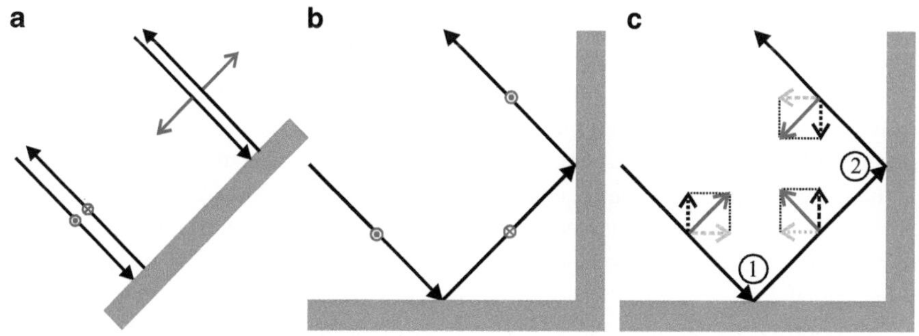

Fig. 1.5 Reflection at metal planes: (**a**) zero incidence angle leads to 180° phase shift jump for any polarisation, because entire E field is tangential, (**b, c**) double-bounce reflection at dihedral structure, in case of polarization direction perpendicular to the image plane (**b**) again the entire E field is tangential resulting in two phase jumps of 180° that sum up to 360°, and for a wave that is polarized parallel to the image plane (**c**) only the field component tangential to the metal plane flips, while the normal component remains unchanged, after both reflections the wave is shifted by 180°

are in phase. Interesting effects are observed when double-bounce reflection occurs at dihedral structures. If the polarization direction is perpendicular to the image plane, again the entire E-field is tangential resulting in two phase jumps of 180° that sum up to 360°. For the configuration shown in Fig. 1.5b this coincides with matrix element S_{HH}. But for a wave that is polarized parallel to the image plane (Fig. 1.5c), only the field component tangential to the metal plane flips, while the normal component remains unchanged. After both reflections the wave is shifted by 180°. As a result, the obtained matrix elements S_{HH} and S_{VV} are shifted by 180°, too.

For Earth observation purposes mostly a single SAR system transmits the signal and collects the backscatter during receive mode, which is referred to as monostatic sensor configuration. In this case, the two cross-polarized matrix components are considered to be equal for the vast majority of targets ($S_{HH} = S_{VV} = S_{XX}$) and the scattering matrix is simplified to:

$$[S] = \frac{e^{jkr} \cdot e^{j\varphi_{HH}}}{\sqrt{4\pi} r} \begin{bmatrix} |S_{HH}| & |S_{XX}|\, e^{j(\varphi_{XX} - \varphi_{HH})} \\ |S_{XX}|\, e^{j(\varphi_{XX} - \varphi_{HH})} & |S_{VV}|\, e^{j(\varphi_{VV} - \varphi_{HH})} \end{bmatrix}.$$

The common multiplicative term outside the matrix is of no interest, useful information is carried by five quantities: three amplitudes and two phase differences.

A variety of methods have been proposed to decompose the matrix **S** optimally to derive information for a given purpose (Boerner et al. 1998). The most common ones are the lexicographic (k_L) and the Pauli (k_P) decompositions, which transform the matrix into 3d vectors:

$$\begin{aligned} k_L &= \left(S_{HH}, \sqrt{2} \cdot S_{XX}, S_{VV}\right)^T, \\ k_P &= \frac{1}{\sqrt{2}} \left(S_{HH} + S_{VV}, S_{HH} - S_{VV}, 2 \cdot S_{XX}\right)^T \end{aligned} \tag{1.7}$$

The Pauli decomposition is useful to discriminate signal of different canonical objects. A dominating first component indicates an odd number of reflections, for example, direct reflection at a plate like in Fig. 1.5a, whereas a large second term is observed for even numbers of reflection like double bounce shown in Fig. 1.5b, c. If the third component is large, either double-bounce at a dihedral object (i.e., consisting of two orthogonal intersecting planes) rotated by 45° is the cause or reflection at multiple objects of arbitrary orientation increases the probability of large cross-polar signal.

As opposed to canonical targets like man-made objects distributed targets like natural land cover have to modeled statistically for PolSAR analysis. For such purpose, the expectation values of the covariance matrix **C** and/or the coherence matrix **T** are often used. These 3 × 3 matrices are derived from the dyadic product of the lexicographic and the Pauli decomposition, respectively:

$$[C_3] = \left\langle k_L \otimes k_L^H \right\rangle, \quad [T_3] = \left\langle k_P \otimes k_P^H \right\rangle, \tag{1.8}$$

where H denotes complex conjugate transpose and the brackets the expectation value. For distributed targets, the two matrices contain the complete scattering information in form of second order statistics. Due to the spatial averaging, they are in general of full rank. The covariance matrix is Wishart distributed (Lee et al. 1994).

Cloude and Pottier (1996) propose an eigenvalue decomposition of matrix T from which they deduce useful features for land cover classification, for example, entropy (H), anisotropy (A), and an angle α. The entropy is a measure of the randomness of the scattering medium, the anisotropy provides insight about secondary scattering processes, and α about the number of reflections.

1.3.6.2 SAR Polarimetry for Urban Analysis

Cloude and Pottier (1997) use the features entropy and angle α extracted by the eigenvalue decomposition of matrix **T** to classify land cover. The authors demonstrate the suitability of the H/α-space for discrimination of nine different object classes using airborne multi-look L-Band SAR data of San Francisco (10 m resolution). The same data were investigated by Lee et al. (1999): building blocks inside the city can clearly be separated from vegetated areas. Chen et al. (2003) apply a fuzzy neural classifier to these data to distinguish the four classes urban areas, ocean, trees, and grass. They achieve very good classification performance based on a statistical distance measure derived from the complex Wishart distribution.

Reigber et al. (2007) suggest applying several state-of-the-art SAR image processing methods for detection and classification of urban structures in high-resolution PolSAR data. They demonstrate these strategies using E-SAR L-band data of 1 m spatial resolution.

The first step of those approaches is sub-aperture decomposition: during SAR image formation many low-resolution real aperture echoes collected along the carrier flight path are integrated to process the full resolution image. As explained above in the context of multi-look processing, connected sub-sequences of pulses cover a smaller aspect angle range with respect to the azimuth direction. The synthesized complex single-look image can be decomposed again into sub-aperture images of lower azimuth resolution by Short-Time-Fourier-Transform. By analysis of the sequence of sub-aperture images, isotropic and anisotropic backscatter can be told apart. An object causing isotropic reflection (e.g., a vertical dipole-like structure) will show up the same in the images, but anisotropic backscatter (e.g., double-bounce at buildings) will appear only at certain aspect angles.

In order to determine isotropic or anisotropic behaviour from PolSAR data, it is convenient to compare the covariance matrices $\mathbf{C}_i$ of the sub-aperture images, which are Wishart distributed: for stationary (i.e., isotropic) backscattering they should be locally equal or at least very similar. This hypothesis is validated using a maximum-likelihood-ratio test Λ, based on the covariance matrices.

A similar technique that does not necessarily require PolSAR data is based on the coherence of the sub-aperture images (Schneider et al. 2006). In contrast to distributed targets governed by speckle, point-like coherent scatterers coincide with high correlation between sub-bands. By applying an indicator called internal coherence Y, Reigber et al. (2007) manage to extract many building boundaries independently of their orientation.

The third investigated feature Ψ is deployed to extract the image texture, which is analyzed using a speckle filter proposed by Lee (1980). Finally, the authors discuss some possibilities to use the features Λ, Ψ, and Y as input for further processing, for example, based on first order statistics or segmentation using a distance transform.

An approach for recognition of urban objects is described in Chapter 5 of this book (Hänsch and Hellwich 2009). The authors give an overview of SAR Polarimetry, discuss features and operators feasible for PolSAR, and go into details of methodologies to recognize objects.

1.3.7 Fusion of SAR Images with Complementing Data

The term fusion as used here relates to supporting the analysis of SAR images by complementary data. Supplementary data sources in this sense are either remote sensing images of different sensor types taken approximately at the same time or GIS content. Fusion can be conducted on different levels of abstraction; in general, approaches are grouped into three classes: pixel or image level (iconic) fusion, feature level (symbolic) fusion, and decision level fusion (Ehlers and Tomowski 2008). Although exceptions from the following rule exist, iconic fusion is rather applied to improve land cover classification based on imagery of medium or coarse resolution, whereas particularly the feature level fusion is more appropriate for images of fine spatial grid.

1.3.7.1 Image Registration

The advent of high-resolution SAR data comes along with the necessity of co-registration of complementary imagery of high quality. As a rule of thumb, the co-registration accuracy should match the spatial resolution of the data. Hence, an average accuracy of 20 m, sufficient in case of Landsat Thematic Mapper (TM) and ERS data, is not acceptable any more for the fusion of TSX or airborne SAR images with complementary data of comparable geometric resolution. Registration of high-resolution images requires suitable similarity measures, which may be based either on distinct image features (e.g., edges or lines) or the local signal distribution (Tupin 2009).

Hong and Schowengerdt (2005) propose to use edges to co-register SAR and optical satellite images of urban scenes precisely which have already roughly been aligned with an accuracy of some tens of pixels. They use ERS SAR data that have been subject to speckle filtering and register those to TM data. Dare and Dowman (2001) suggest a similar approach, but before the final registration is achieved based on edges they conduct a pre-processing step that matches homogeneous image regions of similar shape.

Inglada and Giros (2004) discuss the applicability of a variety of statistical quantities, for example mutual information, for the task of fine-registration of SPOT-4 and ERS-2 images. After resampling of the slave image to the master image grid remaining residuals are probably caused by effects of the unknown true scene topography. Especially urban 3d objects like buildings appear locally shifted in the images according to their height over ground and the different sensor positions and mapping principles. Hence, these residuals may be exploited to generate an improved DEM of the scene. This issue was investigated also in Wegner and Soergel (2008), who determine the elevation over ground of bridges from airborne SAR data and aerial images.

1.3.7.2 Fusion for Land Cover Classification

With respect to iconic image fusion, the problem of different spatial resolution of the data arises. This challenge is well known from multi-spectral satellite sensors like SPOT or TM. Such sensors usually provide at the same time one high-resolution gray value image (i.e., the panchromatic channel that integrates the radiation of a large part of the visible spectrum plus the near infrared, depending on the device) and several multi-spectral channels (representing the radiation of smaller spectral bands) of reduced resolution by a factor 2–4. A large body of literature deals with so-called pan-sharpening, which means to transform as much information as possible from the panchromatic and the spectral images into the 3d RGB space used for computer displays. Klonus et al. (2008) propose a method to adapt such approach to the multi-sensor case: they use a high-resolution TSX image providing object geometry to foster the analysis of multi-spectral images of lower resolution; the test site is a rural area. Their algorithm performs well compared to other approaches. The authors conclude that the benefit of fusion SAR and multi-spectral data with respect to classification performance becomes evident in the case of a resolution ratio better than one to ten.

Multi-spectral satellite sensors provide useful data to complement single or time-series of SAR images in particular in order to classify agricultural crops (Ban 2003). Data fusion was also applied to separate urban areas from other kinds of land cover. Solberg et al. (1996) apply a MRF fusion method to discriminate urban areas from water, forest, and two agricultural classes (ploughed and unploughed) based on ERS and TM images as well as GIS data providing field borders. The authors conclude that fusion significantly improves the classification performance. Haack et al. (2002) investigated the suitability of Radarsat-1 and TM data for urban delineation, the best results were achieved by consideration of texture features derived from the SAR images.

Waske and Benediktsson (2007) use a SVM approach to classify seven natural classes plus the urban class by a dense time series of ERS 2 and ASAR images spanning 2 years supplemented by one multi-spectral satellite image per year. Since the SVM is a binary classifier, the problem of discriminating more than two classes arises. In addition, information from different sensors may be combined in different ways. The authors propose a hierarchical scheme as solution. Each data source is classified separately by a SVM. The final classification result is based on decision fusion of the different outputs using another SVM. In later work of part of the authors (Waske and Van der Linden 2008) besides the SVM also the Random Forests classification scheme is applied to a similar multi-sensor data set.

1.3.7.3 Feature-Based Fusion of High-Resolution Data

Tupin and Roux (2003) propose an algorithm to detect buildings and reconstruct their outlines from one airborne SAR image and one aerial photo of about 50 cm geometric resolution. The images show an industrial area with large halls.

They first extract bright lines in SAR data that probably arise from double-bounce reflection. Then those lines are projected into the optical data. According to the sensor geometry, a buffer is defined around each projected line in which lines in the optical image are searched that are later assembled to closed rectangular building boundary polygons. In later work (Tupin and Roux 2005; Tupin 2009), this method was extended to a 3d approach, which is discussed in Section 1.4.1 in more detail.

Wegner et al. (2009b) propose a method for building detection in residential areas using one high-resolution SAR image and one aerial image. Similar to Tupin and Roux (2003) bright lines are considered as indicators to buildings in SAR images. In addition to the double-bounce line also the parallel line caused by specular reflection are considered. Those features are merged with potential building regions that are extracted independently in the optical image. The segmentation is fully carried out in the original image geometry (i.e., the slant range/azimuth plane in case of SAR) in order to avoid artifacts introduced by image geocoding and only the symbolic representations of building hypotheses are transformed into a common world coordinate system, where the fusion step takes place. The fusion leads to a considerable improvement of the detection completeness compared to results achieved from the SAR image or the optical data alone.

1.4 3d Approaches

The 3d structure of the scene can be extracted from SAR data by various techniques, Toutin and Gray (2000) give an excellent and elaborate overview. We distinguish here Radargrammetry that is based on the pixel magnitude and Interferometry that uses the signal phase. Both techniques can be further subdivided, which is described in the following Sections 1.4.1 and 1.4.2 in more detail.

1.4.1 *Radargrammetry*

The term Radargrammetry suggests the analogy to well-known Photogrammetry applied to optical images to extract 3d information. In fact, the height of objects can be inferred from a single SAR image or a couple of SAR images similar to photogrammetric techniques. For instance, the shadow cast behind a 3d object is useful to determine its elevation over ground. Additionally, the disparity of the same target observed from two different aspects can be exploited in order to extract its height according to stereo concepts similar to those of Photogrammetry. An extensive introduction into Radargrammetry is given in the groundbreaking book of Franz Leberl (1990) that still is among the most important references today. In contrast to Interferometry, Radargrammetry is restricted to the magnitude of the SAR imagery, the phase is not considered.

1.4.1.1 Single Image

The extraction of 3d information from single images is summarized by Toutin and Gray (2000) under the genus clinometry. Such approaches are particularly appropriate if no redundant coverage of the terrain is possible, which is and was often the case for extraterrestrial missions, for example, the Magellan probe to planet Venus (Leberl 1990).

There are two main kinds of useful features for single image analysis: radar shadow and shading. The former is in any case useful for human operators to get a 3d-impression of the scene. In case of flat terrain, it is straightforward to determine an object's height from the length of the cast shadow according to Eq. (1.6). This works well for detached buildings and the shape of the shadow may allow to deduce the type of roof (Bolter 2000; Bennett and Blacknell 2003).

Wegner et al. (2009b) estimate the height over ground of a tall bridge from the cast shadow. Since the bridge body is usually smooth compared to the ground, undulations of the terrain might be inferred from the variation of the shadow length.

Due to the aspect dependence of the shadow, multi-aspect images are generally required to extract a building completely. However, in built-up areas radar shadow is often hard to distinguish from other dark parts in the image like roads or parking lots covered with asphalt. Furthermore, layover from other buildings or trees impairs the value of the shadow feature to extract the height of objects.

Shading (change of grey value) is useful to derive the local 3d structure from the image brightness particularly in case of extended areas of homogeneous land cover on Earth (e.g., deserts) and other planets. It works well if two requirements are met: the reflection of the soil is Lambertian and the position of the signal source is known. Then, the observed gray value of a smooth surface solely depends on the local incidence angle. Since the illumination angle is given from navigation data of the SAR sensor carrier, the incidence angle and finally the local terrain slope can be deduced from the acquired image. Due to the inherent heterogeneity of man-made objects, shading is generally not appropriate for urban terrain.

Kirscht and Rinke (1998) combine both approaches to extract 3d objects. They assume that forests and building roofs would appear brighter in the amplitude image than the shadow they cast on the ground. They screen the image in range direction for ordered couples of bright areas followed by a dark region. The approach works for the test image showing the DLR site located in Oberpfaffenhofen, which is characterized by few detached large buildings and forest. However, for scenes that are more complex this approach seems not to be appropriate.

Quartulli and Datcu (2004) propose a stochastic geometrical modeling for building recognition from a high-resolution SAR image. They mainly model the bright appearance of the layover area followed by salient linear or L-shaped double-bounce signal and finally a shadow region. They consider flat and gable roof buildings. The footprint size tends to be overestimated, problems occur for complex buildings.

1.4.1.2 Stereo

The equivalent radar sensor configurations to the optical standard case of stereo are referred to as same-side and opposite-side SAR stereo (Leberl 1990). Same side means the images have been acquired from parallel flight tracks and the scene was mapped from the same aspect under different viewing angles. Analogous, opposite-side images are taken from antiparallel tracks. The search for matches is a 1d problem, the equivalent of the epipolar lines known from optical stereo are the range lines of the SAR images. Both types of configurations have their pros and cons. On the one hand, the opposite-side case leads to a large disparity, which is advantageous for the height estimate. On the other hand, the similarity of the images drops with increasing viewing angle difference; as a consequence, the number of image patches that can be matched declines. Due to the orbit inclination, both types of configuration are rare for space-borne sensors and more common for airborne data (Toutin and Gray 2000).

Simonetto et al. (2005) investigate same-side SAR stereo using three high-resolution images of the airborne sensor RAMSES taken with 30°, 40°, and 60° viewing angle θ in the image center. The scene shows an industrial zone with large halls. Bright L-shaped angular structures, which are often caused by double-bounce at buildings, are used as features for matching. Two stereo pairs are investigated: P1 with $\Delta\theta$ of 10° and P2 with 30° viewing angle difference. In both cases, large buildings are detected. Problems occur at small buildings, often because of lack of suitable features. As expected, the mean error in altimetry is smaller for the P2 configuration, but fewer buildings are recognized compared to P1.

SAR stereo is not limited to same-side or opposite-side images. Soergel et al. (2009) determine the height of buildings from a pair of high-resolution airborne SAR images taken from orthogonal flight paths. Of course, the search lines for potential matches do not coincide with the range direction anymore. Despite the quite different aspects, enough corresponding features can be matched at least for large buildings. The authors use a production system to group bright lines to rectangular

2d angle objects (Michaelsen et al. 2006), which are matched in 3d to built 3d angular objects. In addition, symmetry axes of the buildings are extracted from the set of chosen 2d angle objects.

Xu and Jin (2007) present an approach for automatic building reconstruction from multi-aspect SAR images of grid size of about one meter. They mainly exploit the layover induced shift of the buildings that are observed as bright parallelograms of varying location and orientation from four aspects. Hough transform is used to identify parallel lines that are further analysed in a probabilistic framework. The method performs well for detached buildings.

1.4.1.3 Image Fusion

In the previous section, it was shown that the 3d structure of the topography and especially the height of buildings can be deduced to some extent from single images and more complete from an image pair by stereo techniques. This is true for both SAR and optical images. Hence, a combination of two images of both sensor types can be considered special cases of clinometry or stereo techniques. The only complication is that two different sensor principles have to be taken into account in terms of mapping geometry and the appearance of object features.

Tupin and Roux (2003) detect building outlines based on the fusion of SAR and optical features. They analyze the same industrial scene as Simonetto et al. (2005), a single SAR image acquired by the RAMSES sensor is complemented by an aerial photo. A line detection operator proposed by Tupin et al. (1998) is applied to segment bright lines in the SAR image. As described previously, those line primitives are projected to the optical data to determine expectation areas for building features detected in the photo. Those hints to buildings are edges, which have been extracted with the Canny operator (Canny 1986). First, an edge in the photo is searched that is oriented parallel and situated closely to the SAR line. Then, sets of quadrangular search areas are defined, which are assessed based on the number of supportive edges. In a subsequent step, more complex building footprints are extracted by closed polygons, whose vertices are calculated from intersections of the segmented edges. In later work (Tupin and Roux 2005; Tupin 2009), this method was extended to a full 3d approach, which is based on a region adjacency graph of an elevation field that is regularized by a MRF. One purpose of this regularization is to achieve consistent heights for several wings of large buildings (prismatic building model). More details are given in Chapter 6 of this book.

1.4.2 SAR Interferometry

1.4.2.1 InSAR Principle

As discussed previously, a drawback of SAR is its diffraction-limited resolution in elevation direction. Similar to stereo, the SAR Interferometry (InSAR) technique

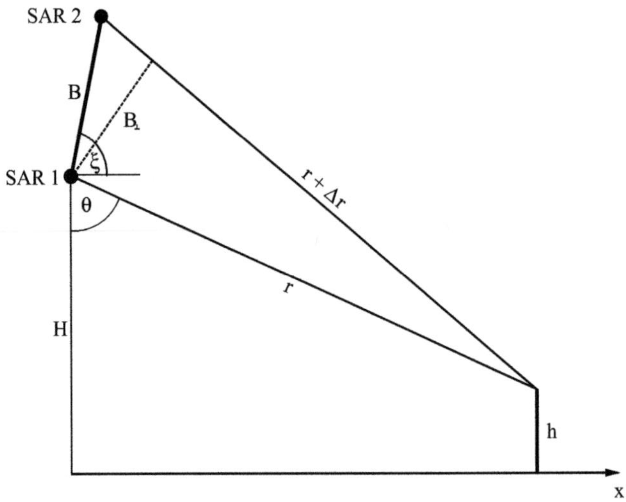

Fig. 1.6 Principle of SAR interferometry

uses more than one image to determine the height of objects over ground (Zebker and Goldstein 1986). However, the principle of information extraction is quite different: In contrast to stereo that relies on the magnitude image, Interferometry is based on the signal phase.

In order to measure elevation, two complex SAR images are required that have been taken from locations separated by a baseline B perpendicular to the sensor paths. The relative orientation of the two antennas is further given by the angle ξ (Fig. 1.6). This sensor set-up is often referred to as Across-Track Interferometry.

Preprocessing of the images usually comprises over-sampling, co-registration, and spectral filtering:

- Over-sampling is required to avoid aliasing: the complex multiplication in space domain carried out later to calculate the interferogram coincides with convolution of the image spectra.
- In order to maintain the phase information, co-registration and interpolation have to be conducted with sub-pixel accuracy of about 0.1 pixel or better.
- Spectral filtering is necessary to suppress non-overlapping parts of the image spectra; only the intersection of the spectra carries useful data for Interferometry.

The interferogram s is calculated by a pixel-by-pixel complex multiplication of the master image u_1 with the complex conjugated slave image u_2. Due to baseline B, the distances from the antennas to the scene differ by Δr, which results in a phase difference $\Delta\varphi$ in the interferogram:

$$s = u_1 \cdot u_2^* = |u_1| \cdot e^{j\varphi_1} \cdot |u_2| \cdot e^{-j\varphi_2} = |u_1| \cdot |u_2| \cdot e^{j\Delta\varphi}$$

$$\textit{with} \quad \Delta\varphi = W\left\{\varphi_{fE} + \varphi_{Topo} + \varphi_{Defo} + \varphi_{Error}\right\} \approx \frac{2\pi \cdot p}{\lambda} \cdot \Delta r \qquad (1.9)$$

The factor p is either 1 or 2 for single-pass or repeat-pass measurements, respectively. In the former case, the data are gathered simultaneously (usually airborne missions and SRTM (Rabus et al. 2003)) and in the latter case, the images are taken at different times, for example, at repeated orbits of a satellite. Phase $\Delta\varphi$ consists mainly of four parts: the term φ_{Topo} carries the height information that has to be isolated from the rest. The so-called phase of the flat Earth φ_{fE} depends only on the variation of the angle θ over swath and can easily be subtracted from $\Delta\varphi$. Error term φ_{Error} consists of several parts, the most important to be discussed here is component φ_{Atmo} that models atmospheric signal delay. The other parts of φ_{Error} and the term φ_{Defo} are neglected for the moment; they will be considered in Section 1.5 dealing with surface motion.

The phase difference $\Delta\varphi$ is only unambiguous in range $[-\pi, \pi]$, indicated by the wrapping operator W in Eq. (1.9). Thus, a phase-unwrapping step is often required before further processing. Thereafter, the elevation differences Δh in the scene depend approximately linearly on $\Delta\varphi$:

$$\Delta h \approx \frac{\lambda}{2\pi \cdot p} \cdot \frac{r \cdot \sin(\theta)}{B_\perp} \cdot \Delta\varphi, \quad B_\perp = B \cdot \cos(\theta - \xi). \tag{1.10}$$

The term $B_\perp$ is called normal baseline. It has to be larger than zero to enable the height measurement. At first glance, it seems to be advantageous to choose the normal baseline as large as possible to achieve a high sensitivity of the height measurement, because a 2π cycle (fringe) would coincide with a small rise in elevation. However, there is an upper limit for $B_\perp$ referred to as critical baseline: the larger the baseline becomes, the smaller the overlapping part of the object spectra gets and the critical value coincides with total loss of overlap. For ERS/Envisat the critical baseline amounts to about 1.1 km, whereas it increases to a few km for TSX, depending, besides other parameters, on signal bandwidth and incidence angle. In addition, a small unambiguous elevation span due to a large baseline leads to a sequence of many phase cycles in undulated terrain or mountainous areas, which have to be unwrapped perfectly in order to follow the terrain. The performance of phase-unwrapping methods very much depends on the signal to noise ratio (SNR). Hence, the quality of a given InSAR DEM may be heterogeneous depending on the local reflection properties of the scene especially for large baseline Interferometry.

To some degree the local DEM accuracy can be assessed a priory from the coherence of the given SAR data. The term coherence is defined as the complex cross-correlation coefficient of the SAR images, for many applications only its magnitude (range [0 ... 1]) is of interest. Coherence is usually estimated from the data by spatial averaging over a suitable area covering N pixels:

$$\gamma = \frac{E\left[u_1 \cdot u_2^*\right]}{\sqrt{E\left[|u_1|^2\right] \cdot E\left[|u_2|^2\right]}} = |\gamma| \cdot e^{j\Phi_0}, \quad |\gamma| \approx \frac{\left|\sum\limits_{n=1}^{N} u_1^{(n)} \cdot u_2^{(n)*}\right|}{\sqrt{\sum\limits_{n=1}^{N} \left|u_1^{(n)}\right|^2 \cdot \sum\limits_{n=1}^{N} \left|u_2^{(n)}\right|^2}} \tag{1.11}$$

Low coherence magnitude values indicate poor quality of the height derived by InSAR, whereas values close to one coincide with accurate DEM data. Several factors may cause loss of coherence (Hanssen 2001): non-overlapping spectral components in range (γ_{geom}) and azimuth (Doppler Centroid decorrelation, γ_{DC}), volume decorrelation (γ_{vol}), thermal noise ($\gamma_{thermal}$), temporal decorrelation ($\gamma_{temporal}$), and imperfect image processing ($\gamma_{processing}$, e.g., co-registration and interpolation errors). Usually those factors are modeled to influence the overall coherence in a multiplicative way:

$$\gamma = \gamma_{geom} \cdot \gamma_{DC} \cdot \gamma_{vol} \cdot \gamma_{thermal} \cdot \gamma_{temporal} \cdot \gamma_{processing}$$

Temporal decorrelation is an important limitation of repeat-pass Interferometry. Particularly in vegetation areas, coherence may be lost entirely after one satellite repeat cycle. However, as previously discussed, temporal decorrelation is useful for time-series analysis aiming at land cover classification and for change detection.

There is a second limitation attached to repeat-pass Interferometry: atmospheric conditions may vary significantly between both data takes leading to a large difference in φ_{Atmo} perturbing the measurement of surface heights. In ERS Interferograms phase disturbances in the order of half a fringe cycle frequently occur (Bamler and Hartl 1998).

In case of single-pass Interferometry neither atmospheric delay nor scene decorrelation have to be taken into account, because both images are acquired at the same time. The quality of such DEM is mostly governed by the impact of thermal noise, which is modeled to be additive, that is, the two images u_i consist of a common deterministic part c plus a random noise component n_i. Then, the coherence is modeled to approximately be a function of the local SNR:

$$|\gamma| \approx \frac{1}{1 + \frac{1}{SNR}}, \quad with \quad SNR = \frac{|c|^2}{|n|^2}.$$

1.4.2.2 Analysis of a Single SAR Interferogram

The opportunity to extract buildings from InSAR data has attracted the attention of many scientists who developed a number of different approaches. We can present only a few here. Tison and Tupin (2009) provide an overview in Chapter 7 of this book.

Gamba et al. (2000) adapt an approach originally developed for segmentation of planar objects in depth images (Jiang and Bunke 1994) for building extraction. The InSAR DEM is scanned along range lines; the data are piecewise approximated by straights. In order to segment 2d regions, homogeneous regions are segmented from sets of adjacent patches of similar range extent and gradient. The test data consists of a 5 m grid InSAR DEM of an urban scene containing large and tall buildings. Due to lack of detail, the buildings are reconstructed as prismatic objects of arbitrary footprint shape. The main buildings were detected; the footprints are approximated by rectangles. However, the footprint sizes are systematically underestimated; problems arise especially due to layover and shadowing issues.

Piater and Riseman (1996) apply a split-and-merge region segmentation approach to InSAR DEM for roof plane extraction. Elevated objects are separated from ground according to the plane equations. In a similar approach, Hoepfner (1999) uses region growing for the segmentation. He explicitly models the far end of a building in the InSAR DEM, which he expects to appear darker (i.e., at lower elevation level) in the image. The test data features a spatial grid better than half a meter and the scene shows a village. Twelve from 15 buildings are detected; under-segmentation occurs particularly where buildings stand together closely.

Up to now in this section, only methods that merely make use of the InSAR DEM have been discussed. However, the magnitude and the coherence images of the interferogram contain also useful data for building extraction. For example, Quartulli and Datcu (2003) propose a MRF approach for scene classification and subsequent building extraction. Burkhart et al. (1996) exploit as well all three kinds of images. They use diffusion-based filtering to de-noise the InSAR data and segment bright areas in the magnitude image that might coincide with layover. In this paper, the term front-porch-effect for characterization of the layover area in front of a building was coined.

Soergel et al. (2003a) also process the entire InSAR data set. They look for bright lines marking the start of a building hypothesis and two kinds of shadow edges at the other end: the first is the boundary between building and shadow and the second is the boundary between shadow and ground. Quadrangular building candidate objects are assembled from those primitives. The building height is calculated from two independent data sources: the InSAR DEM and the length of the shadow. From the InSAR DEM values enclosed by the building candidate region, the average height is calculated. In this step, the coinciding coherence values serve as weights in order to increase the relative impact of the most reliable data. Since some building candidate objects might contradict each other and inconsistencies may occur, processing is done iteratively. In this way, adjustments according to the underlying model, for example, rectangularity and linear alignment of neighboring buildings, are enforced, too. The method is tested for a built-up area showing some large buildings located in close proximity. Most of the buildings are detected and the mayor structures can be recognized. However, the authors recommend multi-aspect analysis to mitigate remaining layover and occlusion issues.

Tison et al. (2007) extent their MRF approach originally developed for high-resolution SAR amplitude images to InSAR data of comparable grid. Unfortunately, the standard deviation of the InSAR DEM is about 2–3 m. The limited quality of the DEM enables to extract mainly large buildings, while small ones cannot be detected. However, the configuration of the MRF seems to be sound. Therefore, better results are expected for data that are more appropriate.

1.4.2.3 Multi-image SAR Interferometry

Luckman and Grey (2003) used a stack of 20 ERS images to infer the height variance of an urban area by analysis of the coherence. This is possible because two factors influencing the coherence are the normal baseline and the vertical distribution of the scatterers. By inverting a simplified coherence model, the authors are able to discriminate residential areas from multistory buildings in the inner city of Cardiff, UK.

One possibility to overcome the layover problem is multi-baseline processing of sets of SAR images of suitable tracks; the key idea is to establish a second synthetic aperture orthogonal to the flight path and to achieve a real 3d imaging of the scene in this manner. This technique, which is referred to as SAR tomography (TomoSAR) as well, was already demonstrated for airborne (Reigber and Moreira 2000) and space borne scenarios (Fornaro et al. 2005). In order to maintain sufficient spectral overlap, the viewing angles of the SAR images of the evaluated stack vary only slightly. Compared to SAR image focusing, SAR Tomography deals with sparse and irregularly spaced samples, because of the limited number of suitable SAR orbits that may deviate arbitrarily from a reference by tens or hundreds of meters. Therefore, special sophisticated digital signal processing techniques have to be applied for resolving different scatterers in elevation direction. This resolution is given by Eq. (1.1) and replacing aperture D by two times the range of normal baselines B_{range}. Zhu et al. (2008) show some very interesting first results achieved by processing of 16 TSX images covering the Wynn hotel, a skyscraper in Las Vegas. However, the special feature of TSX that repeated orbit cycles lay inside a tube of about 300 m diameter in space limits the TomoSAR resolution. In this case, B_{range} is 270 m, which results in an elevation resolution of about 40 m. Nevertheless, this is sufficient to clearly resolve signal contributions from ground and building. The authors suggest to combine TomoSAR with techniques to determine object motion (such approaches are discussed in Section 1.5), that means to add a forth dimension (i.e., time) to information extraction.

1.4.2.4 Multi-aspect InSAR

Multi-image Interferometry from the same aspect may solve layover problems to some extent. However, occlusion behind buildings still is an issue. In order to overcome this, multi-aspect data are useful for InSAR, too.

Xiao et al. (1998) study a village scene of 15 buildings of different roof type and orientation that was mapped from the four cardinal directions by a high-resolution airborne InSAR sensor. This ground range data set was investigated elsewhere also (Piater and Riseman 1996; Bolter 2000, 2001); it is worthwhile to mention that no trees or bushes are present, because it is an artificial scene built for military training purposes. In addition, a multi-spectral image is available. In a first step, a classification of both InSAR and multi-spectral data was conducted in order to separate buildings from the rest. However, the most important part of the approach consists of applying image processing techniques to the InSAR data. The authors fuse the four InSAR DEMs, always choosing the height value of the DEM that shows maximal coherence at the pixel of interest. Gaps due to occlusion vanish since occluded areas are replaced by data from other aspects. A digital terrain model (DTM) is calculated from the fused DEM applying morphologic filtering. Subtraction of the DTM from the DEM yields a normalized DEM (nDEM). In the latter, connected components of adequate areas are segmented. Minimum size bounding rectangles are fit to the contours of those elevated structures. If the majority of pixels inside those rectangular polygons are classified to belong to the building class, the hint is accepted as building object. Finally, 14 from 15 buildings have been successfully detected; the roof structure is not considered.

The same data set was also thoroughly examined by Bolter (2000, 2001). She combines the analysis of the magnitude and the height data by introducing the shadow analysis as alternative way to measure the building elevation over ground. In addition, the position of the part of the building footprint that is facing away from the sensor can be determined. Fusion of the InSAR DEMs is accomplished by always choosing the maximum height, no matter its coherence. One of the most valuable achievements of this paper was to apply simulations to improve SAR image understanding and to study the appearance of buildings in SAR and InSAR data. Balz (2009) discusses techniques and applications of SAR simulation in more detail in Chapter 9 of this book. Based on simulations, the benefit of different kinds of features can be investigated systematically for a large number of arbitrary sensor and scene configurations. All 15 buildings are detected and 12 roofs are reconstructed correctly taking into account two building models: flat-roofed and gable-roofed buildings.

Soergel et al. (2003b) provide a summary of the geometrical constraints attached to the size of SAR layover and occlusion areas of certain individual buildings and building configurations. Furthermore, the authors apply a production system for building detection and recognition that models geometrical and topological constraints accordingly. Fusion is not conducted on the iconic raster, but at object level. All objects found in the slant range InSAR data of the different aspects are transformed to the common world coordinate system according to range and azimuth coordinates of their vertices and the InSAR height. The set union of the objects constructed so far acts as a pool to assemble more complex objects step by step. The approach run iteratively in an analysis-by-synthesis manner. This means intermediate results are used to simulate InSAR data and to predict location and size of building features. Simulated and real data are compared and deviations are minimized in subsequent cycles. The investigated test data covers a small rural scene that was illuminated three times from two opposite aspects, resulting in three full InSAR data sets. All buildings are detected, the fusion improves the completeness of detection and the reconstruction of the roofs (buildings with flat or gable roof are considered).

Thiele et al. (2007), who focus on built-up areas, further developed the previous approach. The test data consist of four pairs of complex SAR images, which were taken in single-pass mode by the AeS Sensor of Intermap Company from two orthogonal aspects. The spatial resolution is 38 cm in range and 18 cm in azimuth. Two procedures are proposed: one is tailored for residential buildings and the other for large buildings (e.g., halls). After interferogramm generation, the magnitude images are transferred to Decibel. Co-registered magnitude images are fused by choosing the maximum value in order to achieve better segmentation results. The operators proposed by Steger and Canny, respectively, detect bright line and edge objects. Primitives are projected to the world coordinate system where further processing takes place. L-structures are built from the set union of the primitives and thereafter the building outlines. Depending on the building class of interest, the higher-level reasoning steps of the two approaches are adapted. The main buildings are found, whereas small buildings are missed during the detection phase and tall vegetation causes problems, too. The authors conclude that both approaches should be merged in order to address areas of mixed architecture.

In later work, a method for the extraction of gable-roofed buildings is proposed (Thiele et al. 2009a). The most promising feature of this kind of buildings is the parallel bright line pair visible for buildings that are oriented in azimuth direction: the line situated closer to the senor is caused by direct reflection, while the other one is due to double bounce (Fig. 1.3e). The appearance of these features is discussed comprehensively using range profiles of the magnitude and phase images for real and simulated data. In addition, geometric constraints for roofs of different steepness are derived.

In orthogonal views only from one aspect the double-line feature may appear, whereas in the other aspect again a bright line or a L-structure should be visible. The line caused by direct reflection from the roof coincides with higher InSAR DEM values than the double-bounce line that represents terrain level. Hence, the height is used to select and project only the double-bounce lines into the scene to be fused with the other hints in order to reconstruct the building footprint.

1.4.3 Fusion of InSAR Data and Other Remote Sensing Imagery

As discussed above, one key problem that burdens 3d recognition of urban areas from InSAR data is the similarity of buildings and trees in the radar data. One solution to compensate the lack of spectral information provided by SAR is to incorporate co-registered multi-spectral or hyperspectral data.

Hepner et al. (1998) use hyperspectral data to improve building extraction from an InSAR DEM. First, potential building locations are extracted from the DEM by thresholding. Buildings and groups of trees are often hard to tell apart from the SAR data alone and thus hyperspectral data come into play, in which both classes can be separated easily.

Jaynes et al. (1996) assemble rectangular structures from lines detected in aerial images that are potential building hypotheses. The building elevation over ground is derived from an InSAR DEM co-registered to the optical data. If the average height is above the threshold, a prismatic building object is reconstructed. As opposed to this procedure, Huertas et al. (2000) look for building hints in the InSAR data to narrow down possible building locations in aerial photos, in which reconstruction is conducted. They assume correspondence of buildings with bright image regions in the InSAR amplitude and height images. First, regions of poor coherence are excluded from further processing. Then, the amplitude and height images are filtered with the Laplacian-of-Gaussian operator. Connected components of coinciding positive filter response are considered building hints. Finally, edge primitives are grouped to building outlines at the corresponding locations in the optical data.

Wegner et al. (2009a, b) developed an approach for building extraction in dense urban areas based on single-aspect aerial InSAR data and one aerial image. Fusion is conducted on object level. In the SAR data, again bright lines serve as building primitives. From the set of all such lines only those are chosen whose InSAR height is approximately at terrain level, that is lines caused by roof structures are rejected. Potential building areas are segmented in the optical data using a constrained region growing approach. Building hypotheses are assessed in the range [0 ... 1], value 1 indicates optimum. For fusion, the objects found in the SAR data are weighted by 0.33, those from the photo by 0.67, and the sum of both values gives a final figure of merit that again can reach value 1 as maximum. A threshold was set to 0.6 to filter only the best building hypothesis objects. The fusion step leads to a significant rise in terms of both completeness and correctness compared to results achieved without fusion.

1.4.4 SAR Polarimetry and Interferometry

The combination of SAR Polarimetry and Interferometry enables information extraction concerning the type of reflection and the 3d location of its source even for multiple objects inside a single resolution cell. Lee et al. (1994) investigated the intensity of phase statistics of multi-look PolSAR and InSAR images. In a seminal paper, Cloude and Papathanassiou (1998) proposed a method to supplement SAR Polarimetry with SAR Interferometry (PolInSAR). The basic idea is to use the concatenated vectors of the Pauli decomposition (Eq. 1.7) of both PolSAR image sets to calculate a 6×6 coherency matrix:

$$\begin{aligned}
k_{P1} &= \frac{1}{\sqrt{2}}\left(S_{HH1}+S_{VV1}, S_{HH1}-S_{VV1}, 2\cdot S_{XX1}\right)^T,\\
k_{P2} &= \frac{1}{\sqrt{2}}\left(S_{HH2}+S_{VV2}, S_{HH2}-S_{VV2}, 2\cdot S_{XX2}\right)^T,\\
k &= k_{P1}k_{P2}\\
[T_6] &= \left\langle k\otimes k^H\right\rangle = \begin{bmatrix} \left\langle k_{P1}\otimes k_{P1}^H\right\rangle & \left\langle k_{P1}\otimes k_{P2}^H\right\rangle \\ \left\langle k_{P1}\otimes k_{P2}^H\right\rangle^H & \left\langle k_{P2}\otimes k_{P2}^H\right\rangle \end{bmatrix} = \begin{bmatrix} [T_{11}] & [\Omega_{12}] \\ [\Omega_{12}]^H & [T_{22}] \end{bmatrix}.
\end{aligned}$$

The matrices $\mathbf{T}_{11}$ and $\mathbf{T}_{22}$ represent the conventional PolSAR coherency matrices, while $\boldsymbol{\Omega}_{12}$ contains also InSAR information.

The opportunity to combine the benefits of PolSAR and InSAR is of course of vital interest for urban analysis, for example, to discriminate different kinds of signal from several objects inside layover areas. Guillaso et al. (2005) propose an algorithm for building characterization in L-Band data of 1.5 m resolution. The first step consists of unsupervised Wishart H-A-α classification and segmentation in the PolSAR data. The result is a partitioning of the scene into the three classes single-bounce, double-bounce, and volume scattering. In order to improve the separation of buildings from vegetation in the volume class, an additional classification is carried out that combines polarimetric and interferometric features. Furthermore, a sophisticated signal processing approach from literature called ESPRIT (estimation of signal parameters via rotational invariant techniques) is applied to remove noise from the InSAR phase signal. Finally, the height of the buildings is reconstructed. Results are in good agreement with ground truth. In later work, almost the same group of authors also proposes an approach capable to cope with multi-baseline PolInSAR data (Sauer et al. 2009).

1.5 Surface Motion

Surface deformation can be triggered by various kinds of anthropogenic or natural processes, for example, on the one hand mining activities or ground water removal and on the other hand earthquake, volcanic activity, or landslide. The magnitude of such deformation process may amount to only some centimeters per year. Depending on the type of deformation process, the motion may proceed slowly with constant velocity or abruptly (e.g., earthquake). In any case, it is hard or even impossible to monitor such subtle change by means of optical sensors. The potential of radar remote sensing to detect and monitor small magnitude soil movement by Interferometry was investigated quite early (Rosen et al. 2000). The basic idea is to isolate the term related to terrain deformation φ_{Defo} from the InSAR phase difference (Eq. 1.9). Two main techniques have been developed called Differential SAR Interferometry (dInSAR) and Persistent Scatterer Interferometry (PSI), which both rely on InSAR processing as described in Section 1.4.2.1. Their basic principles are discussed briefly in the following and in more detail in the Chapter 10 of this book written by Crosetto and Monserrat (2009).

1.5.1 Differential SAR Interferometry

The interferogram is calculated in the usual way. A key issue is to remove the topographic phase term φ_{Topo} in Eq. (1.9). This is done by incorporating a DEM, which is either given as reference or derived by InSAR. In the latter case, three SAR images are required: one interferogram delivers the DEM, the other the deformation pattern. From the DEM the phase term induced by topography is simulated in agreement with the baseline configuration of the interferogram chosen for deformation extraction. The simulated topographic phase φ_{Topo_sim} and the geometry dependent term of the flat Earth φ_{fE} are subtracted from the measured phase difference:

$$\Delta\varphi - \varphi_{Topo} - \varphi_{fE} \approx \varphi_{Defo} + \varphi_{Error} \approx \frac{4\pi}{\lambda}\left(\vec{n}_{LOS} \cdot \vec{v}\Delta t\right). \qquad (1.12)$$

The unit vector n points to the line-of-sight (LOS) of the master SAR sensor, which means only the radial component of surface motion of velocity v in arbitrary direction can be measured. Hence, we observe a 1d projection of an unknown 3d movement. Therefore, geophysical models are usually incorporated, which provide insight whether the soil moves vertically or horizontally. By combination of ascending and descending SAR imagery, two 1d components of the velocity pattern are retrieved.

The dInSAR technique has already been successfully applied to various surface deformations. Massonnet et al. (1993), who used a pre-strike and a post-strike SAR image pair to determine the displacement field of the Landers earthquake, gave a famous example. However, there exist important limitations of this technique that are linked to the error phase term φ_{Error} which can be further subdivided into:

$$\varphi_{Error} = \varphi_{Orbit} + \varphi_{Topo\_sim} + \varphi_{Noise} + \varphi_{Atmo} + \varphi_{Decorrelation} \qquad (1.13)$$

The first two components model deficiencies of the accuracy of orbit estimates and the used DEM, while the third term refers to thermal noise. Proper signal processing and choice of ground truth can minimize those issues. More severe are the remaining two terms dealing with atmospheric conditions during data takes and real changes of the scene in-between SAR image acquisition. The water vapor density in the atmosphere has significant impact on the velocity of light and consequently on the phase measurement. Unfortunately, this effect varies over the area usually mapped by a space borne SAR image. Therefore, a deformation pattern might be severely obscured by atmospheric signal delay leading to large phase difference component φ_{Atmo}, which handicaps the analysis or even makes it impossible. The term $\varphi_{Decorrelation}$ is an important issue in particular for vegetated areas. Due to phenological processes or farming activities, the signal can fully decorrelate in-between repeat cycles of the satellite; in such areas the detection of surface motion is impossible. However, signal from urban areas and non-vegetated mountains may maintain coherence for many years.

1.5.2 Persistent Scatterer Interferometry

This technique was invented to overcome some drawbacks of conventional dInSAR discussed in the last section. Ferretti et al. (2000, 2001) from Politecnico di Milano developed the basic principles of the method. They coined the term permanent scatterers, which is dedicated to their algorithm and the spin-off company TRE. Other research groups have developed similar techniques, today most people use the umbrella term Persistent Scatterer Interferometry (PSI).

In this method, two basic concepts are applied to overcome the problems related to atmospheric delay and temporal decorrelation. The first idea is to use stacks of as many suitable SAR images as possible. Since the spatial correlation of water vapor is large compared to the resolution cell of a SAR image, the related phase component of a given SAR acquisition is in general spatially correlated as well. On the other hand, the temporal correlation of φ_{Atmo} is in general in the scale of hours or days. Hence, the same vapor distribution will never influence two SAR acquisitions taken systematically according to the repeat cycle regime of the satellite spanning many days. In summary, the atmospheric phase screen (APS) is modeled to add spatial low-pass and temporal high-pass signal components. Some authors explicitly model the APS in the mathematical framework to estimate surface motion (Ferretti et al. 2001).

The second concept explains the name of the method: the surface movement cannot be reconstructed gapless for the entire scene. Instead, the analysis relies on pixels whose signal is stable or persistent over time. One method to identify those PS is the dispersion index D_A, which is the ratio of the amplitude standard deviation and the mean value of a pixel over the stack. Alternatively, high signal-to-clutter ratio between a pixel and its surrounding indicates that the pixel might contain a PS (Adam et al. 2004). The PS density very much depends on the type of land cover and may vary significantly over a scene of interest. Since buildings are usually present for long times in the scene and made of planar facets, the highest number of PS is found in settlement areas. Hence, PSI is especially useful to monitor urban subsidence or uplift.

However, Hooper et al. (2004) successfully developed a PSI method for measuring deformation of volcanoes. This is possible because rocks also may cause signal of sufficient strength and stability. Source code of a version of Andrew Hooper's software is available in the internet (StaMPS 2009).

PS density also depends on the spatial resolution of the SAR data. The better the resolution gets, the higher the probability becomes that merely a single strong scatterer is located inside the cell. Bamler et al. (2009) report a significant rise of PS density found in TSX stacks over urban scenes compared to Envisat or ERS. This offers the opportunity to monitor urban surface motion at finer scales (e.g., on building level) in the future.

1.6 Moving Object Detection

SAR focusing relies on stationary scenes. As soon as objects move during data acquisition, this assumption is violated. If the movement occurs parallel to the sensor track, the object appears blurred. In the case of radial motion, an additional Doppler frequency shift takes place. Since the Doppler history is used to focus the SAR image in azimuth, a wrong azimuth position is the consequence. Depending on the object velocity, this shift can reach significantly amounts (train-of-the-track effect). If it is possible to observe the shifted object and to match it with its correct position (e.g., road, track), its radial (i.e., in LOS) velocity v_{LOS} can be determined:

$$\Delta az = -R\frac{v_{LOS}}{v_{Sat}},$$

with satellite speed v_{Sat}, azimuth shift Δaz, and range of minimum distance R. However, often such match is hardly feasible and ambiguities may particularly occur in urban scenes. In addition, acceleration of objects may induce further effects. Meyer et al. (2006) review source and consequences of those phenomena in more detail.

SAR Interferometry is capable to determine radial velocity, too. For such purpose, the antenna set-up has to be adapted such that the baseline is oriented along-track instead of across-track as for DEM extraction. The antennas whose phase centers are separated by Δl pass the point of minimum distance to the target after time t. Meanwhile the object has slightly moved resulting in a velocity dependent phase difference:

$$\Delta\varphi = \frac{4\pi}{\lambda}v_{LOS}\frac{\Delta l}{v_{Sat}} = \frac{4\pi}{\lambda}v_{LOS}t \quad (1.14)$$

Modern agile sensors like TSX are capable of Along-Track Interferometry. Hinz et al. (2009, Chapter 4 of this book) discuss this interesting topic in more detail.

Acknowledgement I want to thank my colleague Jan Dirk Wegner for proofreading the paper.

References

Adam N, Kampes B, Eineder M (2004) Development of a scientific permanent scatterer system: modifications for mixed ERS/ENVISAT time series. Proceedings of Envisat Symposium, Salzburg

Bamler R, Eineder M, Adam N, Zhu X, Gernhardt S (2009): Interferometric potential of high resolution spaceborne SAR. Photogrammetrie Fernerkundung Geoinformation 5/2009:407–420

Bamler R, Hartl P (1998) Synthetic aperture radar interferometry. Inverse Probl 14(4):R1–R54

Bajcsy R, Tavakoli M (1976) Computer recognition of roads from satellite pictures. IEEE Trans Syst Man Cybern 6(9):623–637

Balz T (2009) SAR simulation of urban areas: techniques and applications. Chapter 9 of this book

Ban Y (2003) Synergy of multitemporal ERS-1 SAR and landsat TM data for classification of agricultural crops. Can J Remote Sens 29(4):518–526

Ban Y, Wu Q (2005) RADARSAT SAR data for landuse/land-cover classification in the rural-urban fringe of the greater Toronto area. AGILE 2005, 8th Conference on Geographic Information Science, pp 43–50

Baumgartner A, Steger C, Mayer H, Eckstein W, Ebner H (1999) Automatic road extraction based on multi-scale, grouping, and context. Photogramm Eng Remote Sens 65(7):777–785

Bennett AJ, Blacknell D (2003) The extraction of building dimensions from high-resolution SAR imagery. IEEE Proceedings of the International Radar Conference, pp 182–187

Boerner WM, Mott H, Lüneburg E, Livingston C, Brisco B, Brown RJ, Paterson JS (1998) Polarimetry in radar remote sensing: basic and applied concepts, Chapter 5. In: Henderson FM, Lewis AJ (eds) Principles and applications of imaging radar, vol. 2 of manual of remote sensing (ed: Reyerson RA), 3rd edn. Wiley, New York

Bolter R (2000) Reconstruction of man-made objects from high-resolution SAR images. Proceedings of IEEE Aerospace Conference, Paper No. 6.0305, CD

Bolter R (2001) Buildings from SAR: detection and reconstruction of buildings from multiple view high-resolution interferometric SAR data. Dissertation, University of Graz, Austria.

Bruzzone L, Marconcini M, Wegmuller U, Wiesmann A (2004) An advanced system for the automatic classification of multitemporal SAR images. IEEE Trans Geosci Remote Sens 42(6):1321–1334

Burkhart GR, Bergen Z, Carande R (1996) Elevation correction and building extraction from interferometric SAR imagery. Proceedings of IGARSS, pp 659–661

Chen CT, Chen KS, Lee JS (2003) The use of fully polarimetric information for the fuzzy neural classification of SAR images. IEEE Trans Geosci Remote Sens 41(9):2089–2100

Canny J (1986) A computational approach to edge detection. IEEE Trans Pattern Anal Mach Intell 8(6):679–698

Cloude SR, Papathanassiou KP (1998) Polarimetric SAR interferometry. IEEE Trans Geosci Remote Sens 36(5):1551–1565

Cloude SR, Pottier E (1996) A review of target decomposition theorems in radar polarimetry. IEEE Trans Geosci Remote Sens 34(2):498–518

Cloude SR, Pottier E (1997) An entropy based classification scheme for land applications of polarimetric SAR. IEEE Trans Geosci Remote Sens 35(1):68–78

Crosetto M, Monserrat O (2009) Urban applications of Persistent Scatterer Interferometry. Chapter 10 of this book

Curlander JC, McDonough RN (1991) Synthetic aperture radar: systems and signal processing. Wiley, New York

Dare P, Dowman I (2001) An improved model for automatic feature-based registration of SAR and SPOT images. ISPRS J Photogramm Remote Sens 56(1):13–28

Dekker RJ (2003) Texture analysis and classification of SAR images of urban areas. Proceedings of 2nd GRSS/ISPRS Joint Workshop on Remote Sensing and Data Fusion on Urban Area, pp 258–262

Dell'Acqua F, Gamba P (2001) Detection of urban structures in SAR images by robust fuzzy clustering algorithms: The example of street tracking. IEEE Trans Geosci Remote Sens 39(10):2287–2297

Dell'Acqua F, Gamba P, Lisini G (2003) Road map extraction by multiple detectors in fine spatial resolution SAR data. Can J Remote Sens 29(4):481–490

Dell'Acqua F, Gamba P, Lisini G (2009) Rapid mapping of high-resolution SAR scenes. ISPRS J Photogramm Remote Sens 64(5):482–489

Dell'Acqua F, Gamba P (2009) Rapid mapping using airborne and satellite SAR images. Chapter 2 of this book

Duda RO, Hart PE, Stork DG (2001) Pattern classification, 2nd edn. Wiley, New York

Dong Y, Forster B, Ticehurst C (1997) Radar backscatter analysis for urban environments. Int J Remote Sens 18(6):1351–1364

Ehlers M, Tomowski D (2008) On segment based image fusion. In: Blaschke T, LANG S, Hay G (eds) Object-based image analysis spatial concepts for knowledge-driven remote sensing applications. Lecture notes in geoinformation and cartography. Springer, New York, pp 735–754

Esch T, Roth A, Dech S (2005) Robust approach towards an automated detection of built-up areas from high-resolution radar imagery. Proceedings of 2nd GRSS/ISPRS Joint Workshop on Remote Sensing and Data Fusion on Urban Area, CD, 6 p

Essen H (2009) Airborne remote sensing at millimeter wave frequencies. Chapter 11 of this book

Ferretti A, Prati C, Rocca F (2000) Nonlinear subsidence rate estimation using permanent scatterers in differential SAR interferometry. IEEE Trans Geosci Remote Sens 38(5):2202–2212

Ferretti A, Prati C, Rocca F (2001) Permanent scatterers in SAR interferometry. IEEE Trans Geosci Remote Sens 39(1):8–20

Fornaro G, Lombardini F, Serafino F (2005) Three-dimensional multipass SAR focusing: experiments with long-term spaceborne data. IEEE Trans Geosci Remote Sens 43(4):702–714

Guillaso S, Ferro-Famil L, Reigber A, Pottier E (2005) Building characterisation using L-band polarimetric interferometric SAR data. IEEE Geosci Remote Sens Lett 2(3):347–351

Gamba P, Houshmand B, Saccani M (2000) Detection and extraction of buildings from interferometric SAR data. IEEE Trans Geosci Remote Sens 38(1):611–618

Gamba P, Dell'Acqua F (2003) Improved multiband urban classification using a neuro-fuzzy classifier. Int J Remote Sens 24(4):827–834

Gouinaud G, Tupin F (1996) Potential and use of radar images for characterization and detection of urban areas. Proceedings of IGARSS, pp 474–476

Goodman JW (1985) Statistical optics. Wiley, New York

Haack BN, Solomon EK, Bechdol MA, Herold ND (2002) Radar and optical data comparison/integration for urban delineation: a case study. Photogramm Eng Remote Sens 68: 1289–1296

Hänsch H, Hellwich O (2009) Object recognition from polarimetric SAR images. Chapter 5 of this book

Hanssen R (2001) Radar interferometry: data interpretation and error analysis. Kluwer, Dordrecht, The Netherlands

He C, Xia G-S, Sun H (2006) An adaptive and iterative method of urban area extraction from SAR images. IEEE Geosci Remote Sens Lett 3(4):504–507

Hedman K, Wessel B, Stilla U (2005) A fusion strategy for extracted road networks from multi-aspect SAR images. In: Stilla U, Rottensteiner F, Hinz S (eds) CMRT05. International archives of photogrammetry and remote sensing 36(Part 3 W24), pp 185–190

Hedman K, Stilla U (2009) Feature fusion based on bayesian network theory for automatic road extraction. Chapter 3 of this book

Hellwich O, Mayer H (1996) Extraction line features from Synthetic Aperture Radar (SAR) scenes using a Markov random field model. IEEE International Conference on Image Processing (ICIP), pp 883–886

Henderson FM, Mogilski KA (1987) Urban land use separability as a function of radar polarization. Int J Remote Sens 8(3):441–448

Henderson FM, Xia Z-G (1997) SAR applications in human settlement detection, population estimation and urban land use pattern analysis: a status report. IEEE Trans Geosci Remote Sens 35(1):79–85

Henderson FM, Xia Z-G (1998) Radar applications in urban analysis, settlement detection and population analysis. In: Henderson FM, Lewis AJ (eds) Principles and applications of imaging radar, Chapter 15. Wiley, New York, pp 733–768

Hepner GF, Houshmand B, Kulikov I., Bryant N (1998) Investigation of the potential for the integration of AVIRIS and IFSAR for urban analysis. Photogramm Eng Remote Sens 64(8):813–820

Hinz S, Suchand S, Weihing D, Kurz F (2009) Traffic data collection with TerraSAR-X and performance evaluation. Chapter 4 of this book

Hoepfner KB (1999) Recovery of building structure from IFSAR-Derived elevation maps. Technical Report 99–16, Computer Science Department, University of Massachusetts, Amherst

Hong TD, Schowengerdt RA (2005) A robust technique for precise registration of radar and optical satellite images. Photogramm Eng Remote Sens 71(5):585–593

Hooper A, Zebker H, Segall P, Kampes B (2004) A new method for measuring deformation on volcanoes and other natural terrains using InSAR persistent scatterers. Geophys Res Lett 31(23):611–615

Huertas A, Kim Z, Nevatia R (2000) Multisensor integration for building modeling. IEEE Proceedings of Conference on Computer Vision and Pattern Recognition, pp 203–210

Inglada J, Giros A (2004) On the possibility of automatic multisensor image registration. IEEE Trans Geosci Remote Sens 42(10):2104–2120
Jiang X, Bunke H (1994) Fast segmentation of range images into planar regions by scan line grouping. Mach Vis Appl 7(2):115–122
Jaynes CO, Stolle FR, Schultz H, Collins RT, Hanson AR, Riseman EM (1996) Three-dimensional grouping and information fusion for site modeling from aerial images. ARPA Image Understanding Workshop, Morgan Kaufmann, New Orleans, LA
Kirscht M, Rinke C (1998) 3D-reconstruction of buildings and vegetation from Synthetic Aperture Radar (SAR) images. Proceedings of IAPR Workshop on Machine Vision Applications, pp 228–231
Klare J, Weiss M, Peters O, Brenner A, Ender J (2006) ARTINO: a new high-resolution 3d imaging radar system on an autonomous airborne platform. Geoscience and Remote Sensing Symposium, pp 3842–3845
Klonus S, Rosso P, Ehlers M (2008) Image fusion of high-resolution TerraSAR-X and multispectral electro-optical data for improved spatial resolution. Remote sensing – new challenges of high resolution. Proceedings of the EARSeL Joint Workshop, E-Proceedings
Levine MD, Shaheen SI (1981) A modular computer vision system for picture segmentation and interpretation. Trans Pattern Anal Mach Intell 3(5):540–554
Leberl F (1990) Radargrammetric image processing. Artech House, Boston, MA
Lee JS (1980) Digital image enhancements and noise filtering by use of local statistics. IEEE Trans Pattern Anal Mach Intell 2:165–168
Lee JS, Grunes MR, Ainsworth TL, Du L, Schuler DL, Cloude SR (1999) Unsupervised classification of polarimetric SAR images by applying target decomposition and complex Wishart distribution. IEEE Trans Geosci Remote Sens 37(5):2249–2258
Lee JS, Hoppel KW, Mango SM, Miller AR (1994) Intensity and phase statistics of multilook polarimetric and interferometric SAR imagery. IEEE Trans Geosci Remote Sens 32(5): 1017–1028.
Liao MS, Zhang L, Balz T (2009) Post-earthquake landslide detection and early detection of landslide prone areas using SAR. Proceedings of 5th GRSS/ISPRS Joint Workshop on Remote Sensing and Data Fusion on Urban Area. URBAN 2009, CD, 5 p
Lisini G, Tison C, Tupin F, Gamba P (2006) Feature fusion to improve road network extraction in high-resolution SAR images. IEEE Geosci Remote Sens Lett 3(2):217–221
Lopes A, Nezry E, Touzi R, Laur H (1993) Structure detection and statistical adaptive speckle filtering in SAR images. Int J Remote Sens 1366–5901 14(9):1735–1758
Lu D, Weng Q (2007) A survey of image classification methods and techniques for improving classification performance. Int J Remote Sens 28(5):823–870
Luckman A, Grey W (2003) Urban building height variance from multibaseline ERS coherence. IEEE Trans Geosci Remote Sens 41(9):2022–2025
Massonnet D, Rossi M, Carmona C, Adragna F, Peltzer G, Feigl K, Rabaute T (1993) The displacement field of the landers earthquake mapped by radar interferometry. Nature 364(8):138–142
Meyer F, Hinz S, Laika, A, Weihing D, Bamler R (2006) Performance analysis of the TerraSAR-X traffic monitoring concept. ISPRS J Photogramm Remote Sens 61(3–4):225–242
Michaelsen E, Soergel U, Thoennessen U (2005) Potential of building extraction from multi-aspect high-resolution amplitude SAR data. In: Stilla U, Rottensteiner F, Hinz S (eds) CMRT05, IAPRS 2005 XXXVI(Part 3/W24), pp 149–154
Michaelsen E, Soergel U, Thoennessen U (2006) Perceptual grouping for automatic detection of man-made structures in high-resolution SAR data. Pattern Recognit Lett (Elsevier B.V., Special Issue Pattern Recognit Remote Sens) 27(4):218–225
Moreira, A (2000) Radar mit synthetischer Apertur – Grundlagen und Signalverarbeitung. Habilitation. University of Karlsruhe, Germany
Piater JH, Riseman EM (1996) Finding planar regions in 3-D grid point data. Technical Report UM-CS-1996–047, University of Massachusetts, Amherst, Computer Science

Quartulli M, Datcu M (2004) Stochastic geometrical modeling for built-up area understanding from a single SAR intensity image with meter resolution. IEEE Trans Geosci Remote Sens 42(9):1996–2003

Quartulli M, Datcu M (2003) Information fusion for scene understanding from interferometric SAR data in urban environments. IEEE Trans Geosci Remote Sens 41(9):1976–1985

Rabus B, Eineder M, Roth A, Bamler R (2003) The shuttle radar topography mission – a new class of digital elevation models acquired by spaceborne radar. ISPRS J Photogramm Remote Sens 57(4):241–262

Reigber A, Moreira A (2000) First demonstration of airborne SAR tomography using multibaseline L-band data. IEEE Trans Geosci Remote Sens 38(5, Part 1):2142–2152

Reigber A, Jäger M, He W, Ferro-Famil L, Hellwich O (2007) Detection and classification of urban structures based on high-resolution SAR imagery. Proceedings of 4th GRSS/ISPRS Joint Workshop on Remote Sensing and Data Fusion on Urban Area. URBAN 2007, CD, 6 p

Rosen PA, Hensley S, Joughin IR, Li FK, Madsen SN, Rodríguez E, Goldstein RM (2000) Synthetic aperture radar interferometry. Proc IEEE 88(3):333–382

Schneider RZ, Papathanassiou KP, Hajnsek I, Moreira A (2006) Polarimetric and interferometric characterization of coherent scatterers in urban areas. IEEE Trans Geosci Remote Sens 44(4):971–984

Schreier G (1993) Geometrical properties of SAR images. In: Schreier G (ed) SAR geocoding: data and systems. Karlsruhe, Wichmann, pp 103–134

Sauer S, Ferro-Famil L, Reigber A, Pottier E (2009) Polarimetric dual-baseline InSAR building height estimation at L-band. IEEE Geosci Remote Sens Lett 6(3):408–412

Simard M, Saatchi S, DeGrandi G (2000) The use of decision tree and multiscale texture for classification of JERS-1 SAR data over tropical forest. IEEE Trans Geosci Remote Sens 38(5):2310–2321

Simonetto E, Oriot H, Garello R (2005) Rectangular building extraction from stereoscopic airborne radar images. IEEE Trans Geosci Remote Sens 43(10):2386–2395

Smits PC, Dellepiane SG, Schowengerdt RA (1999) Quality assessment of image classification algorithms for land-cover mapping: a review and proposal for a cost-based approach. Int J Remote Sens 20:1461–1486

Soergel U, Michaelsen E, Thiele A, Cadario E, Thoennessen U (2009) Stereo Analysis of high-resolution SAR images for building height estimation in case of orthogonal aspect directions. ISPRS J Photogramm Remote Sens, Elsevier B.V. 64(5):490–500

Soergel U, Schulz K, Thoennessen U, Stilla U (2005) Integration of 3d data in SAR mission planning and image interpretation in urban areas. Info Fus (Elsevier B.V.) 6(4):301–310

Soergel U, Thoennessen U, Brenner A, Stilla U (2006) High-resolution SAR data: new opportunities and challenges for the analysis of urban areas. IEE Proc Radar Sonar Navig 153(3): 294–300

Soergel U, Thoennessen U, Stilla U (2003a) Reconstruction of buildings from interferometric SAR data of built-up areas. In: Ebner H, Heipke C, Mayer H, Pakzad K (eds) Photogrammetric Image Analysis PIA'03, international archives of photogrammetry and remote sensing 34(Part 3/W8):59–64

Soergel U, Thoennessen U, Stilla U (2003b) Iterative building reconstruction in multi-aspect InSAR data. In: Maas HG, Vosselman G, Streilein A (eds) 3-D Reconstruction from airborne laserscanner and InSAR data, IntArchPhRS 34(Part 3/W13):186–192

Solberg AHS, Taxt T, Jain AK (1996) A Markov random field model for classification of multi-source satellite imagery. IEEE Trans Geosci Remote Sens 34(1):100–112

StaMPS (2009) http://enterprise.lr.tudelft.nl/∼ahooper/stamps/index.html

Steger C (1998) An unbiased detector of curvilinear structures. IEEE Trans Pattern Anal Mach Intell 20:113–125

Strozzi T, Dammert PBG, Wegmuller U, Martinez J-M, Askne JIH, Beaudoin A, Hallikainen NT (2000) Landuse mapping with ERS SAR interferometry. IEEE Trans Geosci Remote Sens 38(2):766–775

Takeuchi S, Suga Y, Yonezawa C, Chen CH (2000) Detection of urban disaster using InSAR – a case study for the 1999 great Taiwan earthquake. Proceedings of IGARSS, on CD
Thiele A, Cadario E, Schulz K, Thoennessen U, Soergel U (2007) Building recognition from multi-aspect high-resolution InSAR data in urban area. IEEE Trans Geosci Remote Sens 45(11):3583–3593
Thiele A, Cadario E, Schulz K, Soergel U (2009a) Analysis of gable-roofed building signatures in multiaspect InSAR data. IEEE Geoscience and Remote Sensing Letters, Digital Object Identifier: 10.1109/LGRS.2009.2023476, online available
Thiele A, Wegner J, Soergel U (2009b) Building reconstruction from multi-aspect InSAR data. Chapter 8 of this book
Tison C, Nicolas JM, Tupin F, Maitre H (2004) A new statistical model for Markovian classification of urban areas in high-resolution SAR images. IEEE Trans Geosci Remote Sens 42(10): 2046–2057
Tison C, Tupin F, Maitre H (2007) A fusion scheme for joint retrieval of urban height map and classification from high-resolution interferometric SAR images. IEEE Trans Geosci Remote Sens 45(2):496–505
Tison C, Tupin F (2009) Estimation of urban DSM from mono-aspect InSAR images. Chapter 7 of this book
Tupin F (2009) Fusion of optical and SAR images. Chapter 6 of this book
Tupin F (2000) Radar cross-views for road detection in dense urban areas. Proceedings of the European Conference on Synthetic Aperture Radar, pp 617–620
Tupin F, Roux M (2003) Detection of building outlines based on the fusion of SAR and optical features. ISPRS J Photogramm Remote Sens 58:71–82
Tupin F, Roux M (2005) Markov random field on region adjacency graph for the fusion of SAR and opitcal data in radargrammetric applications. IEEE Trans Geosci Remote Sens 42(8): 1920–1928
Tupin F, Maitre H, Mangin J-F, Nicolas J-M, Pechersky E (1998) Detection of linear features in SAR images: application to road network extraction. IEEE Trans Geosci Remote Sens 36(2):434–453
Tupin F, Houshmand B, Datcu M (2002) Road detection in dense urban areas using SAR imagery and the usefulness of multiple views. IEEE Trans Geosci Remote Sens 40(11):2405–2414
Touzi R, Lopes A, Bousquet P (1988) A statistical and geometrical edge detector for SAR images. IEEE Trans Geosci Remote Sens 26(6):764–773
Toutin T, Gray L (2000) State-of-the-art of elevation extraction from satellite SAR data. ISPRS J Photogramm Remote Sens 55(1):13–33
Tzeng YC, Chen KS (1998) A fuzzy neural network to SAR image classification. IEEE Trans Geosci Remote Sens 36(11):301–307
Vincent L, Soille P (1991) Watersheds in digital spaces: an efficient algorithm based on immersion simulations. IEEE Trans Pattern Anal Mach Intell 13(6):583–598
Voigt S, Riedlinger T, Reinartz P, Künzer C, Kiefl R, Kemper T, Mehl H (2005) Experience and perspective of providing satellite based crisis information, emergency mapping & disaster monitoring information to decision makers and relief workers. In: van Oosterom P, Zlatanova S, Fendel E (eds) Geoinformation for disaster management. Springer, Berlin, pp 519–531
Walessa M, Datcu M (2000) Model-based despeckling and information extraction from SAR images. IEEE Trans Geosci Remote Sens 38(5):2258–2269
Waske B, Benediktsson JA (2007) Fusion of support vector machines for classification of multisensor data. IEEE Trans Geosci Remote Sens 45(12):3858–3866
Waske B, Van der Linden S (2008) Classifying multilevel imagery from SAR and optical sensors by decision fusion. IEEE Trans Geosci Remote Sens 46(5):1457–1466
Wegner JD, Soergel U (2008) Bridge height estimation from combined high-resolution optical and SAR imagery. Int Arch Photogramm Remote Sens Spat Info Sci 37(Part B7–3):1071–1076
Wegner JD, Thiele A, Soergel U (2009a) Building extraction in urban scenes from high-resolution InSAR data and optical imagery. Proceedings of 5th GRSS/ISPRS Joint Workshop on Remote Sensing and Data Fusion on Urban Area. URBAN 2009, 6 p, CD

Wegner JD, Auer S, Thiele A, Soergel U (2009b) Analysis of urban areas combining high-resolution optical and SAR imagery. 29th EARSeL Symposium 8 p, CD

Wessel B, Wiedemann C, Hellwich O, Arndt WC (2002) Evaluation of automatic road extraction results from SAR imagery. Int Arch Photogramm Remote Sens Spat Info Sci 34(Part 4/IV):786–791

Wessel B (2004) Road network extraction from SAR imagery supported by context information. Int Arch Photogramm Remote Sens 35(Part 3B):360–365

Xiao R, Lesher C, Wilson B (1998) Building detection and localization using a fusion of interferometric synthetic aperture radar and multispectral images. ARPA Image Understanding Workshop, Morgan Kaufmann, pp 583–588

Xu F, Jin YQ (2007) Automatic reconstruction of building objects from multiaspect meter-resolution SAR images. IEEE Trans Geosci Remote Sens 45(7):2336–2353

Zebker HA, Goldstein RM (1986). Topographic mapping from interferometric synthetic aperture radar observations. J Geophys Res 91:4993–4999

Zhu X, Adam N, Bamler R (2008) First demonstration of spaceborne high-resolution SAR tomography in urban environment using TerraSAR-X data. CEOS SAR Workshop 2008, CD

Chapter 2
Rapid Mapping Using Airborne and Satellite SAR Images

Fabio Dell'Acqua and Paolo Gamba

2.1 Introduction

Historically, Synthetic Aperture Radar (SAR) data was made available later than optical data for the purpose of land cover classification (Landsat Legacy Project Website, http://library01.gsfc.nasa.gov/landsat/; NASA Jet Propulsion Laboratory: Missions, http://jpl.nasa.gov/missions/missiondetails.cfm?mission=Seasat); in more recent times, the milestone of spaceborne meter resolution was reached by multispectral optical data first (Ikonos; GEOEye Imagery Sources, http://www.geoeye.com/CorpSite/products/imagery-sources/Default.aspx#ikonos), followed a few years later by radar data (COSMO/SkyMed [Caltagirone et al. 2001] and TerraSAR-X [Werninghaus et al. 2004]). As a consequence, more experience has been accumulated on the extraction of cartographic features from optical rather than SAR data, although in some cases radar data is highly recommendable because of frequent cloud cover (Attema et al. 1998) or because the information of interest is better visible at the microwave frequencies rather than at the optical ones (Kurosu et al. 1995).

Unfortunately, though, SAR data cannot provide complete scene information because radar systems operate on a single band of acquisition, a limitation which is partly compensated, and only in specific cases, by their increasingly available polarimetric capabilities (Treitz et al. 1996).

Nonetheless, the launch of new generation, Very High Resolution (VHR) SAR satellites, with the consequent perspective availability of repeated acquisitions over the entire Earth, do push towards the definition of novel methodologies for exploiting these data even for extraction of cartographic features. This does not mean that a replacement is in progress over the traditional way of cartographic mapping, based on airborne and, more recently, spaceborne sensors in the optical and near infrared regions. There is instead the possibility for VHR SAR to provide basic and complementary information.

F. Dell'Acqua (✉) and P. Gamba
Department of Electronics, University of Pavia. Via Ferrata, 1 - I-27100 Pavia
e-mail: fabio.dellacqua@unipv.it; paolo.gamba@unipv.it

U. Soergel (ed.), *Radar Remote Sensing of Urban Areas*, Remote Sensing and Digital Image Processing 15, DOI 10.1007/978-90-481-3751-0_2,

It has been indeed proven that SAR data is capable of identifying some of the features reputed to be among the most complex to be detected in remotely sensed images (e.g. buildings, bridges, ships, and other complex-shaped objects); semi-automatic procedures are already available providing outputs at a commercially acceptable level. Some examples include the definition of urban extent (He et al. 2006), discrimination of water bodies (Hall et al. 2005), vegetation monitoring (Askne et al. 2003), road element extraction (Lisini et al. 2006), entire road network depiction (Bentabet et al. 2003), and so on. Moreover, the interferometric capabilities of SAR, where available, allow the exploitation of terrain and object height to improve the cartographic mapping process (Gamba and Houshmand 1999).

In terms of cost and possibility of covering large areas, SAR is indeed widely exploited for three-dimensional characterization of the landscape. This can be used to characterize road networks (Gamba and Houshmand 1999), buildings (Stilla et al. 2003) and, more generally, to discriminate between different kinds of cartographic features.

The main obstacle on the way of these processes towards real-world, commercial applications is probably their specialisation on just one among the possible features of cartographic interest. Although a number of approaches intended for SAR image analysis have appeared in technical literature, no single one is expected to cover all the spatial and spectral features needed for a complete process of cartographic feature extraction starting from scratch.

Road extraction, for instance, is treated in many papers (Mena 2003), but this is seldom connected to urban area extraction and the use of different strategies according to the urban or non-urban areas (see Tupin et al. 2002 or Wessel 2004). The same holds for the reverse approach.

In the following it will be shown an example of how an effective procedure can be assembled starting from some of the above cited or similar algorithms, and thus exploiting as much as possible the full range of information available in a SAR scene acquired at high spatial resolution.

The final goal of the research in progress is a comprehensive approach to SAR scene characterization, attentive to the multiple elements in the same scene. It is thus based on multiple feature extraction and various combination/fusion algorithms. Through an analysis of many different details of the scene, either spectral or spatial, a quick yet sufficiently accurate interpretation of a SAR scene can be obtained, useful for focusing further extraction work or as a first step in more precise feature extraction steps.

The stress in this chapter is placed onto the so-called "rapid mapping" which summarizes the above concept: a fast procedure to collect basic information on the contents of a SAR scene, useful in those cases where the limited amount of information needed does not justify the use of complex, computationally heavy procedures or algorithms.

2.2 An Example Procedure

We illustrate the concept of rapid mapping and the choices and technical solutions behind it by making reference to a procedure proposed by the authors of this chapter, and described more in detail in (Dell'Acqua et al. 2008).

In most cases scene interpretation starts from a segmentation of the image based on some sort of knowledge embedded into the algorithms and then proceeds to analyse each single segment more in detail, possibly further partitioning it. The reference procedure also uses this approach which is commonly termed "top-down", meaning that the interpretation starts from the "top" level objects (biggest objects, largest partitions, widest categories) successively moving "down" (to smaller objects, ...) better specifying and possibly also perfecting the recognition and analysis of the objects found. The procedure in (Dell'Acqua et al. 2008) features also contemporary exploitation of spatial (texture analysis, extraction and combination of linear elements) and radiometric (mostly local intensity) features.

The general information flow is visible in Fig. 2.1, while the proposed standard structure is presented in Fig. 2.2. The next subchapters describe how the basic information can be extracted from a given high-resolution SAR image.

2.2.1 *Pre-processing of the SAR Images*

The example procedure, as shown in Fig. 2.2, is performed stepwise starting from the easiest-to-extract land cover moving on to categories requiring more complicated processing or obtainable by exclusion of the former –already assigned- land covers. It is worthwhile mentioning that the entire procedure can be realized relying also on other algorithms than those cited in the present chapter, provided that those can guarantee a comparable accuracy and quality of results. The procedure is not particularly demanding in terms of the data characteristics: input data are single-polarisation, single-date amplitude images. Results may benefit from fusion with multi-polarisation and/or multitemporal data; research is underway to fuse information coming from more images or more polarisation, but results are not yet assessed and not worthwhile being presented here.

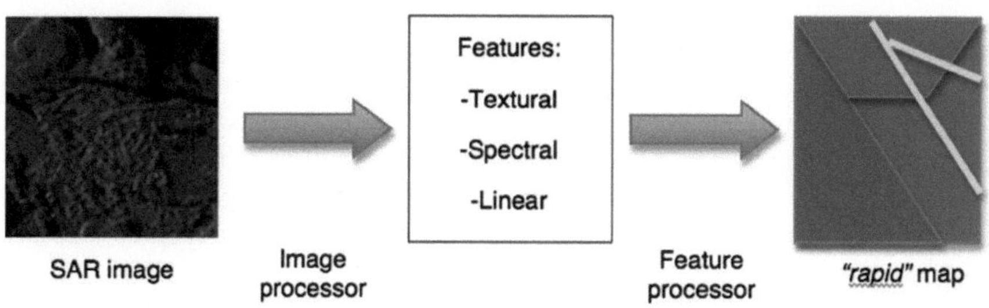

Fig. 2.1 The information flow

Fig. 2.2 Processing steps

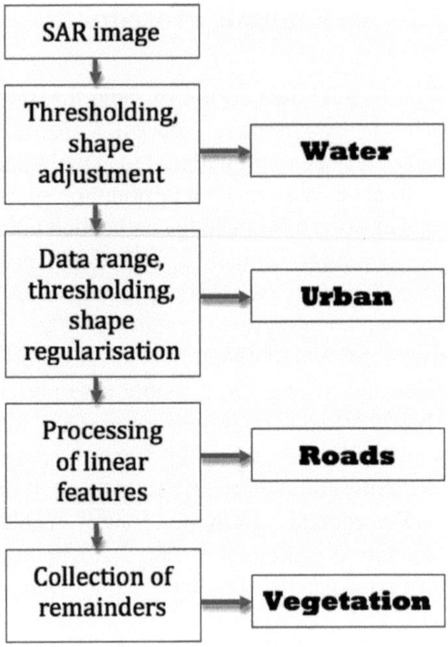

One of the first steps is speckle removal, which is in general useful but has shown to be not really indispensable for rapid mapping. Probably thanks to the extraction of geometric features, which is nearly independent from the single pixel value, experiments have shown indeed that even when the speckle-filtering step is completely skipped, the worsening in the quality of final results is not significant. In our experiments we have used the classical Lee filter and performed filtering on all the images, as the tiny saving in computation time does not justify working on unfiltered images.

Let us now consider the various land cover/element extraction steps in the respective order: water bodies, human settlements, road network, vegetated areas.

2.2.2 *Extraction of Water Bodies*

It is commonly acknowledged that internal water bodies are one of the easiest land covers to detect in SAR images, as calm water surfaces cause mirror-like reflection to reflect the incident electromagnetic wave away from the sensor. This results in a particularly low backscatter (Hess et al. 1990; Horritt et al. 2003) which in turn– thanks to the noise being multiplicative- translates into a homogeneous, nearly featureless and textureless (Ulaby et al. 1986) region in the water-covered area in a SAR image.

Moreover, inner water bodies cover areas several pixel wide and form shapes which can be considered smooth and regular even at high spatial resolution. Therefore a thresholding of the image can be used and followed by a procedure like the one described in (Gamba et al. 2007). There, the reasoning behind regularization is applied to building, but the same considerations may be easily found to be applicable also to water bodies. The procedure is split into two steps, the first one devoted to better delineate edges and separate elements, while the second step aims instead at filling possible small holes and gaps inside an object, generally resulting from local classification errors. The reader is referred to (Gamba et al. 2007) for more details on the procedure. Alternative procedures for regularisation may be considered such as (Heremans et al. 2005), based on an adjustment of a procedure (Chesnaud et al. 1999) conceived for general object extraction in images. As mentioned in the introduction, it is not crucial to choose one or the other method for extraction of the single object as far as a reasonable accuracy can be achieved, even more so with easy-to-extract water bodies.

2.2.3 Extraction of Human Settlements

Several methods have been proposed so far for detecting human settlements in radar remotely sensed images, most of them being efficient in detecting the sheer presence of human settlements but generally showing poor performances in precisely delineating the extent of the urban areas (Gouinaud and Tupin 1996). Methods relying on a priori knowledge (Yu et al. 1999) to improve classification are not usable for rapid mapping purposes and one should rather attempt to make the extraction more precise by exploiting textural information (Duskunovic et al. 2000; Dekker 2003; Dell'Acqua and Gamba 2003) and even spatial proximity information based, for example, on Markov Random Fields (Tison et al. 2004). An important issue is however the scale of the texture considered, and this is becoming especially relevant with the increasing availability of VHR SAR images. Such issue is discussed in (Dell'acqua et al. 2006), where an approach combining co-occurrence matrix and semivariogram analysis was tested for mapping urban density in satellite SAR data. Results show that, in terms of final classification accuracy, the joint use of those two features to optimize the texture window size can be nearly as effective as an exhaustive search. A methodology is thus introduced to compute the co-occurrence features with a window consistent with the local scale, provided by the semivariogram analysis. Orientation is the second important issue after scale, and for a discussion of texture orientation the reader is referred to (Pesaresi et al. 2007) where optical images are considered but some geometric considerations may be extended to SAR images as well.

We will illustrate here the approach proposed in (Dell'Acqua et al. 2008) which relies on a simple, isotropic occurrence measure, namely data range or the difference between the maximum and the minimum pixel intensity value in the considered local window. Three steps compose the procedure.

In the first step a pre-scaling of the image to a pixel size of 5 m is performed, according to the considerations expressed in (Pesaresi et al. 2007), and a 5×5 pixel window is used to compute data range, resulting in a $25 \times 25\,\mathrm{m}^2$ area to be analysed for the local texture measure computation.

The second step consists of a threshold operation over the computed occurrence map. The threshold value is generally determined heuristically, and a value of 100 was found to provide acceptable results in most cases after a radiometric rescaling of the texture image values to the range 0–255 has been performed. Criteria for a suitable, automatic choice of the threshold value are under investigation. This step is the one where previously performed speckle filtering can make some difference to the accuracy of the results, although the next step is intended also to suppress pixel-wise errors due to local speckle peaks.

The third and last step consists of spatial homogenisation in the form of morphological closing. Again, based on the considerations in (Pesaresi et al. 2007), a size of 5×5 pixels has been used for the morphological operator, which is applied twice as in our experience this produces better results. More refined techniques can be found in (Soille and Pesaresi 2002); however, a reasonable balance between accuracy and complexity should be made before using more sophisticated algorithms where rapid mapping is the context at hand.

A typical, critical pattern for the algorithm outlined above consists of tall trees when they are sufficiently sparse to cause isolated reflection peaks. Some improvement can however be obtained by exploiting relationship with formerly extracted water bodies: it is quite uncommon to find small clusters of urban pixels at a 5 m scale aside a water body, and a buffer area around this latter can be cleared of all urban area pixels as assigned by the texture thresholding step. A further refinement may rely on the exclusion of strongly textured areas, which are likely to be caused by sparse trees, although this implies computation of other texture measures and thus a heavier computational burden. An active research line in this direction is to exploit local extraction of linear features in very high-resolution images to better distinguish urban areas, characterised by man-made features, which are expected to contain several straight lines visible in the images (Aldrighi et al. 2009).

2.2.4 Extraction of the Road Network

The next step consists of extracting another very important piece of information for mapping purposes, that is the main road network. In order to differentiate the problem between two very different contexts, the road network is extracted in non-urban areas first and then within urban areas.

In non-urban areas, that is outside the areas which have been assigned to the "urban" class in the previous steps, a trivial simplification consists of discarding all the areas recognised as belonging to other extracted land-cover classes, that is water and tall trees. In the remaining area, many approaches can be used for road extraction. This problem has been indeed considered for quite a long time

(Bajcsy and Tavakoli 1976) and many different algorithms have been proposed over the years. Naturally, at an initial stage the work concentrated on aerial, optical images. Fischler et al. (1981) used two types of detectors: one optimised against false alarms, and another optimised against misdetections, and combined their responses using dynamic programming. McKeown and Denlinger (McKeown and Denlinger 1988) proposed a road-tracking algorithm for aerial images, which relied on road-texture correlation and road-edge following.

At the time when satellite SAR images started becoming widely available, methods focussed on this type of data made an appearance. Due to the initially coarse resolution of the images, most of such methods exploit a local criterion evaluating radiometric values on some small neighbourhood of a target pixel to start discriminating lines from background, possibly relying on classical edge extractors such as Canny (1986). These segments are eventually connected into a network by introducing larger-scale knowledge about the structures to be detected (Fischler et al. 1981). In an attempt to generalise the approach (Chanussot et al. 1999) extracted roads by combining results from different edge detectors in a fuzzy framework.

Noticeably these approaches refer to the geometrical or structural context of a road, undervaluing its radiometric properties as a region. These are instead considered in Dell'Acqua and Gamba (2001) and Dell'Acqua et al. (2002), where the authors propose clustering of pixels that a classifier has assigned to the "road" class. In the cited papers the authors try and discriminate roads by grouping "road" pixels into linear or curvilinear segments using modified Hough transforms or dynamic programming. The dual approach is proposed in (Borghys et al. 2000), where segmentation is used to skip uniform areas and concentrate the extraction of edges where statistical homogeneity is lower.

Tupin et al. (1998), proposed an automatic extraction methodology for the main axes of road networks. They presented two local line detectors and a method for fusing the information obtained from these detectors to obtain segments. The real roads were identified among the segment candidates by defining a Markov random field for the set of segments. Jeon et al. (1999), proposed an automatic road detection algorithm for radar satellite images. They presented a map-based method based on a coarse-to-fine, two-step matching process. The roads were finally detected by applying snakes to the potential field, which was constructed by considering the characteristics and the structures of roads. As opposed to simple straight-line element detection, in (Jeon et al. 2002), the authors propose extraction of curvilinear structures associated to the use of a genetic algorithm to select and group best candidates in the attempt to optimise the overall accuracy of the extracted road network.

With the increasing availability of new generation, very-high-resolution spaceborne SAR data, multiresolution approach are becoming a sensible choice. In (Lisini et al. 2006), the authors propose a method for road network detection from high-resolution SAR data that includes a data fusion procedure in a multiresolution framework. It takes into account the information available by both a line detector and a classification algorithm to improve the road segment selection and the road network reconstruction. This could be used as a support for rapid mapping over HR spaceborne SAR images.

To complement road extraction in rural areas, extraction of urban road network is the next step. In this environment the scale of relevant objects is much smaller and thus the meter-resolution becomes a requirement. Since in VHR SAR images the roads no longer appear as single image edges but rather as dark, elongated areas with side edges generally brighter than the inside, the strategy needs to be slightly changed. Therefore, one may detect roads by searching pairs of parallel edges or dark, elongated, homogeneous areas. What appears to be a promising approach is fusion of results from different detectors, optimised for the different geometric and radiometric characteristics of the road elements, as proposed in (Dell'Acqua et al. 2003).

After the road elements have been detected, a multiscale-feature fusion framework followed by a final alignment (Dell'Acqua et al. 2005), can be made to follow in order to remove false positives and discard repeated, slightly different detection of the same road element.

Finally, if the focus is placed on the extraction of the road network rather than single roads, geometric features contained in the scene (such as junctions, as shown in Negri et al. (2006)) can be used to infer the presence of missed roads and try and complete the extracted road network.

As shown in (Dell'Acqua et al. 2008), a further refinement of the results is possible when SAR and InSAR data are jointly available, this latter producing a DSM (Digital Surface Model) of the observed area. A simple 2-dimensional low-pass filtering of the DSM is used to approximate a DTM (Digital Terrain Model). This allows identifying as buildings the clusters of pixels stemming above the estimated local ground level. The complementary pixel set are potentially parks, places, playgrounds or similar, and roads (urban environment is implied). The first categories can be discriminated thanks to their aspect ratio, expected to be very different with respect to roads. The remaining ground-level pixels are likely to be road pixels, and they may be reused as clues for better road recognition in a fusion process with the initially extracted road network.

2.2.5 *Extraction of Vegetated Areas*

Assuming that only the limited set of classes mentioned at the beginning is to be discriminated (water, human settlements, roads vegetation) for "rapid mapping" to be performed, once all the other classes have been extracted, the remaining pixels should belong to the vegetation class. Within vegetation it seems quite sensible to try and distinguish trees and woods from low-rise vegetation.

Two approaches are possible to such discrimination, and integration of the results from both approaches seems even more promising (Dell'Acqua et al. 2008). The first approach relies on texture information: woods show a remarkably evident texture, not found in most of the other vegetated land covers. In particular, the co-occurrence measure "correlation" is the best option for discriminating woods and other taller cultures from the background, since this measure shows significantly larger values on big windows (30×30 m was used in our experiments) and long displacements (around 10 m).

The second approach involves availability and use of 3D data: a difference operation between the DSM and the DTM will highlight the areas where vegetation is present. Please recall the underlying assumption that urban areas have already been detected and thus removed from the areas considered for vegetation detection; buildings, which may generate similar signatures in the DSM-DTM difference, should have already been masked away at this stage.

Even better results can be achieved by combining results from both approaches. A logical AND operation between the two results has been found by experiment to be advantageous in terms of reduction in false positives vs. increase in missed woods.

2.2.6 Other Scene Elements

As a final remark, we may note that a limited amount of further processing may lead to detection and location of further scene elements not directly addressed in the previous subchapters. Examples are represented by intersections between roads and water bodies, which can be identified as bridges with a good degree of confidence (actually, to a combinations of the degree of confidence with which each supporting element was detected); or lake islands, that is vegetated areas completely surrounded by a "water" region. This issue is not however discussed in depth here as the focus of this chapter is on the extraction of information from the SAR image itself rather than on further stages of processing which may lead to the determination of derived pieces of information.

2.3 Examples on Real Data

To illustrate the usefulness of rapid mapping we will refer to a typical application, that is mapping in the context of disaster management, currently performed by institutions like the International Charter on Space and Major Disasters, SERTIT, UNOOSA and others with methods which imply a massive amount of labour by human experts; the procedures may benefit from the support of tailored tools enlarging the fraction of operations required to produce disaster maps. The Sichuan, China earthquake happened on the 12th of May, 2008, and the extensive rescue operations following this tragic event, proved the value of high-resolution optical and radar remote sensing during the emergency response. While optical data provide a fast and simple way to value "at glance" the damage level, radar sensors have showcased their ability to deliver images independent of weather conditions –which were quite poor at that time in the stricken area- and of time of the day, and demonstrated that in principle they can represent a mean to obtain an up-to-date situation map in the immediate aftermath of an event, which is precious information for intervention planning.

Immediately after the Sichuan earthquake our group did activate immediately two mechanisms to collect data:

- The Italian Civil Protection Department (Dipartimento della Protezione Civile or DPC) was activated to bring help and support to the stricken population; in this framework the European Centre for Training and Research in Earthquake Engineering (EUCENTRE), a foundation of the University of Pavia, as an "expert centre" of DPC was enabled to access COSMO/SkyMed (C/S) data acquired over the affected area
- Our research group is entitled to apply for TerraSAR-X (TSX) data for scientific use following the acceptance of a project proposal connected to urban area mapping submitted in response to a DLR AO

Both the C/S and TSX data covered quite large areas, on the order of ten times 10 km. In order to limit the processing times and avoid dispersing the analysis efforts, the images were cropped to selected subareas. Since the focus of this work is on mapping of significant element rather than damage mapping, in the following we will concentrate only on areas, which reported slight damage or no damage at all, namely:

- C/S sub-image: a village located on the outskirts of Chengdu, around 30°33′ 17.14″N, 104°14′0.18″E; in this area almost no damage was reported. An urban area including a number of wide, well-visible main urban roads aligned to two principal, perpendicular directions, almost no secondary roads. The urban area is surrounded by countryside with low-rise vegetation crossed by a few rural, connecting roads.
- TSX sub-image: Luojiang, no damage, some water surface (TSX) 31°18′27.85″N, 104°29′46.73″E; in this area no damage was reported. The image contains the urban area of Luojiang, crossed by a large river, a big pond, and several urban roads with sparse directions.

These two areas, which reported almost no damage, were chosen to illustrate an application related to disaster mapping, that is "peacetime" extraction of fresh information aimed at keeping maps of the disaster-prone area constantly up-to-date. Other areas of the same images were instead used for damage mapping purposes (Dell'Acqua et al. 2009).

2.3.1 *The Chengdu Case*

As mentioned above, this urban area was chosen because of its large number of urban roads and indeed the rapid mapping procedure focussed on road extraction. The original COSMO/SkyMed image crop is shown in Fig. 2.3, courtesy of the Italian Space Agency and the Italian Civil Protection Department.

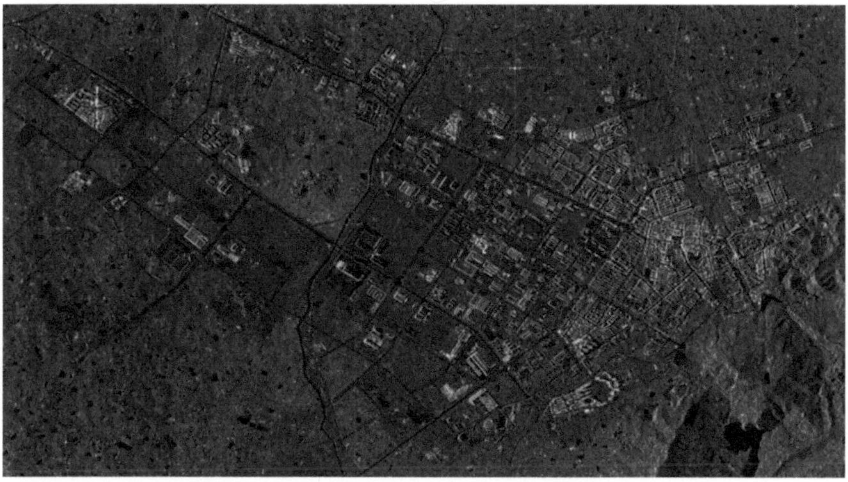

Fig. 2.3 COSMO/SkyMed image of Chengdu outskirts. Italian Space Agency

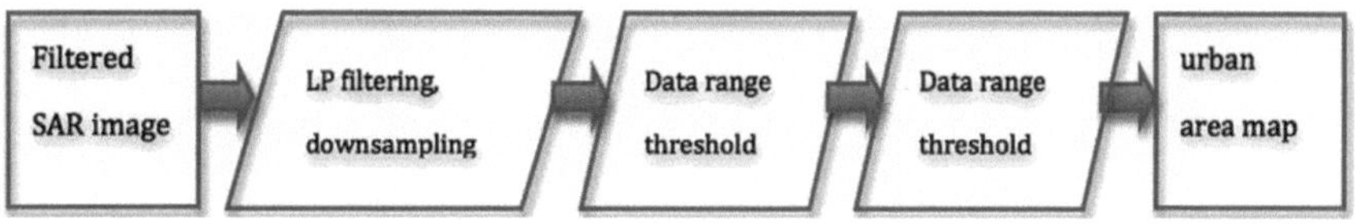

Fig. 2.4 The urban area extraction procedure

After despeckle filtering, the first processing step performed on this image was extraction of the urban area as described in (Dell'Acqua et al. 2008) and briefly outlined in the scheme in Fig. 2.4. Extraction results appear as a red overlay on the original SAR image in Fig. 2.5.

Looking carefully at the image one can note some facts:

- Some small blocks not connected with the main urban area are missed; note the cluster of buildings located at mid height on the far left side of the image. Although it is quite difficult to tell exactly the reason why the co-occurrence measure ended below the fixed threshold, a reasonable guess is the peculiar shape of the building results in smooth transition between double bounce and mirror reflection areas. This translates into a data range measure lower than commonly found in areas containing buildings.
- Remarkably, where urban areas are detected, their contours are defined accurately. Please refer to the bottom central area of Fig. 2.5, where the irregular boundaries of the urban area are followed with a good correctness.
- Thanks to the morphological closing operation, single buildings are not considered, although they locally cause an above-threshold value for the data range texture measure. An example is the strong reflector located on top centre of the

Fig. 2.5 The results of urban area extraction over Chengdu image

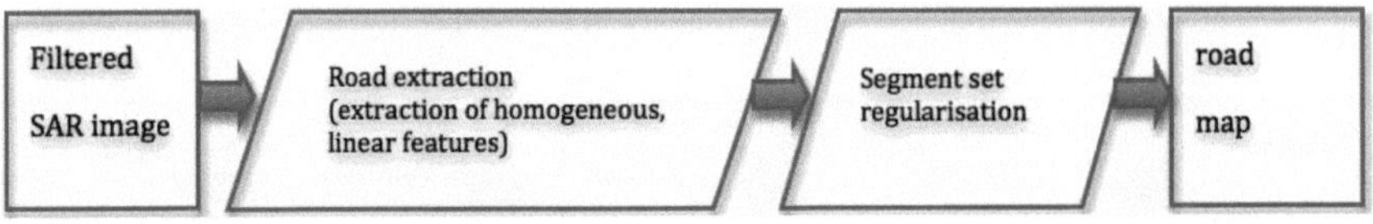

Fig. 2.6 The road extraction procedure

image, which causes the impulse response of the system to appear in the shape of a cross. By inspection of the Google Earth © image of the area, this appears to be a single building probably with a solid metal structure.

The next operation was extraction of the road network (Fig. 2.6), whose results are illustrated in Fig. 2.7. Again, this operation was performed following the procedure described in (Dell'Acqua et al. 2008), and briefly recalled in Fig. 2.6. The urban road system is basically extracted, no important roadway was missed; nonetheless some gaps are visible in a number of roads. The advantage in the context of rapid mapping is that the basic structure of the road network becomes available, including pieces of non linear roads, like for the bend in mid centre left of the image. On the other hand, though, in some cases narrow waterways like the trench flowing vertically across the image are detected as roads. Moreover, the gaps in detected roads prevent the use of the current version of the extractor in an emergency situation where a fast detection of uninterrupted communication lines is required. Nonetheless, the imposition of geometric constraints may be the correct step for completing the road network and keeping maps up-to date.

Fig. 2.7 Street network extracted from Chengdu image

2.3.2 The Luojiang Case

The second area selected for experimenting a rapid mapping procedure is the town of Luojiang, featuring less ordered urban roads, a large river crossed by a series of bridges, and two big ponds on top right. The built-up area is actually quite sparse, with clusters of buildings interspersed among bare areas. The corresponding crop of TerraSAR-X image (courtesy of DLR) is shown in Fig. 2.8.

The same procedure (Dell'Acqua et al. 2008) used for the COSMO/SkyMed image was re-applied to this image, and the results of the urban area extraction are shown in Fig. 2.10, left, as a red overlay on the gray-scale SAR image.

Noticeably, the classified urban area correctly reproduces the sparse pattern of blocks in the town, especially in the southernmost area (the images are geo-coded north upwards). Unfortunately, though, some missed buildings are reported in the eastern part of the image, probably due to the lower contrast found in that area.

In Fig. 2.10, left, a blue area represents the result of extracting water bodies in the same image according to the procedure reported in (Dell'Acqua et al. 2008) and briefly recalled in the scheme in Fig. 2.9. Generally speaking, water bodies are extracted conservatively, and several water pixels are missed. No false positives are reported, and a portion of the lower pond in upper right part of the image was lost. This is a consequence of a particularly strict selection of parameters for the extraction of water pixels, favouring correctness against completeness. Different selections of parameters may result in a more balanced situation between correctness and completeness, however discussing this issue is off the scope of this chapter. As a general consideration, the most appropriate strategy will depend on the purpose of the rapid mapping operation; for example, in the case of a flood where non-submerged pieces of land are sought to move people to a temporary haven,

Fig. 2.8 TerraSAR-X image of Luojiang, Sichuan, China

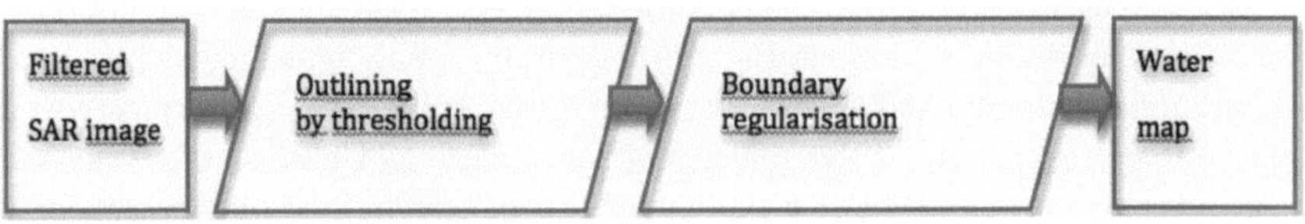

Fig. 2.9 Water land cover extraction procedure

completeness of class should be favoured (fewer pixels reliably classified as land rather than more and unsure), while in case of possible obstruction of a river due to a landslide, correctness is preferable.

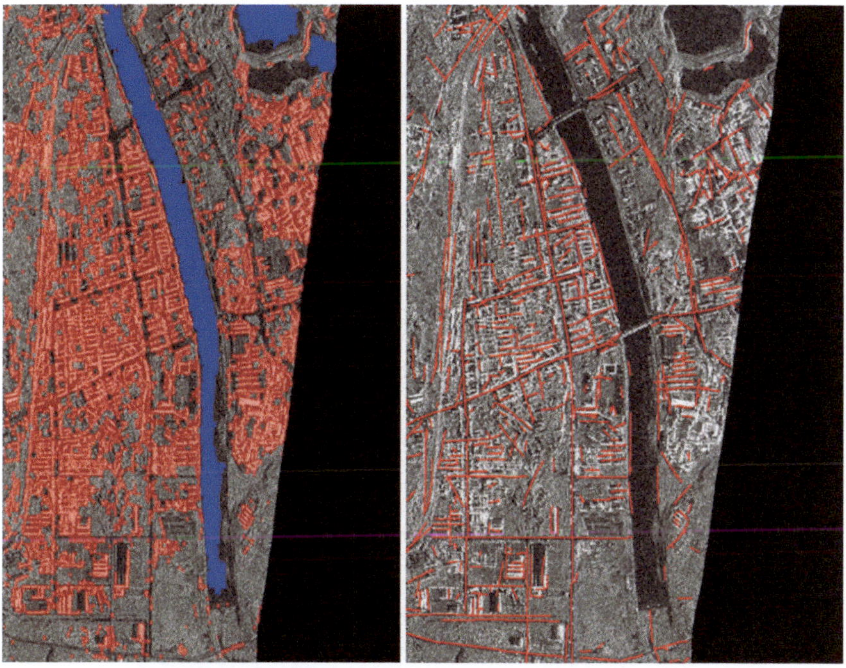

Fig. 2.10 *Left*: Water and urban area extraction; *Right*: Road network extraction on the Luojiang image

Figure 2.10, right, shows the results of road extraction applied to the Luojiang image. As can be seen in the figure, the extracted road network, overlaid in red over the gray-scale image, is quite complete. Just like for the C/S image, boundaries of waterways (in this case, the river) may be confused for roads, but their suppression is achievable by comparison with the water body map. In this sense the extraction of pieces of information from the image can improve the correctness of the following extraction steps, as mentioned in Section 2.2.

Again, a certain number of gaps are reported in the road network, although the overall structure is quite evident from the extraction result. Similar considerations to those made in the previous subchapter apply also to this extraction.

A final step may consist, as anticipated in Section 2.2, of vegetation extraction.

The easiest way to perform such extraction, considering the limited set of land cover classes assumed, is to define vegetation as the remainder of the image once all the other land cover classes have been extracted. Although quite simple, this approach provides acceptable results in a context of rapid mapping, shown for this case in Fig. 2.11, where the "vegetation" class is overlaid to the original gray-scale image. Naturally the accuracy of this result is tied to the accuracy of the former class extraction; if one looks at the missed part of the pond on upper right, it is easy to see that it ended up in the class "vegetation" causing a lower correctness value.

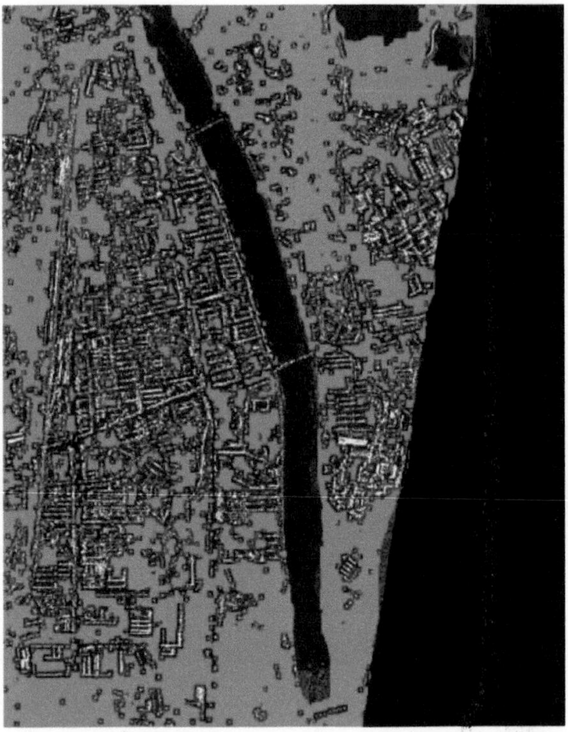

Fig. 2.11 Vegetation extraction over Luojiang image

2.4 Conclusions

The appearance on the scene of the new generation of SAR satellites capable of providing meter and sub-meter resolution SAR scenes potentially over any portion of the Earth surface has overcome the traditional limits connected with airborne acquisition and has boosted research on this alternative source of information in the context of mapping procedures.

Both 2D and, where available, 3D information may profitably be exploited for the so-called "rapid mapping" procedure, that is a fast procedure to collect basic information on the contents of a SAR scene, useful in those cases where the limited amount of information needed does not justify the use of complex, computationally heavy procedures or algorithms.

It has been shown by examples that rapid mapping on HR SAR scenes is feasible once suitable, efficient tools for the extraction of relevant features are available.

Although the proposed results are acceptable for rapid mapping, the usual cartographic applications need accuracy levels that are not achievable with the proposed tools. The two problems are then to be considered as separate items:

- On the one side, "rapid mapping" with its requirement of light computational load and speed in production of results
- On the other side, traditional cartographic applications with much loosed speed requirements but far stricter accuracy constraints

Needless to say, rapid mapping can still be useful to provide a starting point, over which precise cartographic extraction can successively build, in a two-stage approach which is expected to be overall more efficient than addressing directly the precise extraction.

A big advantage of using SAR data for rapid mapping is the availability of 3D interferometric data derived directly through suitable processing of different satellite passes over the site; 3D data is naturally perfectly registered with the underlying 2D radiometric data.

This chapter has presented a general framework for performing rapid mapping based on SAR scenes, but some issues still remain open and deserve further investigation:

- Small clusters of buildings sometimes may not be detected as urban areas and result in the production of false positives for the class "wood".
- The model for roads is a series of linear segments, thus curvilinear roads have to be piecewise approximated, with a consequent loss of accuracy and possibly also completeness. This is a problem especially in higher relief areas where bends are frequent. A curvilinear model for roads should be integrated into the extraction algorithm if this is to be really complete. The trade-off between precision and speed of execution should not however be forgotten.

It is the opinion of the authors that a structure like the one presented in this chapter is a good starting point for the setting up of a "scene interpreter" in a context of rapid mapping over SAR images. The modular structure allows the inclusion of new portions of code or algorithms as needed. Thanks to the increasing availability of very high-resolution spaceborne SAR all over the world, and the capability of those systems to acquire images over a given area within a few days or even hours, will make the topic of rapid mapping to increase its appeal for many applications, especially for those related to disaster monitoring.

Acknowledgements The authors wish to acknowledge the Italian Space Agency and the Italian Civil Protection Department for providing the COSMO/SkyMed image used in the examples of rapid mapping, the German Space Agency (DLR) for providing the TerraSAR-X image, and Dr. Gianni Lisini for performing the processing steps discussed in this chapter.

References

Aldrighi M, Dell'Acqua F, Lisini G (2009) Tile mapping of urban area extent in VHR SAR images. In: Proceedings of the 5th IEEE/ISPRS joint event on remote sensing over urban areas, Shanghai, China, 20–22 May 2009

Askne J, Santoro M, Smith G, Fransson JES (2003) Multitemporal repeat-pass SAR interferometry of boreal forests. IEEE Trans Geosci Remote Sens 41(7):1540–1550

Attema EPW, Duchossois G, Kohlhammer, G (1998) ERS-1/2 SAR land applications: overview and main results. In: Proceedings of IGARSS'08, vol 4, pp 1796–1798

Bajcsy R, Tavakoli M (September 1976) Computer recognition of roads from satellite pictures. IEEE Trans Syst Man Cybern SMC-6:623–637

Bentabet L, Jodouin S, Ziou D, Vaillancourt J (2003) Road vectors update using SAR imagery: a snake-based method. IEEE Trans Geosci Remote Sens 41(8):1785–1803

Borghys D, Perneel C, Acheroy M (2000) A multivariate contour detector for high-resolution polarimetric SAR images. In: Proceedings of the 15th International Conference Pattern Recognition, vol 3, pp 646–651, 3–7 September 2000

Caltagirone F, Spera P, Gallon A, Manoni G, Bianchi L (2001) COSMO-Skymed: a dual use Earth observation constellation. In: Proceedings of the 2nd international workshop on satellite constellation and formation flying, pp 87–94

Canny J (November 1986) A computational approach to edge detection. IEEE Trans Pattern Anal Mach Intell PAMI-8(11):679–698

Chanussot J, Mauris G, Lambert P (May 1999) Fuzzy fusion techniques for linear features detection in multitemporal SAR images. IEEE Trans Geosci Remote Sens 37(3):2287–2297

Chesnaud C, Réfrégier P, Boulet V (November 1999) Statistical region snake-based segmentation adapted to different physical noise models. IEEE Trans Pattern Anal Mach Intell 21(11):1145–1157

Dell'Acqua F, Gamba P (October 2001) Detection of urban structures in SAR images by robust fuzzy clustering algorithms: the example of street tracking. IEEE Trans Geosci Remote Sens 39(10):2287–2297

Dell'Acqua F, Gamba P (January 2003) Texture-based characterization of urban environments on satellite SAR images. IEEE Trans Geosci Remote Sens 41(1):153–159

Dell'Acqua F, Gamba P, Lisini G (2002) Extraction and fusion of street network from fine resolution SAR data. In: Proceedings of IGARSS, vol 1. Toronto, ON, Canada, pp 89–91, June 2002

Dell'Acqua F, Gamba P, Lisini G (2003). Road map extraction by multiple detectors in fine spatial resolution SAR data. Can J Remote Sens 29(4):481–490

Dell'Acqua F, Gamba P, Lisini G (2005) Road extraction aided by adaptive directional filtering and template matching. In: Proceedings of the third GRSS/ISPRS joint workshop on remote sensing over urban areas (URBAN 2005), Tempe, AZ, 14–16 March 2005 (on CD-ROM)

Dell'Acqua F, Gamba P, Trianni G (March 2006) Semi-automatic choice of scale-dependent features for satellite SAR image classification. Pattern Recognit Lett 27(4):244

Dell'Acqua F, Gamba P, Lisini G (2008) Rapid mapping of high-resolution SAR scenes. ISPRS J Photogramm Remote Sens, doi:10.1016/j.isprsjprs.2008.09.006

Dell'Acqua F, Lisini G, Gamba P (2009) Experiences in optical and SAR imagery analysis for damage assessment in the Wuhan, May 2008 earthquake. In: Proceedings of IGARSS 2009, Cape Town, South Africa, 13–17 July 2009

Dekker RJ (September 2003) Texture analysis and classification of ERS SAR images for map updating of urban areas in the Netherlands. IEEE Trans Geosci Remote Sens 41(9):1950–1958

Duskunovic I, Heene G, Philips W, Bruyland I (2000) Urban area selection in SAR imagery using a new speckle reduction technique and Markov random field texture classification. In: Proceedings of IGARSS, vol 2, pp 636–638, July 2000

Fischler MA, Tenenbaum JM, Wolf HC (1981) Detection of roads and linear structures in low resolution aerial imagery using a multisource knowledge integration technique. Comput Graph Image Process 15(3):201–223

Gamba P, Houshmand B (1999) Three-dimensional road network by fusion of polarimetric and interferometric SAR data. In: Proceedings of IGARSS099, vol 1, pp 302–304

Gamba P, Dell'Acqua F, Lisini G (2007) Raster to vector in 2D urban data. In: Proceedings of joint urban remote sensing event 2007, Paris, France, 13–15 April (on CD-ROM)

GEOEye Imagery Sources. http://www.geoeye.com/CorpSite/products/imagery-sources/Default.aspx#ikonos

Gouinaud C, Tupin F (1996) Potential and use of radar images for characterization and detection of urban areas. In: Proceedings of IGARSS, vol 1. Lincoln, NE, pp 474–476, May 1996

Hall O, Falorni G, Bras RL (2005) Characterization and quantification of data voids in the shuttle Radar topography mission data, IEEE Geosci Remote Sens Lett 2(2):177–181

He C, Xia G-S, Sun H (2006) An adaptive and iterative method of urban area extraction from SAR images. IEEE Geosci Remote Sens Lett 3(4):504–507

Heremans R, Willekens A, Borghys D, Verbeeck B, Valckenborgh J, Acheroy M, Perneel C (June 2005) Automatic detection of flooded areas on ENVISAT/ASAR images using an object-oriented classification technique and an active contour algorithm. In: Proceedings of the 31st international symposium on remote sensing of environment. Saint Petersburg, Russia, pp 20–24. http://www.isprs.org/publications/related/ISRSE/html/papers/219.pdf

Hess L, Melack J, Simonett D (1990) Radar detection of flooding areas beneath the forest canopy: a review. Int J Remote Sens 11(5):1313–1325

Horritt M, Mason D, Cobby D, Davenport I, Bates P (2003) Waterline mapping in flooded vegetation from airborne SAR imagery. Remote Sens Environ 85:271–281

Jeon B, Jang J, Hong K (1999) Road detection in spaceborne SAR images based on ridge extraction. In: Proceedings of ICIP, vol. 2. Kobe, Japan, pp 735–739

Jeon B-K, Jang J-H, Hong K-S (January 2002) Road detection in spaceborne SAR images using a genetic algorithm. IEEE Trans Geosci Remote Sens 40(1):22–29

Kurosu T, Fujita M, Chiba K (1995) Monitoring of rice crop growth from space using the ERS-1 C-band SAR. IEEE Trans Geosci Remote Sens 33(4):1092–1096

Landsat Legacy Project Website. http://library01.gsfc.nasa.gov/landsat/

NASA Jet Propulsion Laboratory: Missions. SEASAT. http://jpl.nasa.gov/missions/missiondetails.cfm?mission=Seasat

Lisini G, Tison C, Tupin F, Gamba P (2006) Feature fusion to improve road network extraction in high-resolution SAR images. IEEE Geosci Remote Sens Lett 3(2):217–221

McKeown DM, Denlinger L (1988) Cooperative methods for road tracking in aerial imagery. In: Proceedings of CVPR, Ann Arbor, MI, pp 662–672

Mena JB (December 2003) State of the art on automatic road extraction for GIS update: a novel classification. Pattern Recognit Lett 24(16):3037–3058

Negri M, Gamba P, Lisini G, Tupin F (2006) Junction-aware extraction and regularization of urban road networks in high-resolution SAR images. IEEE Trans Geosci Remote Sens 44(10): 2962–2971

Pesaresi M, Gerhardinger A, Kayitakire F (2007) Monitoring settlement dynamics by anisotropic textural analysis by panchromatic VHR data. In: Proceedings of joint urban remote sensing event 2007, Paris, 11–13 April 2007 (on CD-ROM)

Service Régional de Traitement d'Image et de Télédétection (SERTIT). http://sertit.u-strasbg.fr/

Soille P, Pesaresi M (2002) Advances in mathematical morphology applied to geoscience and remote sensing. IEEE Trans Geosci Remote Sens 40(9):2042–2055

Stilla U, Soergel U, Thoennessen U (2003) Potential and limits of InSAR data for building reconstruction in built-up areas. ISPRS J Photogramm Remote Sens 58(1–2):113–123

The International Charter – Space and Major Disasters. http://www.disasterscharter.org/

Tison C, Nicolas JM, Tupin F, Maitre H (October 2004) A new statistical model for Markovian classification of urban areas in high-resolution SAR images. IEEE Trans Geosci Remote Sens 42(10):2046–2057

Treitz PM, Rotunno OF, Howarth PJ, Soulis ED (1996) Textural processing of multi-polarization SAR for agricultural crop classification. In: Proceedings of IGARSS'96, pp 1986–1988

Tupin F, Maitre H, Mangin J-F, Nicolas J-M, Pechersky E (March 1998) Detection of linear features in SAR images: application to road network extraction. IEEE Trans Geosci Remote Sens 36(2):434–453
Tupin F, Houshmand B, Datcu M (2002) Road detection in dense urban areas using SAR imagery and the usefulness of multiple views. IEEE Trans Geosci Remote Sens 40(11):2405–2414
Ulaby FT, Kouyate F, Brisco B, Williams THL (March 1986) Textural information in SAR images. IEEE Trans Geosci Remote Sens GE-24(2):235–245. ISSN: 0196-2892. Digital Object Identifier: 10.1109/TGRS.1986.289643
UNOSAT is the UN Institute for Training and Research (UNITAR) Operational Satellite Applications Programme. http://unosat.web.cern.ch/unosat/
Werninghaus R, Balzer W, Buckreuss St, Mittermayer J, Mühlbauer P (2004) The TerraSAR-X mission. EUSAR, Ulm, Germany
Wessel B (2004) Context-supported road extraction from SAR imagery: transition from rural to built-up areas. In: Proceedings of the EUSAR, Ulm, Germany, pp 399–402, May 2004
Yu S, Berthod M, Giraudon G (July 1999) Toward robust analysis of satellite images using map information-application to urban area detection. IEEE Trans Geosci Remote Sens 37(4): 1925–1939

Chapter 3
Feature Fusion Based on Bayesian Network Theory for Automatic Road Extraction

Uwe Stilla and Karin Hedman

3.1 Introduction

With the development and launch of new sophisticated Synthetic Aperture Radar (SAR) systems such as Terra SAR-X, Radarsat-2 and COSMO/Skymed, urban remote sensing based on SAR data has reached a new dimension. The new systems deliver data with much higher resolution than previous SAR satellite systems. Interferometric, polarimetric and different imaging modes have paved the way to new urban remote sensing applications. A combination of image data acquired from different imaging modes or even from different sensors is assumed to improve the detection and identification of man-made objects in urban areas. If the extraction fails to detect an object in one SAR view, it might succeed in another view illuminated from a more favorable direction.

Previous research has shown that the utilization of multi-aspect data (i.e. data of the same scene, but acquired from different directions) improves the results. This has been tested both for building recognition and reconstruction (Bolter 2001; Michaelsen et al. 2007; Thiele et al. 2007) and for road extraction (Tupin et al. 2002; Dell'Acqua et al. 2003; Hedman et al. 2005). Multi-aspect images supply the interpreter with both complementary and redundant information. However, due to complexity of the SAR data, the information is also often contradicting. Especially in urban areas, the complexity arises through dominant scattering caused by building structures, traffic signs and metallic objects in cities. Furthermore one has to deal with the imaging characteristics of SAR, such as speckle-affected images,

U. Stilla
Institute of Photogrammetry and Cartography, Technische Universitaet Muenchen, Arcisstrasse 21, 80333 Munich, Germany
e-mail: stilla@bv.tum.de

K. Hedman (✉)
Institute for Astronomical and Physical Geodesy, Technische Universitaet Muenchen, Arcisstrasse 21, 80333 Munich, Germany
e-mail: karin.hedman@bv.tum.de

U. Soergel (ed.), *Radar Remote Sensing of Urban Areas*, Remote Sensing and Digital Image Processing 15, DOI 10.1007/978-90-481-3751-0_3,

foreshortening, layover, and shadow. A correct fusion step has the ability to combine information from different sources, which in the end is more accurate and better than the information acquired from one sensor alone.

In general, better accuracy is obtained by fusing information closer to the source and working on the signal level. But in contrary to multi-spectral optical images, a fusion of multi-aspect SAR data on pixel-level hardly makes any sense. SAR data is far too complex. Instead of fusing pixel-information, features (line primitives) shall be fused. Decision-level fusion means that an estimate (decision) is made based on the information from each sensor alone and these estimates are subsequently combined in a fusion process. Techniques for decision-level fusion worthy of mention are fuzzy-theory, Dempster-Shafer's method and Bayesian theory. Fuzzy-fusion techniques especially for automatic road extraction from SAR images have already been developed (Chanussot et al. 1999; Hedman et al. 2005; Lisini et al. 2006). Tupin et al. (1999) proposed an evidential fusion process of several structure detectors in a framework based on Dempster-Shafer theory. Bayesian network theory has been successfully tested for feature fusion for 3D building description (Kim and Nevatia 2003). Data fusion based on Bayesian network theory has been applied in numerous other applications such as vehicle classification (Junghans and Jentschel 2007), acoustic signals (Larkin 1998) and landmine detection (Ferarri and Vaghi 2006).

One advantage of Bayesian network theory is the possibility of dealing with relations rather than dealing with signals or objects. Contrary to Markov random field, directions of the dependencies are stated which allow top-down or bottom-up combinations of evidence.

In this chapter, high-level fusion, that is fusion of objects and modelling of relations is addressed. A fusion module developed for automatic road extraction from multi-aspect SAR data is presented. The chapter is organized as follows: Section 3.2 gives a general introduction to Bayesian network theory. Section 3.3 formulates first the problem and then presents a Bayesian network fusion model for automatic road extraction. Section 3.3 also focuses on the estimation of conditional probabilities, both continuous and discrete. Finally (Section 3.4), we will test the performance and present some results of the implementation of a fusion module into an automatic road extraction system.

3.2 Bayesian Network Theory

The advantage of a Bayesian network representation is that it allows the user to map causal relationships among all relevant variables. By means of Bayesian probability theory conflicting hypotheses can be discriminated based on the evidence available on hand. Hypotheses with high support can be regarded as true while hypotheses with low support are considered false. Another advantage is that the systems are flexible and allow changing directions between the causal relations, depending on the flow of new evidence.

The equations of interest are Bayes' theorem:

$$P(Y|X,I) = \frac{P(X|Y,I) \cdot P(Y|I)}{P(X|I)} \tag{3.1}$$

and marginalisation:

$$P(X|I) = \int_{-\infty}^{+\infty} P(X,Y|I)\, dY \tag{3.2}$$

where $p(X|Y,I)$ is called the conditional probability or likelihood function, which specifies the belief in X under the assumption that Y is true. $P(Y|I)$ is called the prior probability of Y that was known before the evidence X became available. $P(Y|X,I)$ is often referred to as the posterior probability. The denominator $p(X|I)$ is called the marginal probability, that is the belief in the evidence X. This is merely a normalization constant, which nevertheless is important in Bayesian network theory.

Bayes' theorem follows directly from the product rule.

$$P(X,Y|I) = P(X|Y,I) \cdot P(Y|I) \tag{3.3}$$

The strength of Bayes' theorem is that it relates to the probability that the hypothesis Y is true given the data X to the probability that we have observed the measured data X if the hypothesis Y is true. The latter term is much easier to estimate. All probabilities are conditional on I, which is made to denote the relevant background information at hand.

Bayesian networks expound Bayes' theorem into a directed acyclic graph (DAG) (Jensen 1996; Pearl 1998). The *nodes* in a Bayesian network represent the *variables*, such as temperature of a device, gender of a patient or feature of an object. The *links*, or in other words the *arrows*, represents the informational or causal dependencies between the nodes. If there is an arrow from node Y to node X; this means that Y has an influence on X. Y is called the parental node and X is called the child node. X is assumed to have n states $x_1, \ldots, x_n$ and $P(X = x_i)$ is the probability of each certain state x_i.

The mathematical definition of Bayesian networks is as follows (Jensen 1996; Pearl 1998). The Bayesian network U is a set of nodes $U = \{X_1, \ldots, X_n\}$, which are connected by a set of arrows $\mathbf{A} = \{(X_i, X_j) \,|X_i, X_j \in \mathbf{X}, i \neq j\}$. Let $P(U) = P(x_1, \ldots, x_n)$ be the joint probability distribution of the space of all possible state values x. For being a Bayesian network, U has to satisfy the Markov condition, which means that a variable must be conditionally independent of its nondescendents given its parents. $P(x)$ can therefore be defined as,

$$P(x) = \prod_{x_i \in \mathbf{x}} P(x_i \,|\mathbf{pa}\,(X_i)), \tag{3.4}$$

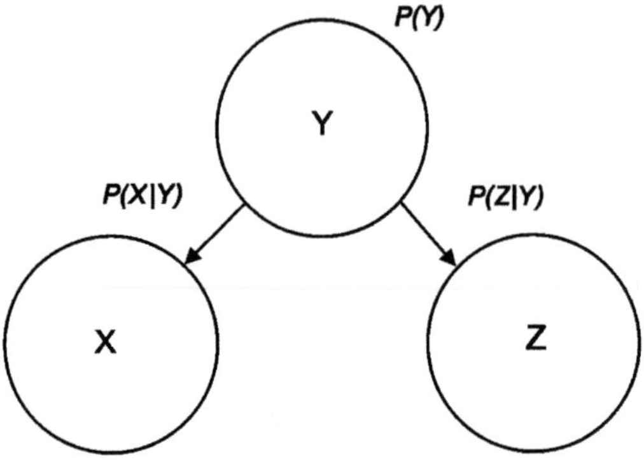

Fig. 3.1 A Bayesian network with one parental node (Y) and its two child nodes (X and Z) and their corresponding conditional probabilities

where **pa** (X_i) represents the parents states of node X_i. If this node has no parents, the prior probability $P(x_i)$ must be specified.

Assume a Bayesian network is composed of two child nodes, X and Z, and one parental node, Y (Fig. 3.1). Since X and Z are considered to be independent given the variable Y, the joint probability distribution $P(y,x,z)$ can be expressed as

$$P(y,x,z) = P(y)P(x\,|y)P(z\,|y). \tag{3.5}$$

Probability distributions in a Bayesian network can have a countable (discrete) or a continuous set of states. Conditional probabilities for discrete states are usually realized by conditional probability tables. Conditional probabilities for continuous states can be estimated by probability density functions.

More detailed information on Bayesian network theory can be found in Jensen (1996) and Pearl (1998).

3.3 Structure of a Bayesian Network

The Bayesian network fusion shall be implemented into an already existing road extraction approach (Wessel and Wiedemann 2003; Stilla et al. 2007). The approach was originally designed for optical images with a ground pixel of about 2 m (Wiedemann and Hinz 1999). The first step consists of line extraction using Steger's differential geometry approach (Steger 1998), which is followed by a smoothening and splitting step. Afterwards specific attributes (i.e. intensity, straightness and length) are specified for each line primitive. A weighted graph of the evaluated road primitives is constructed. For the extraction of the roads from the graph, supplementary road segments are introduced and seed points are defined. Best-valued road candidates serve as seed points, which are connected by an optimal path search through the graph.

A line extraction from SAR images often delivers partly fragmented and erroneous results. Over-segmentation occurs especially frequently in forestry and in

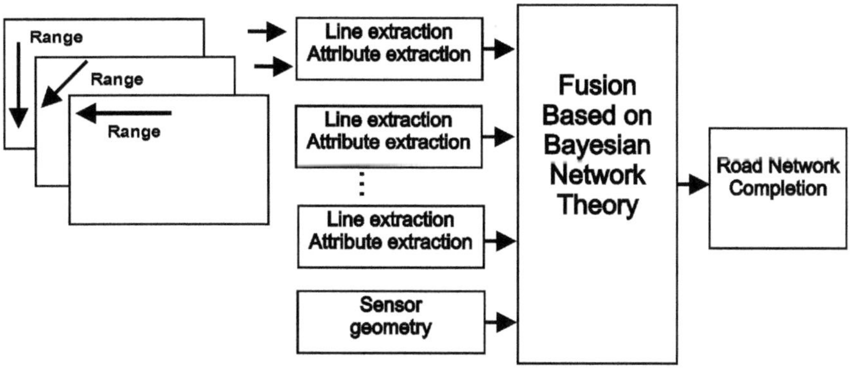

Fig. 3.2 The road extraction approach and its implementation of the fusion module

urban areas. Attributes describing geometrical and radiometric properties of the line features can be helpful for selection and especially for sorting out the most probable false alarms. However, these attributes may be ambiguous and are not considered to be reliable enough when used alone. Furthermore, occlusions due to surrounding objects may cause gaps, which are hard to compensate. On one hand multi-aspect images supply the interpreter with both complementary and redundant information. But on the other hand, the information is often contradicting, due to the over-segmented line extraction. The seed point selection for the optimal path search is the most sensitive parameter. The idea is that the fusion (Fig. 3.2) should contribute to a more complete intermediate result and help to get reliable weights for lines.

The main feature involved in the road extraction process is the line primitive. The line extraction detects not only roads, but also linear shadow regions (*shadows*) and relatively bright line extractions mainly occurring in forest areas (*false alarms*), caused by volume scattering. The first step is to classify these linear features by means of their characteristic attributes *(intensity, length, etc.)*, a set of *n variables* $X_1, \ldots, X_n$. The *variable L* (Fig. 3.3a) is assumed to have the following *states*:

- l_1 = *An extracted line primitive belongs to a ROAD*
- l_2 = *An extracted line primitive belongs to a FALSE ALARM*
- l_3 = *An extracted line primitive belong to a SHADOW*

If relevant, the hypotheses above can be extended with more states $l_4, \ldots, l_n$ (e.g. river, etc.). The flow of evidence may come from the top (state of Y is known) or from the bottom (state of X is known). On one hand, if a shadow is present, one expects that the linear primitive has low intensity. On the other hand, if a linear primitive has the same low intensity, one can assume that a shadow region has been extracted.

If two or more images are available, we shall combine line primitives extracted from two or more images. We need to add a fourth state to our variable L; *the*

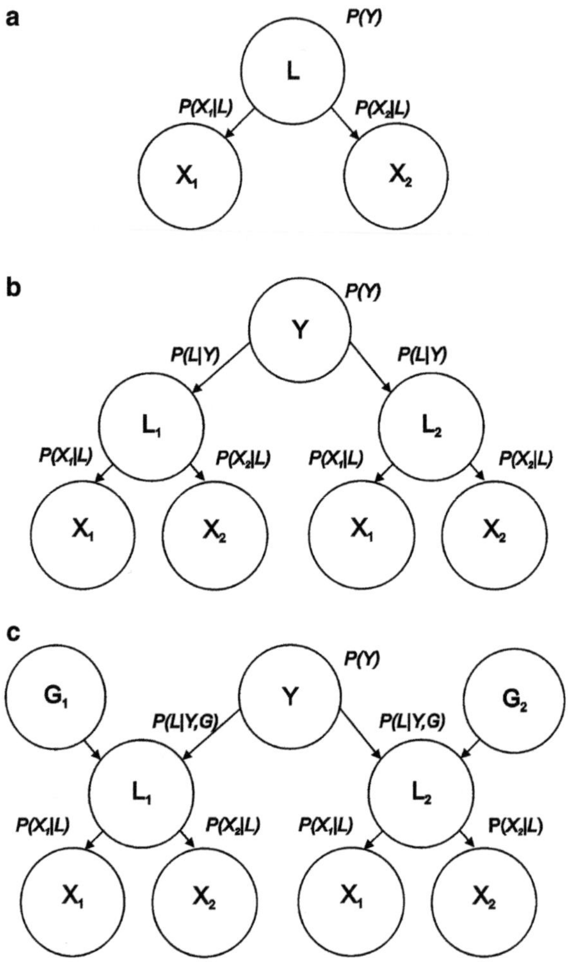

Fig. 3.3 A Bayesian network of (**a**) three nodes: parental node L (linear primitives) and two child nodes, X_1 and X_2 (the attributes) (**b**) two linear features, L_1 and L_2, extracted from two different SAR scenes, (**c**) with different sensor geometries, G_1 and G_2

fact that a line primitive has not been extracted in that scene, l_4. By introducing this state, we also consider the case that the road might not be detected by the line extraction in all processed SAR scenes.

Exploiting sensor geometry information relates to the observation that road primitives in range direction are less affected by shadows or layover of neighbouring elevated objects. A road beside an alley, for instance, can be extracted at its true position when oriented in range direction. However, when oriented in azimuth direction, usually only the parallel layover and shadow areas of the alley are imaged but not the road itself (Fig. 3.4). Hence a third variable is incorporated into the Bayesian network, the sensor geometry, G, which considers the look and incidence angle of the

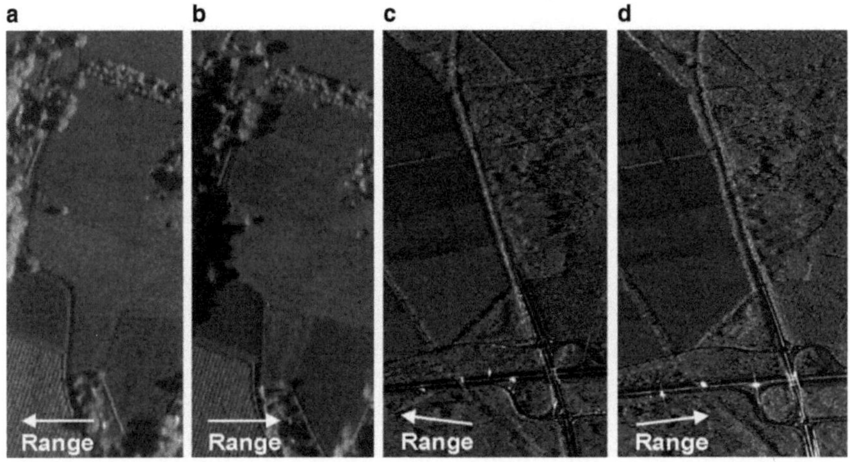

Fig. 3.4 The anti-parallel SAR views exemplify the problem of roads with trees nearby. Depending on the position of the sensor shadow effects occlude the roads. (**a, b**) Sensor: MEMPHIS (FGAN-FHR), (**c, d**) Sensor: TerraSAR-X

sensor in relation to the direction of the detected linear feature (Fig. 3.3c). Bayesian network theory allows us to incorporate a reasoning step which is able to model the relation of linear primitives. These primitives are detected and classified differently in separate SAR scenes. Instead of discussing hypotheses such as the classification of detected linear features, we now deal with the hypothesis whether a road exist or not in the scene. A fourth variable Y with the following four states is included:

- y_1 = *A road exists in the scene*
- y_2 = *A road with high objects, such as houses, trees or crash barriers, nearby exist in the scene*
- y_3 = *High objects, such as houses, trees or crash barriers*
- y_4 = *Clutter*

Since roads surrounded by fields and no objects nearby and roads with high objects nearby appear differently, these are treated as different states. If relevant, the variable Y can easily be extended with further states $y_5, .., y_n$, and makes it possible to describe roads with buildings and roads with trees as separate states.

Instead of dealing with the hypothesis; "whether a line primitive belongs to road or not", the variables Y and G enable us to deal with the hypothesis; "whether a road exist or not". It is possible to support the assumption that a road exists given that two line primitives, one belonging to a road and one belonging to a shadow, are detected. Modeling such hypothesis is much easier using Bayesian network theory compared to a fusion based on classical Bayesian theory.

Writing the chain rule formula, we can express the Bayesian Network (Fig. 3.3b) as

$$P(Y, L_1, L_2, X_1, X_2) = P(Y)P(L_1|Y)P(L_2|Y)P(X_1|L_1)P(X_2|L_2) \quad (3.6)$$

and the Bayesian Network (Fig. 3.3c) as

$$P(Y, G_1, G_2, L_1, L_2, X_1, X_2) = P(Y)P(L_1|Y, G_1)P(L_2|Y, G_2) \times P(X_1|L_1)P(X_2|L_2). \tag{3.7}$$

As soon as the Bayesian Network and their conditional probabilities are defined, knowledge can propagate from the observable variables to the unknown. The only information variable in this specific case is the extraction of the linear segment and their attributes, X. The remaining conditional probabilities to specify are $P(l|y, g)$ and $P(x|l)$. We will discuss the process of defining these in the following two subchapters.

3.3.1 Estimating Continuous Conditional Probability Density Functions

The selection of attributes of the line primitives is based on the knowledge of roads. Radiometric attributes such as mean and constant intensity, and contrast of a line as well as geometrical attributes such as length and straightness are all good examples. It should be pointed out that more attributes do not necessarily yield better results, instead rather the opposite occurs. A selection including few, but significant attributes is recommended. In this work, we have decided to concentrate on three attributes, length of the line primitive, straightness and intensity.

The joint conditional probability that the variable L belongs to the state l_i under the condition that its attributes $\boldsymbol{x}$ (an attribute vector) are known is estimated by the following equation:

$$P(l_i|\mathbf{x}) = \frac{P(\mathbf{x}|l_i)\,P(l_i)}{P(\mathbf{x})}. \tag{3.8}$$

If there is no correlation between the attributes, the likelihood $P(\boldsymbol{x}|l_i)$ can be assumed to be equal to the product of the separate likelihoods for each attribute

$$P(\mathbf{x}|l_i) = P(x_1, x_2, .., x_n|l_i) = P(x_1|l_i)\,P(x_2|l_i)\ldots P(x_n|l_i) \tag{3.9}$$

A final decision on the variable of L can be achieved by the solution, which yields the greatest value for the probability of the observed attributes, usually referred to as the Maximum-A-Posteriori estimation;

$$\hat{l}_M = \arg\max_l p(l|x) \tag{3.10}$$

Each separate likelihood $P(x_i|l_j)$ can be approximated by a probability density function learned from training data. Learning from training data means that the extracted line segments are sorted manually into three groups, roads, shadows and false alarms. Attributes of the line primitives are dependent not only on a range of factors such as characteristics of the SAR scene (rural, urban, etc.), but also on the parameter settings by the line extraction. The aim is to achieve probability density functions which represent a degree of belief of a human interpreter rather than a frequency of the behaviour of the training data. For this reason, different training data sets have been used and for each set the line primitives have been selected carefully.

Histograms are one of the most common tools for visualizing and estimating the frequency distribution of a data set. The Gaussian distribution

$$p(l_i \mid x) = \frac{1}{\sigma\sqrt{2\pi}} e^{\left(-\frac{(x-\mu)^2}{2\sigma^2}\right)} \tag{3.11}$$

is most often assumed to describe random variation that occurs in data used in most scientific disciplines. However, if the data shows a more skewed distribution, has a low mean value, large variance and values cannot be negative, as in this case, the distribution fits better to a log-normal distribution (Limpert et al. 2001). A random variable X is said to be log-normally distributed if $log(X)$ is normally distributed. The rather high skewness and remarkable high variance of the data indicated that the histograms might follow a lognormal distribution, that is

$$p(l_i \mid x) = \frac{1}{S\sqrt{2\pi}\,x} e^{-\frac{(\ln x - M)^2}{2S^2}}. \tag{3.12}$$

The shape of a histogram is highly dependent on the choice of the bin size. Larger bin width normally yields histograms with a lower resolution and as a result the shape of the underlying distribution cannot be represented correctly. Smaller bin widths produce on the other hand irregular histograms with bin heights having great statistical fluctuations. Several formulas for finding the optimum bin width are well-known, such as *Sturges' Rule* or the *Scott's rule*. However most of them are based on the assumption that the data is normally distributed. Since the histograms show a large skewness, a method, which estimates the optimal bin size out of the data directly (Shimazaki and Shinomoto 2007), is used instead. The probability density functions have been fitted to the histograms by a least square adjustment of S and M since it allows the introducing of a-priori variances. Figure 3.5a and b show the histogram of the attribute *length* and its fitted lognormal distributed curve. A fitting carried out in a histogram with one dimension is relatively uncomplicated, but as soon as the dimensions increase, the task of fitting becomes more complicated. As soon as attributes tend to be correlated, they cannot be treated as independent. A fitting of a multivariate lognormal distribution shall then be carried out. The independence condition can be proved by a correlation test.

The obtained probability assessment shall correspond to our knowledge about roads. At a first glance, the histograms in Figs. 3.5a and b seem to overlap. However, Fig. 3.5c exemplifies for the attribute *length* that the discriminant function

$$g(x) = \ln(p(x \mid l_1)) - \ln(p(x \mid l_2)) \tag{3.13}$$

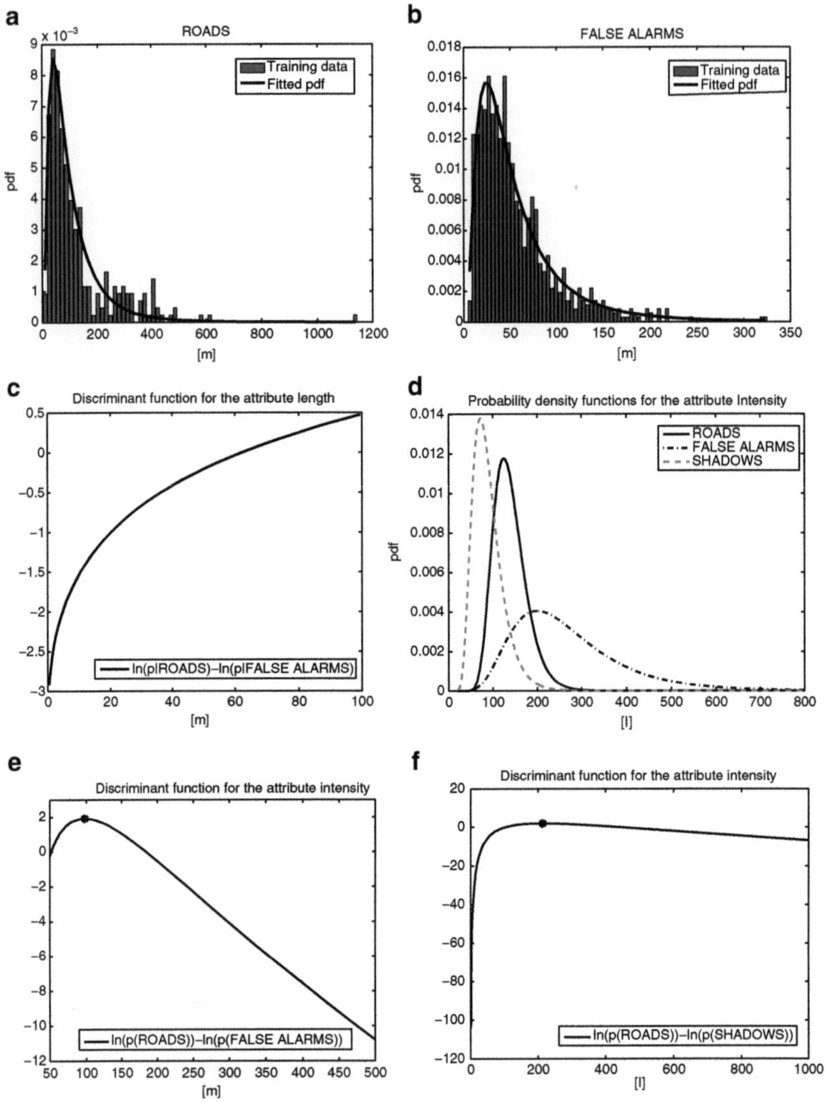

Fig. 3.5 A lognormal distribution is fitted to a histogram of the attribute length (**a**) roads (l_1), (**b**) false alarms (l_2). (**c**) The discriminant function for the attributes length (roads and false alarms). (**d**) Fitted probability density functions for the three states, roads (l_2), false alarms (l_2) and shadows (l_3), (**e, f**) Discriminant function for the attribute intensity, $l_1 - l_2$, and $l_1 - l_3$

increases as the length of the line segment increases. The behaviour of the discriminant function corresponds to the belief of a human interpreter. The behaviour of the discriminant function was tested for all attributes. The discriminant functions seen in Fig. 3.5d–f certainly correspond to the frequency behaviour of the training data but hardly to the belief of a human interpreter.

Irrespective of the data we can make the following conclusions:

1. Line primitives belonging to a shadow have most likely a low intensity compared to false alarms and roads.
2. From the definition of false alarms (see Section 3.3) we can make the conclusion that its line primitives have a rather bright intensity.

For the attribute intensity, thresholds are defined,

$$
\begin{aligned}
p(x|l_2) &= 0 && \mathit{for}\ x < x_L \\
p(x|l_2) &= \frac{1}{S\sqrt{2\pi}\,x} e^{-\frac{(\ln x - M)^2}{2S^2}} && \mathit{for}\ x > x_L
\end{aligned}
$$

and

$$
\begin{aligned}
p(x|l_3) &= \frac{1}{S\sqrt{2\pi}\,x} e^{-\frac{(\ln x - M)^2}{2S^2}} && \mathit{for}\ x < x_H \\
p(x|l_3) &= 0 && \mathit{for}\ x > x_H
\end{aligned}
\tag{3.14}
$$

where x_L and x_H are the local maximum points obtained from the discriminant functions.

Whenever possible the same probability density functions should be used for each SAR scene. However, objects in SAR data acquired by different SAR sensors have a naturally different intensity range. Hence, the probability density functions for intensity should preferably be adjusted as soon as new data sets are included.

3.3.2 Discrete Conditional Probabilities

The capacity of estimating conditional probability density functions is dependent on the availability of training data. If one has no access to sufficient training data, one is forced to express the belief by tables consisting of discrete probabilities. At best, probabilities can be numerically estimated directly from training data, but in the worst case they have to be estimated based on subjective estimates.

By such tables' definition, the nodes in a Bayesian network are variables with a finite number of mutually exclusive states. If the variable Y has states $y_1, \ldots, y_n$ and the variable L has states $l_1, \ldots, l_m$, then $P(l|y)$ is an $m\ x\ n$ table containing numbers $P(l_i|y_j)$ such as

$$
p(L = l\,|Y = y) = \begin{bmatrix} p(l_1\,|y_1) & p(l_1\,|y_2) & \cdots & p(l_1\,|y_n) \\ p(l_2\,|y_1) & p(l_2\,|y_2) & & p(l_2\,|y_n) \\ \vdots & \vdots & & \vdots \\ p(l_m\,|y_1) & p(l_m\,|y_2) & \cdots & p(l_m\,|y_n) \end{bmatrix}. \tag{3.15}
$$

The sum of the columns should be one.

The joint conditional probability that the variable Y belongs to the state y_i under the condition that a linear feature L are extracted from one SAR scene is estimated by

$$P(Y = y_j | L = l) = \alpha P(y_j) \sum_{m}^{i=0} (P(l_i | y_j)\, P(l_i)), \tag{3.16}$$

where α is the marginalization term, which is in this case equal to $1/P(l)$. There are m different events for which L is in state l_i, namely the mutually exclusive events $(y_i, l_1), \ldots, (y_i, l_m)$. Therefore $P(l)$ is

$$P(l) = \sum_{n}^{j=0} \sum_{m}^{i=0} (P(l_i | y_j)\, P(l_i)), \tag{3.17}$$

which is called *marginalization*. Each node can be marginalized.

A soon as the attributes X are known, node Y should be updated with this information coming from node X. $P(l_i)$ should be exchanged to $P(x|l_i)$ estimated by Eq. (3.8).

$$P(Y = y_j | L = l, X = x) = \alpha P(y_j) \sum_{m}^{i=0} (P(l_i | y_j)\, P(l_i | x)), \tag{3.18}$$

Once information from p SAR scenes is extracted the belief in node Y can be expressed as

$$P(Y = y_j | L = l, X = x) = \alpha P(y_j) \prod_{p}^{k=0} \sum_{m}^{i=0} (P(l_i | y_j)\, P(l_i | x))_k. \tag{3.19}$$

The child node L is dependent on both parental nodes, Y and G. If G is included, tables for $p(l|y, g)$ have to be estimated, which results in the following expression:

$$P(Y = y_j | G = g, L = l, X = x) = \alpha P(y_j) \prod_{p}^{k=0} \sum_{m}^{i=0} (P(l_i | y_j, g)\, P(l_i | x))_k. \tag{3.20}$$

3.3.3 Estimating the A-Priori Term

Prior certainties are required for the events which are affected by causes outside of the network. Prior information represents the knowledge of the Bayesian network that is already known. If this knowledge is missing, the prior term for each state can be valued equally. For the Bayesian network proposed here, a prior term $p(Y)$ can be introduced. The prior represent the frequency of the states $y_1, \ldots, y_n$ among our line primitives. The frequency of roads is on one hand proportionately low in some context areas, for instance in forestry regions. On the other hand, the frequency of roads in urban areas is rather high. Hence, global context (i.e. urban, rural and forestry regions) can play a significant role in the definition of the priori term. Global context regions are derived from maps or GIS before road extraction, or can be segmented automatically by a texture analysis. The priori probability can be set differently in these areas.

The advantage of Bayesian network theory is that belief can propagate both upwards and downwards. If maps or GIS information are missing, one could certainly derive context information solely based on the extracted roads (i.e. belief update for variable Y).

3.4 Experiments

The Bayesian network fusion was tested on two multiaspect SAR images (X-band, multi-looked, ground range SAR data of a suburban scene located near the airport of DLR in Oberpfaffenhofen, southern Germany (Fig. 3.6). Training data was

Fig. 3.6 The multi-aspect SAR data analyzed in this example. The scene is illuminated once from the bottom and once from the bottom-right corner

Table 3.1 Conditional probabilities $P(l_i|y_j)$

	$Y = y_1$	$Y = y_2$	$Y = y_3$	$Y = y_4$
$L = l_1$	0.544[a]	0.335	0.236[a]	0.157[a]
$L = l_2$	0.013	0.016	0.00	0.414
$L = l_3$	0.212[a]	0.363	0.364	0.029
$L = l_4$	0.231[a]	0.286	0.400	0.400

[a]Means that the data was directly estimated from training data

collected from the data acquired from the same sensor, but tested on a line extraction performed with different parameter settings. A cross correlation was carried out in order to examine if the assessment of node L based on X delivers a correct result. About 70% of the line primitives were correctly classified.

The conditional probability table $P(L|Y)$ (Table 3.1) could be partly estimated from comparison between ground truth and training data and partly by subjective belief from a user.

The performance was tested on two examples; a road surrounded by fields and a road with a row of trees on one side (marked as 1 and 2 in Fig. 3.7). In each scene linear primitives were extracted and assessed by means of Eq. (3.9) (Table 3.2). For each one of the examples the Bayesian fusion was carried out with a final classification of the variable L with and without a-priori information, and with the uncertainties of L, with and without a-priori information. A comparison between the resulting uncertainty (Eq. 3.17) that the remaining fused linear primitive belongs to the states $y_1, \ldots, y_n$ demonstrates that the influence of the prior term is quite high (Figs. 3.7 and 3.8). Prior term is important for a correct classification of clutter. A fact that also becomes clear from Fig. 3.8 is the importance of keeping the uncertainty assessment by node L instead of making a definite classification. Even if two linear primitives such as L_{S1} and L_{S2} are fused, they may in the end be a good indicator that a road truly exists. This can be of particular importance as soon as a conditional probability table also includes the variable representing sensor geometry, G_1 and as soon as global context is incorporated as a-priori information.

3.5 Discussion and Conclusion

In this chapter, we have presented a fusion approach modeled as a Bayesian Network. The fusion combines linear features from multi-aspect SAR data as part of an approach for automatic road extraction from SAR. Starting with a general introduction to Bayesian network theory, we then presented the main aspects. Finally, some results of some fusion situations were shown.

A smaller Bayesian network such as the one proposed in this work is quite easy to model and implement. The model has a flexible architecture, allowing the implementation of nodes representing new information variables (i.e. global context, further features, a-priori information, etc.). The most time-consuming part is the

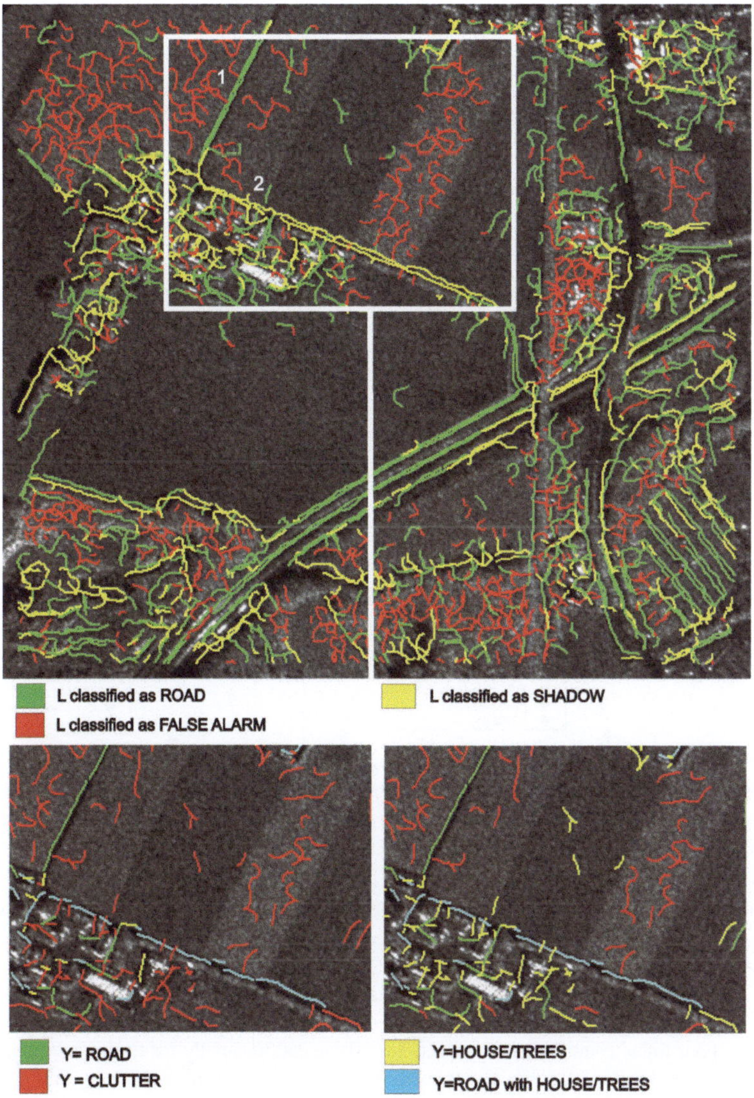

Fig. 3.7 The fusion process was tested on an E-SAR multi-aspect data set (Fig. 3.6). The *upper* image shows node L, which is the classification based on attributes before fusion. The two *lower* images show the end result (node Y) with (to the *left*) and without (to the *right*) prior information. The numbers highlight two specific cases; 1 is a small road surrounded by fields and 2 is a road with trees below. These two cases are further examined in Fig. 3.8

estimation of the conditional probabilities between the nodes. Unfortunately, these need to be updated as soon as data coming from a different SAR sensor is used. In general, different SAR sensors imply different characteristics of the SAR data. The goal is to have a rather small amount of training data which should be enough for an

Table 3.2 Assessment of selected line primitives based on their attributes $P(l_i|x)$

L	$P(l\|x)$		$P(l\|x)$
L_{R1}	$P(l\|x) = [0.749,\ 0.061,\ 0.190,\ 0]$	L_{R1} (classification)	$P(l\|x) = [1,\ 0,\ 0,\ 0]$
L_{R2}	$P(l\|x) = [0.695,\ 0.075,\ 0.230,\ 0]$	L_{R2} (classification)	$P(l\|x) = [1,\ 0,\ 0,\ 0]$
L_{S1}	$P(l\|x) = [0.411,\ 0,\ 0.589,\ 0]$;	L_{S1} (classification)	$P(l\|x) = [0,\ 0,\ 1,\ 0]$
L_{S2}	$P(l\|x) = [0.341,\ 0.158,\ 0.501,\ 0]$;	L_{S2} (classification)	$P(l\|x) = [0,\ 0,\ 1,\ 0]$
L_{No}	$P(l\|x) = [0,\ 0,\ 0,\ 1]$	Priori information $P(Y)$	$P(Y) = [0.20,\ 0.20,\ 0.20,\ 0.40]$

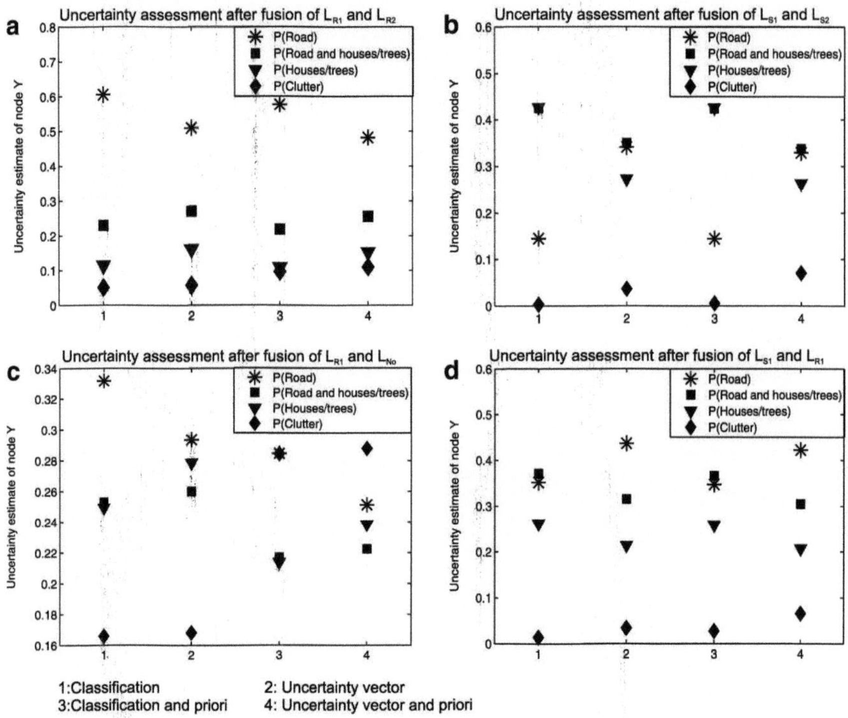

Fig. 3.8 Four linear primitives were selected from the data manually for further investigation of the fusion. The resulting uncertainty assessment of $y_1, \ldots, y_n$ were plotted: (**a**) L_{R1} and L_{R2}, (**c**) L_{RI} and L_{No} (missing line detection), (**b**) L_{S1} and L_{S2}, (**d**) L_{S1} and L_{R1} considering four situations: (1) Classification, (2) Classification and a-priori information, (3) Uncertainty vector, (4) Uncertainty vector and a-priori information. The linear primitives can be seen in Fig. 3.7 and their numerical values are presented in Table 3.2. L_{R1} and L_{R2} are marked with a 1 and L_{S1} and L_{S2} and are marked with a 2

adjustment of the conditional probabilities. Preferably the user would set the main parameters by selecting a couple of linear primitives. Most complicated is the definition of the conditional probability table (Table 3.1), as rather *ad hoc* assumptions need to be made. Nevertheless, the table is important and plays a rather prominent role in the end result. Also, the prior term can be fairly hard to approximate, but should also be implemented for a more reliable result.

One should keep in mind that the performance of fusion processes is highly dependent on the quality of the incoming data. In general, automatic road extraction is a complicated task, merely due to the side-looking geometry. In urban areas, the roads are not even visible due to high surrounding buildings. Furthermore, differentiating between true roads and shadow regions is difficult due to their similar appearance. It is almost impossible to distinguish between roads surrounded by objects (i.e. building rows) and simply the shadow-casting objects with no road nearby. In future work, bright linear features such as layover or strong scatterers could also be included in the Bayesian network for supporting or neglecting these hypotheses.

Nevertheless, this work demonstrated the potential of fusion approaches based on Bayesian networks not only for road extraction but also for various applications within urban remote sensing based on SAR data. Bayesian network fusion could be especially useful for a combination of features extracted from multi-aspect data for building detection.

Acknowledgement The authors would like to thank the Microwaves and Radar Institute, German Aerospace Center (DLR) as well as FGAN-FHR for providing SAR data.

References

Bolter R (2001) Buildings from SAR: detection and reconstruction of buildings from multiple view high-resolution interferometric SAR data. Ph.D. thesis, University of Graz, Austria

Chanussot J, Mauris G, Lambert P (May 1999) Fuzzy fusion techniques for linear features detection in multitemporal SAR images. IEEE Trans Geosci Remote Sens 37:1292–1305

Dell'Acqua F, Gamba P, Lisini G (September 2003) Improvements to urban area characterization using multitemporal and multiangle SAR images. IEEE Trans Geosci Remote Sens 4(9):1996–2004

Ferrari S, Vaghi A (April 2006) Demining sensor modeling and feature-level fusion by Bayesian networks. IEEE Sens J 6:471–483

Hedman K, Wessel B, Stilla U (2005) A fusion strategy for extracted road networks from multi-aspect SAR images. In: Stilla U, Rottensteiner F, Hinz S (eds) CMRT05. Int Arch Photogramm Remote Sens 36(Part 3 W24):185–190

Jensen FV (1996) An introduction to Bayesian networks. UCL Press, London

Junghans M, Jentschel H (2007) Qualification of traffic data by Bayesian network data fusion. In: 10th International Conference on Information Fusion, pp 1–7, July 2007

Kim Z, Nevatia R (June 2003) Expandable Bayesian networks for 3D object description from multiple views and multiple mode inputs. IEEE Trans Pattern Anal Mach Intell 25:769–774

Larkin M (November 1998) Sensor fusion and classification of acoustic signals using Bayesian networks. In: Conference record of the thirty-second Asilomar conference on signals, systems & computers, 1998, vol 2, pp 1359–1362, November 1998

Limpert E, Stahel WA, Abbt M (May 2001) Log-normal distributions across the sciences: keys and clues. BioScience 51:341–352

Lisini G, Tison C, Tupin F, Gamba P (April 2006) Feature fusion to improve road network extraction in high-resolution SAR images. IEEE Geosci Remote Sens Lett 3:217–221

Michaelsen E, Doktorski L, Soergel U, Stilla U (2007) Perceptual grouping for building recognition in high-resolution SAR images using the GESTALT-system. In: 2007 Urban remote sensing joint event: URBAN 2007 – URS 2007 (on CD)

Pearl J (1998) Probabilistic reasoning in intelligent systems: networks of plausible inference. Morgan Kaufmann, San Francisco, CA

Shimazaki H, Shinomoto S (2007) A method for selecting the bin size of a time histogram. Neural Comput 19:1503–1527

Steger C (February 1998) An unbiased detector of curvilinear structures. IEEE Trans Pattern Anal Mach Intell 20:113–125

Stilla U, Hinz S, Hedman K, Wessel B (2007) Road extraction from SAR imagery. In: Weng Q (ed) Remote sensing of impervious surfaces. Tayor & Francis, Boca Raton, FL

Thiele A, Cadario E, Schulz K, Thonnessen U, Soergel U (November 2007) Building recognition from multi-aspect high-resolution InSAR data in urban areas. IEEE Trans Geosci Remote Sens 45:3583–3593

Tupin F, Bloch I, Maitre H (May 1999) A first step toward automatic interpretation of SAR images using evidential fusion of several structure detectors. IEEE Trans Geosci Remote Sens 37:1327–1343

Tupin F, Houshmand B, Datcu M (November 2002) Road detection in dense urban areas using SAR imagery and the usefulness of multiple views. IEEE Trans Geosci Remote Sens 40:2405–2414

Wessel B, Wiedemann C (2003) Analysis of automatic road extraction results from airborne SAR imagery. In: Proceedings of the ISPRS conference "PIA'03", international archives of photogrammetry, remote sensing and spatial information sciences, Munich, vol 32(3–2W5), pp 105–110

Wiedemann C, Hinz S (September 1999) Automatic extraction and evaluation of road networks from satellite imagery. Int Arch Photogramm Remote Sens 32(3–2W5):95–100

Chapter 4
Traffic Data Collection with TerraSAR-X and Performance Evaluation

Stefan Hinz, Steffen Suchandt, Diana Weihing, and Franz Kurz

4.1 Motivation

As the amount of traffic has dramatically increased over the last years, traffic monitoring and traffic data collection have become more and more important. The acquisition of traffic data in almost real-time is essential to immediately react to current traffic situations. Stationary data collectors such as induction loops and video cameras mounted on bridges or traffic lights are matured methods. However, they only provide local data and are not able to observe the traffic situation in a large road network. Hence, traffic monitoring approaches relying on airborne and space-borne remote sensing come into play. Especially space-borne sensors do cover very large areas, even though image acquisition is strictly restricted to certain time slots predetermined by the respective orbit parameters. Space-borne systems thus contribute to the periodic collection of statistical traffic data in order to validate and improve traffic models. On the other hand, the concepts developed for space-borne imagery can be easily transferred to future HALE (High Altitude Long Endurance) systems, which show great potential to meet the demands of both temporal flexibility and spatial coverage.

With the new SAR missions such as TerraSAR-X, Cosmo-Skymed, or Radarsat-2, high-resolution SAR data in the (sub-)meter range are now available. Thanks to this high-resolution, significant steps forward towards space-borne traffic data acquisition are currently made. The concepts basically rely on earlier work on Ground Moving Target Indication (GMTI) and Space-Time Adaptive Processing (STAP) such as Klemm (1998) and Ender (1999), yet as, for example, Livingstone

S. Hinz (✉)
Photogrammetry and Remote Sensing, Karlsruhe Institute of Technology, Germany
e-mail: stefan.hinz@kit.edu

S. Suchandt and F. Kurz
Remote Sensing Technology Institute, German Aerospace Center DLR, Germany
e-mail: steffen.suchandt@dlr.de; franz.kurz@dlr.de

D. Weihing
Remote Sensing Technology, TU Muenchen, Germany
e-mail: diana.weihing@bv.tum.de

U. Soergel (ed.), *Radar Remote Sensing of Urban Areas*, Remote Sensing and Digital Image Processing 15, DOI 10.1007/978-90-481-3751-0_4,

et al. (2002), Chiu and Livingstone (2005), Bethke et al. (2006), and Meyer et al. (2006) show, significant modifications and extensions are necessary when taking the particular sensor and orbit characteristics of a space mission into account.

An extensive overview on current developments and potentials of airborne and space-borne traffic monitoring systems is given in the compilation of Hinz et al. (2006). It shows that civilian SAR is currently not competitive with optical images in terms of detection and false alarm rates, since the SAR image quality is negatively influenced by Speckle as well as layover and shadow effects in case of city areas or rugged terrain. However, in contrast to optical systems, SAR is an active and coherent sensor enabling interferometric and polarimetric analyzes. While the superiority of optical systems for traffic monitoring are in particular evident when illumination conditions are acceptable, SAR has the advantage of operating in the microwave range and thus being illumination and weather independent, which makes it to an attractive alternative for data acquisition in case of natural hazards and crisis situations.

To keep this chapter self-contained, we briefly summarize the SAR imaging process of static and moving objects (Section 4.2), before describing the scheme for detection of moving vehicles in single and multi-temporal SAR interferograms (Section 4.3). The examples are mainly related to the German TerraSAR-X mission but can be easily generalized to other high-resolution SAR missions. Section 4.4 outlines the matching strategy for establishing correspondences between detection result and reference data derived from aerial photogrammetry. Finally, Section 4.5 discusses various quality issues, before Section 4.6 draws conclusions about the current developments and achievements.

4.2 SAR Imaging of Stationary and Moving Objects

In contrast to optical cameras, RADAR is an active sensor technique that emits typically frequency-modulated signals – so-called chirps – with a predefined "pulse repetition frequency" (PRF) in a side-looking, oblique imaging geometry and records the echoes scattered at the objects on the ground; see Fig. 4.1 (left) for illustration of the RADAR imaging geometry. The received echoes are correlated with reference functions eventually yielding a compressed pulse-shaped signal whose width is mainly determined by the chirp's band width (see Fig. 4.2). The travelling time of the signals is proportional to the distance to the objects and defines the image dimension perpendicular to the flight direction, the so-called *range* or *across-track* co-ordinates. The second dimension, *azimuth* or *along-track*, is simply aligned with the flight direction. While the resolution in range direction ρ_R is determined by the chirp band width (cf. Fig. 4.2) and is typically in the (sub-)meter area, the resolution in azimuth direction of the raw data depends on the antenna's real aperture characteristics (antenna length L, carrier wavelength λ, and range R) and is impractically coarse for geospatial applications. Hence, to enhance the azimuth resolution, the well-known Synthetic Aperture Radar (SAR) principle is applied, that is, the

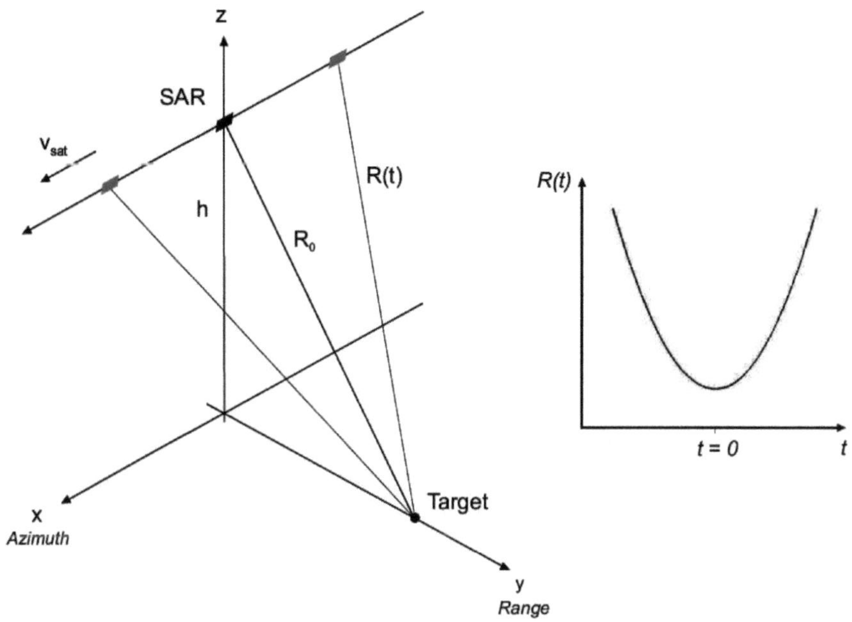

Fig. 4.1 Imaging geometry of a space-borne SAR

Fig. 4.2 Compression of sent chirp into pulse

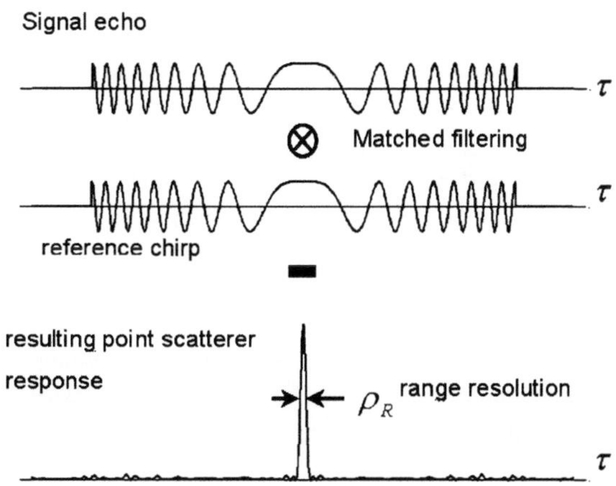

motion of the real antenna is used to construct a very long synthetic antenna by exploiting each point scatterer's range history recorded during a point's entire observation period. Since the length of the synthetic aperture increases proportional with the flying height, the resolution in azimuth direction ρ_{SA} is purely depending on the length of the physical antenna given a sufficiently large PRF to avoid aliasing.

To identify and quantify movements of objects on the ground, a thorough mathematical analysis of this so-called SAR focusing process is necessary:

The position of a Radar transmitter on board a satellite is given by $P_{sat}(t) = [x_{sat}(t); y_{sat}(t); z_{sat}(t)]$ with x being the along-track direction, y the across-track ground range direction and z being the vertical (see Fig. 4.1). An arbitrarily moving and accelerating vehicle is modeled as point scatterer at position $P(t) = [x(t);\ y(t);\ z(t)]$, and the range to it from the radar platform is defined by $R(t) = P_{sat}(t) - P(t)$. Omitting pulse envelope, amplitude, and antenna pattern for simplicity reasons, and approximating the range history $R(t)$ by a parabola (Fig. 4.1, right), the measured echo signal $u_{stat}(t)$ of this point scatterer can be written as $u_{stat}(t) = \exp\{j\ \pi\ FM\ t^2\}$ with FM being the frequency modulation rate of the azimuth chirp:

$$FM = -\frac{2}{\lambda}\frac{d^2}{dt^2}R(t) = -\frac{2}{\lambda R}v_{sat}v_B$$

and v_{sat} and v_B being the platform velocity and the beam velocity on ground, respectively. Azimuth focusing of the SAR image is performed using the matched filter concept (Bamler and Schättler 1993; Cumming and Wong 2005). According to this concept the filter must correspond to $s(t) = \exp\{-j\ \pi\ FM\ t^2\}$.

An optimally focused image is obtained by complex-valued correlation of $u_{stat}(t)$ and $s(t)$. To construct $s(t)$ correctly, the actual range or phase history of each target in the image must be known, which can be inferred from sensor and scatterer position. Usually, time dependence of the scatterer position is ignored yielding $P(t) = P$. This concept is commonly referred to as *stationary-world matched filter* (SWMF). Because of this definition, a SWMF does not correctly represent the phase history of a significantly moving object.

To quantify the impact of a significantly moving object we first assume the point to move with velocity v_{x0} in azimuth direction (along-track, see Fig. 4.3 left). The relative velocity of sensor and scatterer is different for the moving object and the surrounding stationary world. Thus, along track motion changes the frequency modulation rate *FM* of the received scatterer response. The echoed signal of a moving object is compared with the shape of the SWMF in Fig. 4.3 (right). Focusing the signal with a SWMF consequently results in an image of the object burred in azimuth direction. It is unfortunately not possible to express the amount of defocusing exactly in closed form. Yet, when considering the stationary phase approximation of the Fourier-Transform, the width Δt of the focused peak can be approximated by

$$\Delta t \approx 2T_A\frac{v_{x0}}{v_B}[s] \text{ with } T_A \text{ being the synthetic aperture time.}$$

As can be seen, the amount of defocusing depends strongly on the sensor parameters. A car traveling with 80 km/h, for instance, will be blurred by approx. 30 m when inserting TerraSAR-X parameters. However, it has to be kept in mind that this approximation only holds if $v_{x0} >> 0$. It is furthermore of interest, to which extent blurring causes a reduction of the amplitude h at position $t = 0$ (the position of the signal peak) depending on the point's along-track velocity. This can be calculated

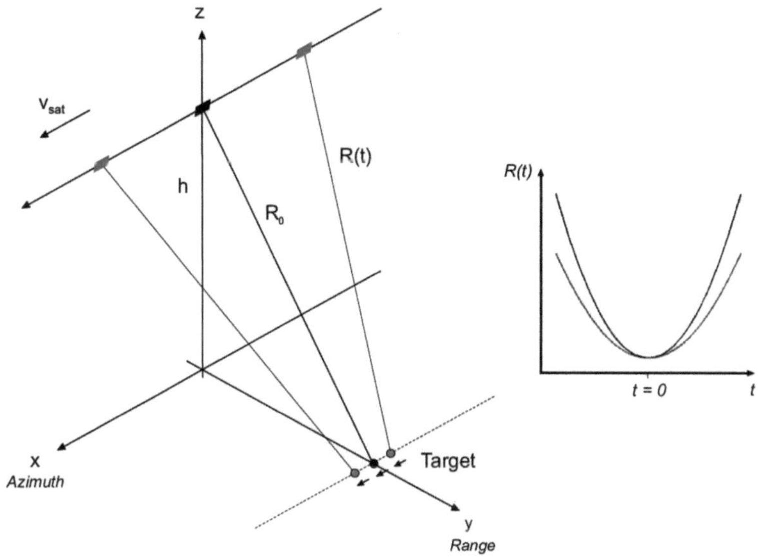

Fig. 4.3 Along-track moving object imaged by a RADAR (*left*) and resulting range history function compared with the shape of the matched filter (*right*)

by integrating the signal spectrum and making again use of the stationary phase approximation:

$$h(t = 0, v_{x0}) \approx \frac{B}{T_A} \frac{v_B}{v_{sat}} \text{ with } B \text{ being the azimuth bandwidth.}$$

When a point scatterer moves with velocity v_{y0} in across-track direction (Fig. 4.4, left), this movement causes a change of the point's range history proportional to the projection of the motion vector into the line-of-sight direction of the sensor $v_{los} = v_{y0} \sin(\theta)$, with θ being the local elevation angle. In case of constant motion during illumination the change of range history is linear and causes an additional linear phase trend in the echo signal, sketched in Fig. 4.4 (right). Correlating such a signal with a SWMF results in a focused point that is shifted in azimuth direction by

$$t_{shift} = \frac{2v_{los}}{\lambda \cdot FM}[s] \text{ in time domain, and by}$$
$$\Delta_{az} = -R\frac{v_{los}}{v_{sat}}[m] \text{ in space domain, respectively.}$$

In other words, across-track motion leads to the fact that moving objects do not appear at their "real-world" position in the SAR image but are displaced in azimuth direction – the so-called "train-off-the-track" effect. Again, when inserting typical TerraSAR-X parameters, the displacement reaches an amount of 1.5 km for a car traveling with 80 km/h in across-track direction. Figure 4.5

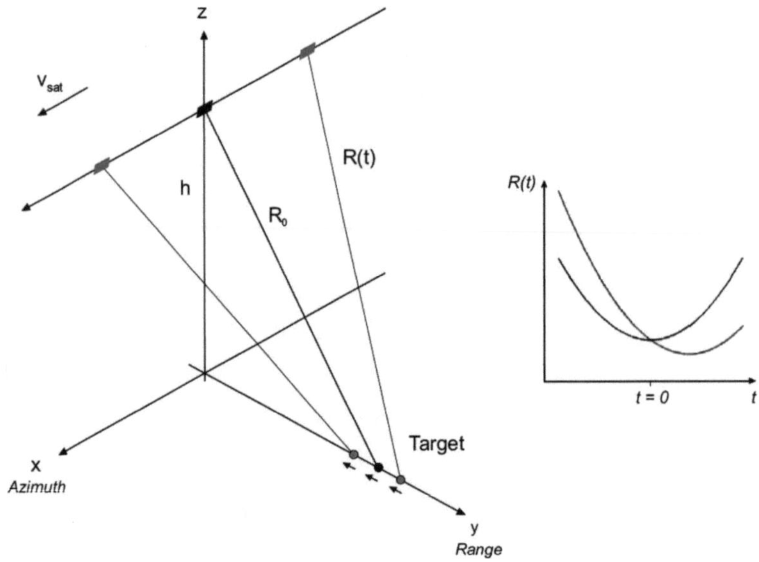

Fig. 4.4 Across-track moving object imaged by a RADAR (*left*) and resulting range history function compared with the shape of the matched filter (*right*)

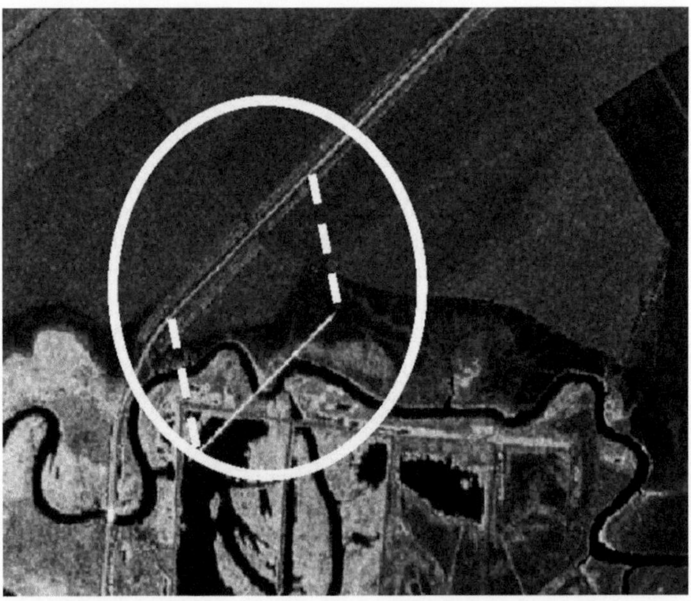

Fig. 4.5 Train off the track imaged by TerraSAR-X (due to across-track motion)

shows a cut-out of the first TerraSAR-X image that, by coincidence, included an example of the displacement effect for a train. Due to the train's across track motion, the image position of the train is displaced from its real-word position on the track.

Across-track motions not only influence the position of an object in the SAR image but also the phase difference between two images in case of an along-track interferometric data acquisition, that is, the acquisition of two SAR images within a short time frame with baseline Δl aligned with the sensor trajectory. The interferometric phase is defined as the phase difference of the two co-registered SAR images $\psi = \varphi_1 - \varphi_2$, which is proportional to motions in line-of-sight direction. Hence, the interferometric phase can also be related to the displacement in space domain:

$$\Delta_{az} = -R\frac{v_{los}}{v_{sat}} = -R\psi\frac{\lambda}{4\pi \cdot \Delta l}[m]$$

In the majority of the literature, it is assumed that vehicles travel with constant velocity and along a straight path. If vehicle traffic on roads and highways is monitored, target acceleration is commonplace and should be considered in any processor or realistic simulation. Acceleration effects do not only appear when drivers physically accelerate or brake but also due to curved roads, since the object's along-track and across-track velocity components vary on a curved trajectory during the Radar illumination. The effects caused by along-track or across-track acceleration have recently been studied in Sharma et al. (2006) and Meyer et al. (2006). These investigations can be summarized such that along-track acceleration a_x results in an asymmetry of the focused point spread function, which leads to a small azimuth-displacement of the scatterer after focusing, whose influence can often be neglected. However, the acceleration in across-track direction a_y causes a spreading of the signal energy in time or space domain. The amount of this defocusing is significant and comparable with that caused by along-track motion. We refer the interested reader to Meyer et al. (2006) where an in-depth study about all the above mentioned influences in TerraSAR-X data can be found.

4.3 Detection of Moving Vehicles

The effects of moving objects hinder the detection of cars in conventionally processed SAR images. On the other hand, these effects are mainly deterministic and thus can be exploited to not only detect vehicles but also measure their velocity. As the new space-borne SAR sensors are equipped with a Dual Receive Antenna (DRA) mode or allow to masking different parts of the antenna on a pulse-by-pulse basis (Aperture Switching, AS [Runge et al. 2006]), two SAR images of the same scene can be recorded within a small timeframe eventually forming an along-track interferogram. In defense related research the problem of detecting moving objects in such images is known as Ground Moving Target Indication (GMTI) and commonly

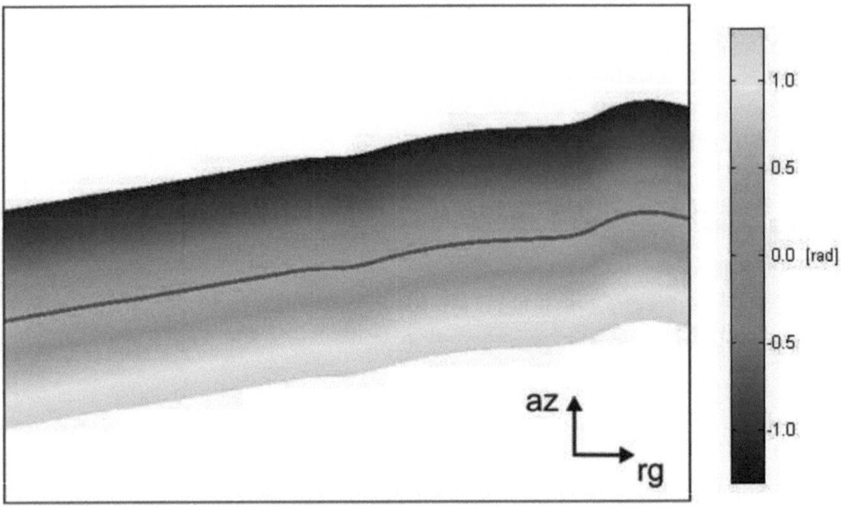

Fig. 4.6 Expected interferometric phase for a particular road depending on the respective displacement

relies on highly specialized multi-channel systems (Klemm 1998; Ender 1999). Even though civilian SAR missions are suboptimal for GMTI, their along-track interferometric data from the DRA or AS mode can be used for the detection of objects moving on ground. Several publications deal with this issue (e.g. Sikaneta and Gierull 2005; Gierull 2002).

To make detection and velocity estimation more robust Meyer et al. (2006), Suchandt et al. (2006), Hinz et al. (2007), and Weihing et al. (2007) include also GIS data from road databases as a priori information. Knowing the positions and directions of roads from GIS data, it is possible to derive a-priori knowledge for the acquired scene. Depending on the distance of a pixel to an associated road segment, which corresponds to the shift Δ_{az}, the expected phase ϕ can be predicted for each pixel. Figure 4.6 illustrates the a-priori phase ϕ for a road section of a TerraSAR-X data take. The phase is only predicted up to a maximum displacement corresponding to a maximum speed.

4.3.1 *Detection Scheme*

Since the signal of a moving vehicle will be displaced or blurred in the image, the signal will superpose with the background signal (clutter), which hampers the detection of ground moving objects. To decide whether a moving vehicle is existent or not, an expected signal hidden in clutter is compared with the actual measurement in the SAR data. Two hypotheses H_0 and H_1 shall be distinguished:

H_0: only clutter and noise are present
H_1: additionally to clutter and noise a vehicle's signal is present

The mathematical framework is derived from statistical detection theory. The optimal test is the likelihood-ratio-test:

$$\Lambda = \frac{f\left(\vec{x}|H_1\right)}{f\left(\vec{x}|H_0\right)}$$

where $$f\left(\vec{x}|H_0\right) = \frac{1}{\pi^2 |C|} \exp\left\{-\vec{X}^H C^{-1} \vec{X}\right\}$$

and $$f\left(\vec{x}|H_1\right) = \frac{1}{\pi^2 |C|} \exp\left\{-\left(\vec{X} - \vec{S}\right)^H C^{-1} \left(\vec{X} - \vec{S}\right)\right\}$$

are the probability density functions. $\vec{S}$ represents the expected signal, $\vec{X}$ stands for the measured signal, and C is the covariance matrix (see, e.g. Bamler and Hartl 1998). From the equations above the decision rule of the log-likelihood test based on threshold α can be derived:

$$\left|\vec{S}^H C^{-1} \vec{X}\right| > \alpha$$

The measured signal $\vec{X}$ consists of the SAR images from the two apertures:

$$\vec{X} = \begin{pmatrix} X_1 \\ X_2 \end{pmatrix},$$

where the indices stand for the respective channel. With the a-priori phase ϕ derived for every pixel (see, e.g., Fig. 4.6) the expected signal $\vec{S}$ can be derived:

$$\vec{S} = \begin{pmatrix} S_1 \\ S_2 \end{pmatrix} = \begin{pmatrix} \exp\left\{j\frac{\phi}{2}\right\} \\ \exp\left\{-j\frac{\phi}{2}\right\} \end{pmatrix}$$

The covariance matrix is defined:

$$C = E\left\{XX^H\right\} = \begin{pmatrix} \bar{I}_1 & \gamma \bar{I} \\ \gamma^* \bar{I} & \bar{I}_2 \end{pmatrix} \approx \begin{pmatrix} 1 & \gamma \\ \gamma & 1 \end{pmatrix}$$

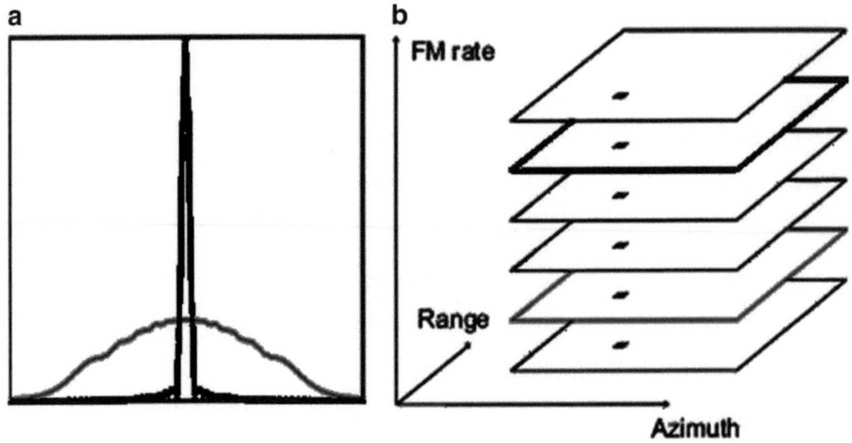

Fig. 4.7 (**a**) Blurred signal of a vehicle focused with the filter for stationary targets (*grey curve*) and the same signal focused with the correct FM rate (*black curve*). (**b**) Stack of images processed with different FM rates

with

$$\bar{I} = \sqrt{\bar{I}_1 \bar{I}_2} = \sqrt{E\left[|u_1|^2\right] E\left[|u_2|^2\right]}$$

being the normalized intensity.

A locally varying threshold α is evaluated for each pixel and decides whether a vehicle is present or not. It thereby depends on a given false alarm rate, which determines the cut-off value for the cumulative function of the log-likelihood test.

It must be considered, however, that this detection scheme assumes well-focused point scatterers. To achieve this also for (constantly) moving objects the amount of motion-induced focusing is predicted in a similar way as the expected interferometric phase based on position and orientation of the corresponding road and the parameters of the matched filter are adjusted accordingly. In addition to this, a slight uncertainty of the predicted value is accommodated by applying the detection scheme to a small stack of images focused with different FM rates, of which the best Λ is selected. Figure 4.7 illustrates this procedure schematically.

4.3.2 *Integration of Multi-temporal Data*

Using a road database and deriving the expected interferometric phase is however not the only way for including a-priori knowledge about the scene. In addition to this, the periodic revisit time of a satellite allows to collecting multi-temporal data about the scene to evaluate. The resulting image stack contains much more information – in particular about the stationary background – which can also be used to enhance the detection process.

Due to the considerable noise in space-borne SAR images, a typical feature of a detection approach as the one described above is to produce false alarms for bright stationary scatterers whose interferometric phase, by coincidence, matches the expected phase value fairly well. Hence, suppressing noise by not loosing spatial resolution is a key-issue for reliable detection. This can be accomplished for stationary objects by averaging an image stack pixel-wise over time. Figure 4.8 gives an impression of this effect. In the same sense, bright stationary spots likely to

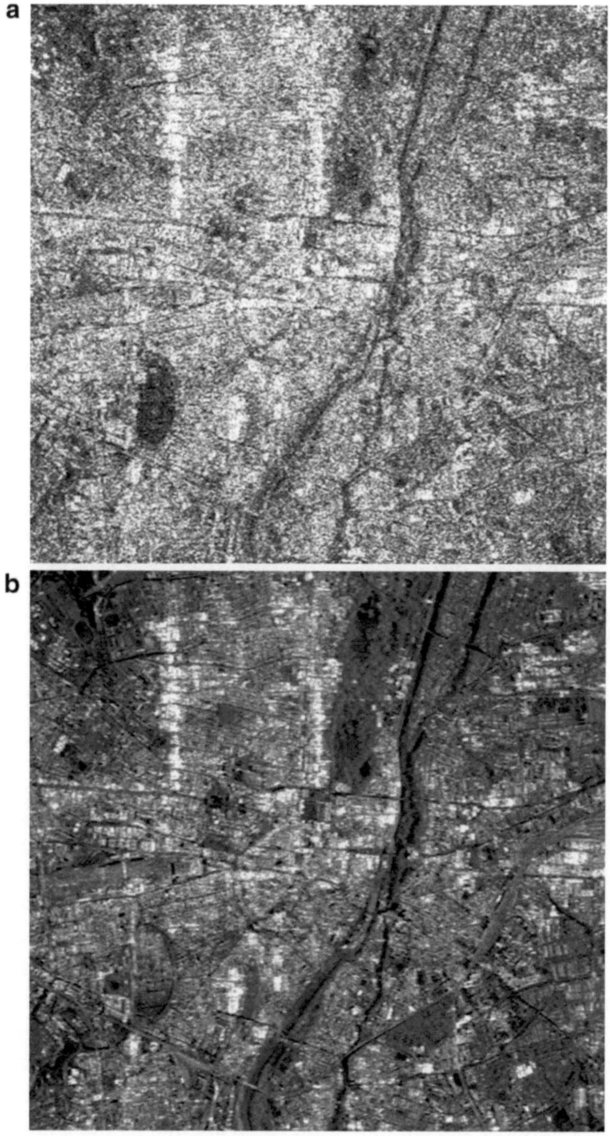

Fig. 4.8 Filtering of multi-temporal SAR data. (**a**) Single SAR amplitude image; (**b**) mean SAR amplitude image after mean filtering of 30 images

be confused with vehicles can be detected and masked before vehicle extraction. To this end we adapted the concept of Persistent Scatterer Interferometry (Ferretti et al. 2001; Adam et al. 2004) and eliminate Persistent Scatterers (PS), which feature a high and time-consistent signal-to-clutter-ratio (SCR).

Before evaluating and discussing the results achieved with the aforementioned approach, we turn to the question of matching moving objects detected in SAR images with reference data derived from optical image sequences.

4.4 Matching Moving Vehicles in SAR and Optical Data

Validating the quality of SAR traffic data acquisition is crucial to estimate the benefits of using SAR in situations motivated in the introduction. In the following, an approach for evaluating the performance of detection and velocity estimation of vehicles in SAR images is presented, which utilizes reference traffic data derived from simultaneously acquired optical image sequences. While the underlying idea of this approach is straightforward, the different sensor concepts imply a number of methodological challenges that need to be solved in order to compare the dynamics of objects in both types of imagery.

Optical image sequences allow to deriving vehicle velocities by vehicle tracking and, when choosing an appropriate focal length, they can also cover the same part of a scene as SAR images. In addition, optical images are rather easy to interpret for a human operator so that reliable reference data of moving objects can be achieved. Yet matching dynamic objects in SAR and optical data remains challenging since the two data sets do not only differ in geometric properties but also in temporal aspects of imaging. Hence, our approach for matching vehicles consists of a geometric part (Section 4.4.1) and a time-dependent part (Section 4.4.2).

4.4.1 Matching Static Scenes

Digital frame images, as used in our approach, imply the well-known central perspective imaging geometry that defines the mapping $[X, Y, Z] => [x_{img}, y_{img}]$ from object to image co-ordinates. As sketched in Fig. 4.9, the spatial resolution on ground (ρ_X) is mainly depending on the flying height H, the camera optics with focal length c, and the size of the CCD elements (ρ_x). On the other side, the geometry of SAR results from time/distance measurements in range direction and parallel scanning in azimuth direction defining a mapping $[X, Y, Z] => [x_{SAR}, R_{SAR}]$. 3D object co-ordinates are thus mapped onto circles of radii R_{SAR} parallel aligned in azimuth direction x_{SAR}. As mentioned above, after SAR focusing, the spatial resolutions (ρ_R, ρ_{SA}) of range and azimuth dimension are mainly depending on the bandwidth of the range chirp and the length of the physical antenna. Please note that the

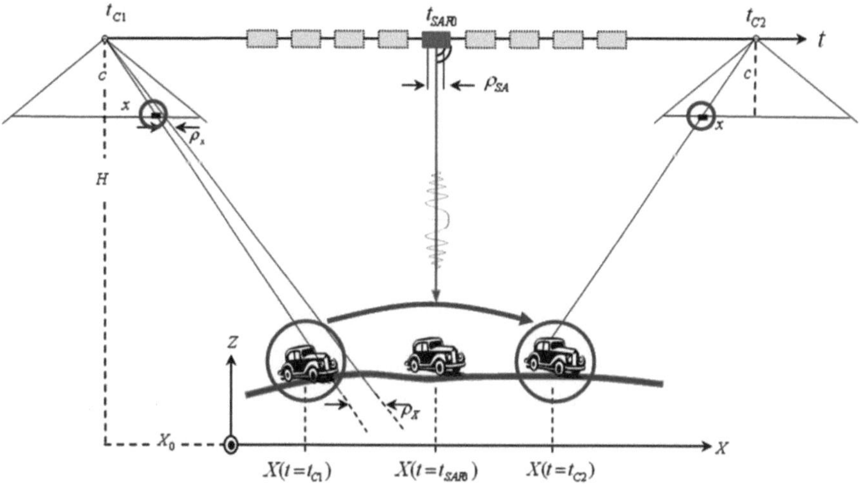

Fig. 4.9 Imaging moving objects in optical image sequences compared to SAR images in azimuth direction

field of view defined by the side-looking viewing angle of a RADAR system is usually too large to derive the 3 D directly so that SAR remains a 2D imaging system.

The different imaging geometries of frame imagery and SAR require the incorporation of differential rectification to assure highly accurate mapping of one data set onto the other. To this end, we employ a Digital Elevation Model (DEM), on which both data sets are projected.[1] Direct georeferencing the data sets is straightforward, if the exterior orientation of both sensors is known precisely. In case the exterior orientation lacks of high accuracy – which is especially commonplace for the sensor attitude – an alternative and effective approach (Müller et al. 2007) is to transform an existing ortho-image into the approximate viewing geometry at sensor position C:

$$[x_C, y_C] = f(\boldsymbol{p}_{ortho}, X_{ortho}, Y_{ortho}, Z_{ortho})$$

where $\boldsymbol{p}_{ortho}$ is the vector of approximate transformation parameters. Refining the exterior orientation reduces then to finding the relative transformation parameters $\boldsymbol{p}_{rel}$ between the given image and the transformed ortho-image, that is

$$[x_{img}, y_{img}] = f(\boldsymbol{p}_{rel}, x_C, y_C),$$

which is accomplished by matching interest points. Due to the large number of interest points, $\boldsymbol{p}_{rel}$ can be determined in a robust manner in most cases. This procedure can be applied to SAR images in a very similar way – with the only modification that, now, $\boldsymbol{p}_{ortho}$ describe the transformation of the ortho-images into the SAR slant range geometry. The result of geometric matching consists of accurately geo-coded

[1] We use an external DEM; though, it could be derived directly from the frame images.

optical and SAR images, so that for each point in the one data set a conjugate point in the other data set can be assigned. However, geometrically conjugate points may have been imaged at different times. This is crucial for matching moving vehicles and has not been considered in the approach outlined so far.

4.4.2 Temporal Matching

The different sensor principles of SAR and optical cameras lead to the fact that the time of imaging a moving object would differ for both sensors – even in the theoretical case of exactly coinciding trajectories of the SAR antenna's phase center and the camera's projection center. Frame cameras take snapshots of a scene at discrete time intervals with a frame rate of, for example, 0.3–3Hz. Due to overlapping images, most moving objects are imaged at multiple times. SAR, in contrast, scans the scene in a quasi-continuous mode with a PRF of 1,000–6,000 Hz, that is each line in range direction gets a different time stamp. Due to the parallel scanning principle, a moving vehicle is imaged only once, however, as outlined above, possibly defocused and at a displaced position.

Figure 4.9 compares the two principles: It shows the overlapping area of two frame images taken at position C_1 at time t_{C1} and position C_2 at t_{C2}, respectively. A car traveling along the sensor trajectory is thus imaged at the time-depending object co-ordinates $X(t = t_{C1})$ and $X(t = t_{C2})$. On the other hand, this car is imaged by the SAR at Doppler-zero position $X(t = t_{SAR0})$, that is when the antenna is closest to the object. Figure 4.9 illustrates that exact matching the car in both data sets is not possible because of the differing acquisition times. Therefore, a temporal interpolation along the trajectory is mandatory and the specific SAR imaging effects must be considered. Hence, our strategy for matching includes following steps:

- Reconstruction of a continuous car trajectory from the optical data by piecewise interpolation (e.g. between control points $X[t = t_{C1}]$ and $X[t = t_{C2}]$ in Fig. 4.9).
- Calculation of a time-continuous velocity profile along the trajectory, again using piecewise interpolation. An uncertainty buffer can be added to this profile to include the measurement and interpolation inaccuracies.
- Transforming the trajectory into the SAR image geometry and adding the displacement due to the across track velocity component. In the same way, the uncertainty buffer is transformed.
- Intersection/matching of cars detected in the SAR image with the trajectory by applying nearest neighbor matching. Cars not being matched are considered to be false alarms.

As result, each car detected in the SAR data (and not labeled as false alarm) is assigned to a trajectory and, thereby, uniquely matched to a car found in the optical data. Figure 4.10 visualizes intermediate steps of matching: a given highway section (maroon line); the corresponding displacement area color coded by an iso-velocity surface; a displaced track of a smoothly decelerating car (green line); and a cut-out

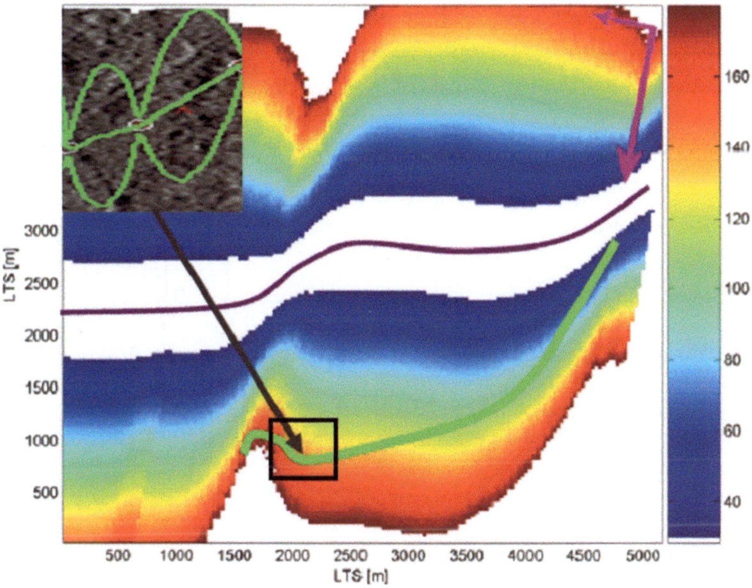

Fig. 4.10 Matching: highway section (magenta line), corresponding displacement area (color-coded iso-velocity surface), displaced track of a decelerating car (green line), local RADAR coordinate system (magenta arrows). Cut-out shows detail of uncertainty buffer. Cars correctly detected in the SAR image are marked by red crosses

of the displaced uncertainty buffer. The car correctly detected in the SAR image and assigned to the trajectory is marked by the red cross in the cut-out. The local RADAR co-ordinate axes are indicated by magenta arrows.

4.5 Assessment

In order to validate the matching and estimate the accuracy, localization and velocity determination have been independently evaluated for optical and SAR imagery.

4.5.1 Accuracy of Reference Data

To determine the accuracy of reference data, theoretically derived accuracies are compared with empirical accuracies measured in aerial images sequences containing reference cars. Under the of constant image scale, the vehicle velocity v_{I2-1} derived from two consecutive co-registered or geo-coded optical images $I1$ and $I2$ is simply calculated by the displacement Δs over the time elapsed Δt:

$$v_{I2-1} = \frac{\Delta s}{\Delta t} = \frac{\sqrt{(X_{I2}-X_{I1})^2+(Y_{I2}-Y_{I1})^2}}{t_{I2}-t_{I1}} = m\frac{\sqrt{(r_{I2}-r_{I1})^2+(c_{I2}-c_{I1})^2}}{t_{I2}-t_{I1}}$$

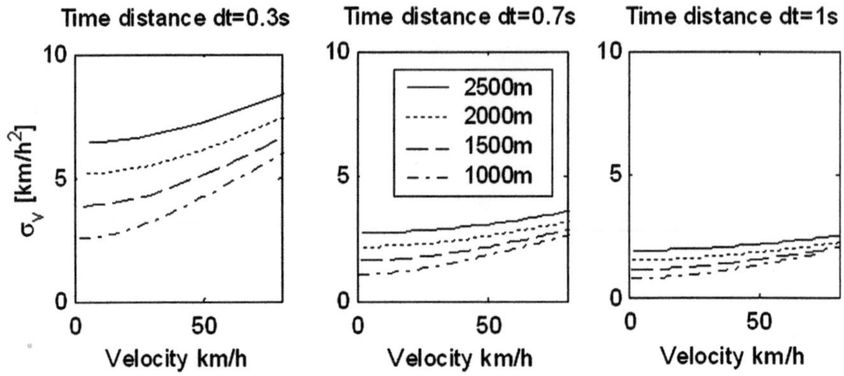

Fig. 4.11 Standard deviation of vehicle velocities (0–80 km/h) derived from vehicle positions in two consecutive frames. Time differences between frames vary (0.3 s, 0.7 s, 1.0 s) as well as flying height (1,000 up to 2,500 m)

where X_{Ii} and Y_{Ii} are object coordinates, r_{Ii} and c_{Ii} the pixel coordinates of moving cars, and t_{Ii} the acquisition times of images $i = 1, 2$. The advantage of the second expression is the separation of the image geo-coding process (represented by factor m) from the process of car measurements, which simplifies the calculation of theoretical accuracies. Thus, three main error sources on the accuracy of car velocity can be identified: the measurement error σ_P in pixel units, the scale error σ_m assumed to be caused mainly by DEM error σ_H, and finally the time error σ_{dt} of the image acquisition time. For the simulations shown in Fig. 4.11 following values have been used: $\sigma_P = 1$, $\sigma_{dt} = 0.02s$, $\sigma_H = 10$ m. It shows decreasing accuracy for greater car velocities and shorter time distances, because the influence of the time distance error gets stronger. On the other hand, the accuracies decrease with higher flight heights as the influence of measurement errors increases. Last is converse to the effect, that with lower flight heights the influence of the DEM error gets stronger.

The theoretical accuracies are assessed with measurements in real airborne images and with data from a reference vehicle equipped with GPS receivers. The time distance between consecutive images was 0.7 s, so that the accuracy of GPS velocity can be compared to the center panel of Fig. 4.11. Exact assignment of the image acquisition time to GPS track times was a prerequisite for this validation and was achieved by connecting the camera flash interface with the flight control unit. Thus, each shoot could be registered with a time error less than 0.02 s. The empirical accuracies derived from the recorded data are slightly worse than theoretical values due to inaccuracies in the GPS/IMU data processing. Yet, it also showed that the empirical standard deviation is below 5 km/h which provides a reasonable hint for defining the velocity uncertainty buffer described above.

Table 4.1 Comparison of velocities from GPS and SAR

Vehicle #	${v_{Tn}}^{GPS}$ (km/h)	${v_{Tn}}^{disp}$ (km/h)	Δv (km/h)
4	5.22	5.47	0.25
5	9.24	9.14	0.1
6	10.03	9.45	0.58
8	2.16	2.33	0.17
9	4.78	4.86	0.08
10	3.00	2.01	0.01
11	6.31	6.28	0.03

4.5.2 Accuracy of Vehicle Measurements in SAR Images

Several flight campaigns have been conducted, to estimate the accuracy of velocity determination from SAR images. To this end, an airborne Radar system has been used with a number of modifications, so that the resulting raw data is comparable with the satellite data. During the campaign eight controlled vehicles moved along the runway of an airfield. All vehicles were equipped with a GPS system with a 10 Hz logging frequency for measuring their position and velocity. Some small vehicles were equipped with corner reflectors to make them visible in the image.

A quantitative estimate of the quality of velocity determination using SAR images can by obtained by comparing the velocity computed from the along-track displacement in the SAR images ${v_{Tn}}^{disp}$ to the GPS velocity ${v_{Tn}}^{GPS}$ (see Table 4.1). The numerical results show that the average difference between the velocity measurements is significantly below 1 km/h. When expressing the accuracy of velocity in form of a positional uncertainty, this implies that the displacement effect influences a vehicle's position in the SAR image only up to a few pixels depending on the respective sensor parameters.

4.5.3 Results of Traffic Data Collection with TerraSAR-X

A modular traffic processor has been developed in prior work at DLR (Suchandt et al. 2006), in which different moving vehicle detection approaches are integrated. The proposed likelihood ratio detector has been included additionally into this environment. The test site near Dresden, Germany, has been used for analyses. The AS data take DT10001 was processed with the traffic processor, while only the likelihood ratio detector described above was used to detect the vehicles in the SAR data. In addition, a mean image was calculated based on the multi-temporal images of this scene, in order to generate a SCR-map and then to determine PS candidates. Candidates were chosen with an SCR greater than 2.0.

During the acquisition of DT10001 by TerraSAR-X a flight campaign over the same scene was conducted. Optical images were acquired with the DLR's 3K optical system mounted on the airplane. Detection and tracking of the vehicles in the optical images delivered reference data to ensure the detections results of the likelihood ratio detector in SAR data.

Figure 4.12 shows a part of the evaluated scene. The temporal mean image is overlaid with the initial detections plotted in green. The blue rectangles mark the displaced positions of the reference data which have been estimated by

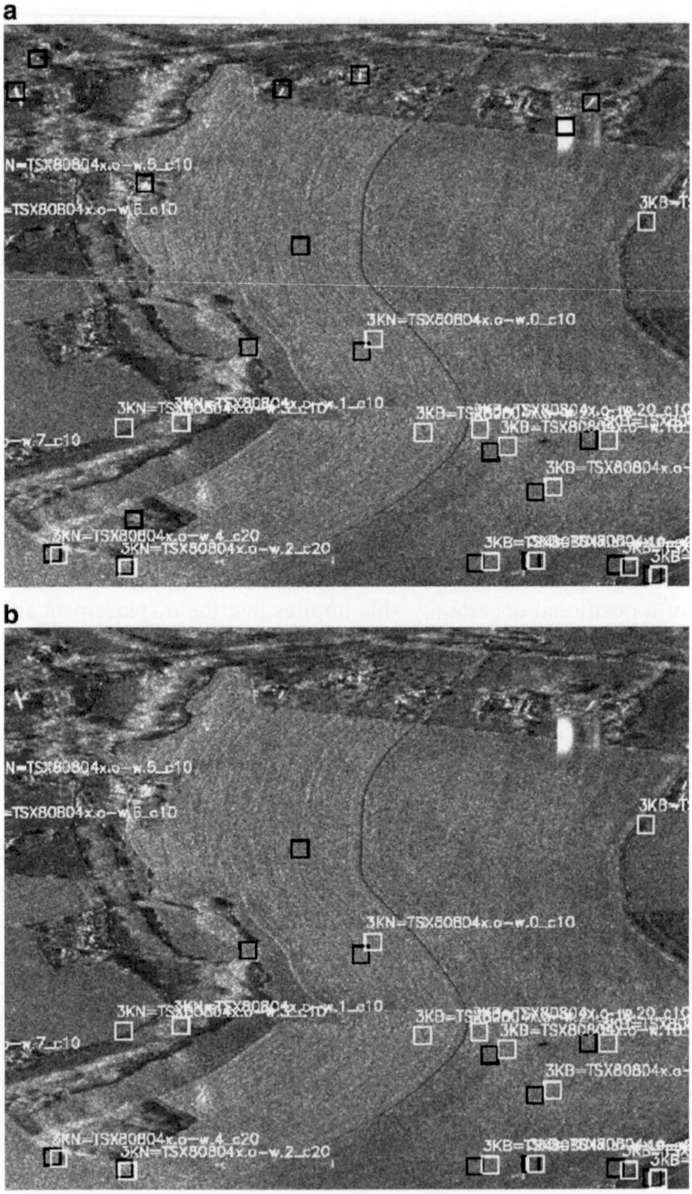

Fig. 4.12 Detections (*green*) and reference data (*blue*) at the displaced positions of the vehicles overlaid on the temporal mean image: (**a**) all initial detections; (**b**) after PS elimination

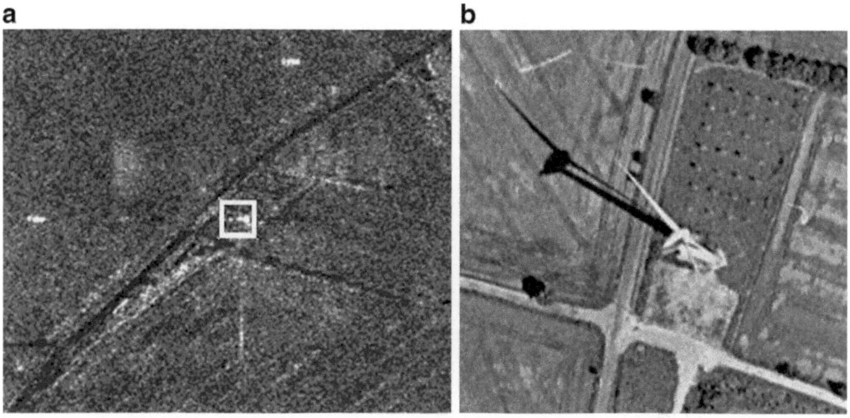

Fig. 4.13 (**a**) Detection in the SAR image; (**b**) optical image of the same area

calculating the displacement according to their measured velocities. Due to measuring inaccuracies described above these positions may differ a bit from those of the detections in the SAR images.

Having analyzed the SCR over time to identify PS candidates, some false detections have been eliminated (compare Fig. 4.12a and b). One example for such a persistent scatterer which was detected wrongly is shown in Fig. 4.13. On the left hand side the position of the detection is marked in the mean SAR image and on the left hand side one can see the same area in an optical image. The false detection is obviously a wind wheel. Figure 4.14 shows the final results for the evaluated data take DT10001, a section of the motorway A4. The final detections results of the traffic processor using the likelihood ratio detector are marked with the red rectangles. The triangles are the positions of these vehicles backprojected to the assigned road. These triangles are color-coded regarding their estimated velocity ranging from red to green (0–250 km/h). Finally 33 detections have been estimated as vehicles. In this figure again the blue rectangles label the estimated positions of the reference data. Eighty-one reference vehicles have been measured in the same section in the optical images.

Comparing the final detections in the SAR data with the reference data, it arises that one detection is a false one. Consequently we have for this example a correctness of 97% and a completeness of 40%. This kind of quality values has been achieved for various scenes. The detection rate is generally quite fair, as expected also from theoretical studies (Meyer et al. 2006). However, the low false alarm rate encourages an investigation of the reliability of more generic traffic parameters like mean of velocity per road segment or traffic flow per road segment etc. To assess the quality of these parameters, Monte Carlo simulations with varying detection rates and false alarm rates have been carried out and compared with reference data, again derived from optical image sequences. The most essential simulation results

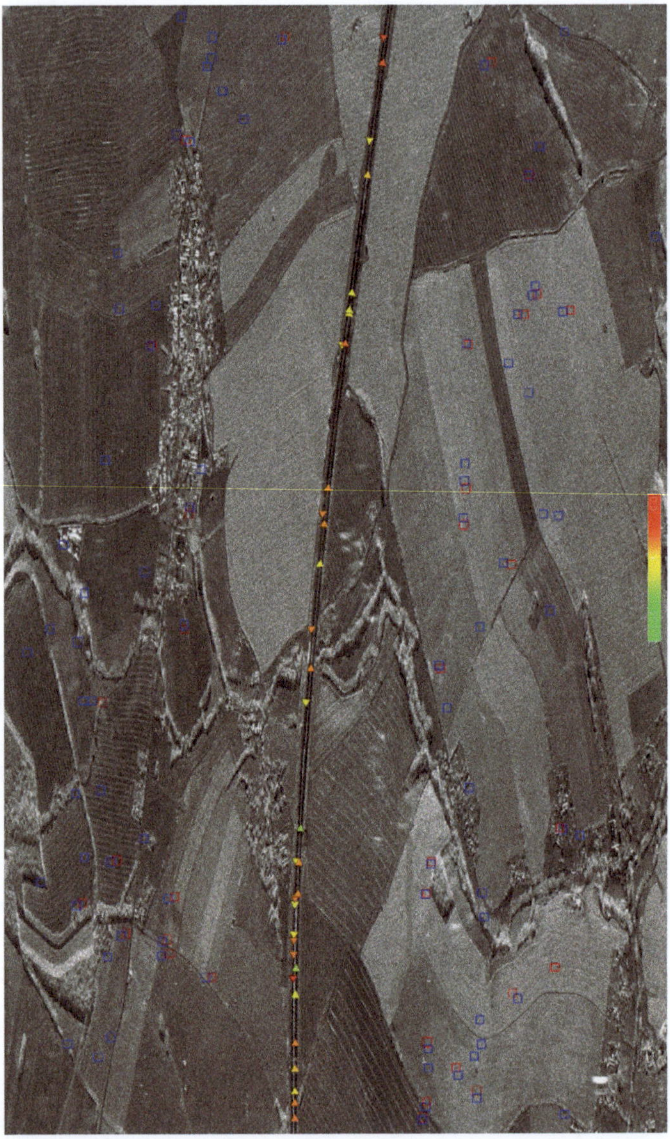

Fig. 4.14 Final detection results (*red*) and reference data (*blue*) at the displaced positions of the vehicles overlaid on the mean SAR image

are listed in Table 4.2. As can be seen, even for a lower percentage of detections in the SAR data, reliable parameters for velocity profiles can be extracted. A detection rate of 50% together with a false alarm rate of 5% still allows to estimating the velocity profile along a road section with a mean accuracy of approximately 5 km/h at a computed standard deviation of the simulation of 2.6 km/h.

Table 4.2 Result of Monte Carlo simulation to estimate the accuracy of reconstructing a velocity profile along a road section depending on different detection and false alarm rates

30% correct/5% false		30% correct/10% false		30% correct/25% false		50% correct/5% false		50% correct/10% false		50% correct/25% false	
RMS (km/h)]	σ (km/h)	RMS (km/h)	σ (km/h)	RMS (km/h)	σ (km/h)	RMS (km/h)	σ (km/h)	RMS (km/h)	σ (km/h)	RMS (km/h)	σ (km/h)
5.97	3.17	8.03	4.66	11.30	6.58	5.22	2.61	7.03	4.01	10.25	6.27

4.6 Summary and Conclusion

This chapter presented an approach for moving vehicle detection in space-borne SAR data and demonstrated its applicability using TerraSAR-X AS data. To evaluate the performance of the approach, a sophisticated scheme for spatio-temporal co-registration of dynamic objects in SAR and optical imagery has been developed. It was used to validate the performance of vehicle detection and velocity estimation from SAR images compared to reference data derived from aerial image sequences. The evaluation showed the limits of the approach in terms of detection rate but also the potential to deliver reliable information about the traffic situation on roads concerning more generic traffic parameters (mean velocity, traffic flow). These were additionally analyzed by Monte Carlo simulations. It should be noted, however, that the approach is limited to open and rural scenes, where layover and radar-shadow rarely appears and the assumption of homogeneous background clutter is approximately fulfilled.

References

Adam N, Kampes B, Eineder M (2004) Development of a scientific permanent scatterer system: modifications for mixed ERS/ENVISAT time series. In: Proceedings of ENVISAT symposium, Salzburg, Austria

Bamler R, Hartl P (1998) Synthetic aperture radar interferometry. Inverse Probl 14:R1–R54

Bamler R, Schättler B (1993) SAR geocoding, Chapter 3. Wichmann, Karlsruhe, pp 53–102

Bethke K-H, Baumgartner S, Gabele M, Hounam D, Kemptner E, Klement D, Krieger G, Erxleben R (2006) Air- and spaceborne monitoring of road traffic using SAR moving target indication – Project TRAMRAD. ISPRS J Photogramm Remote Sens 61(3/4):243–259

Chiu S, Livingstone C (2005) A comparison of displaced phase centre antenna and along-track interferometry techniques for RADARSAT-2 ground moving target indication. Can J Remote Sens 31(1):37–51

Cumming I, Wong F (2005) Digital processing of synthetic aperture radar data. Artech House, Boston, MA

Ender J (1999) Space-time processing for multichannel synthetic aperture radar. Electron Commun Eng J 11(1):29–38

Ferretti A, Prati C, Rocca F (2001) Permanent scatterers in SAR interferometry. IEEE Trans Geosci Remote Sens 39(1):8–20

Gierull C (2002) Moving target detection with along-track SAR interferometry. Technical Report DRDC-OTTAWA-TR-2002–084, Defence Research & Development Canada

Hinz S, Bamler R, Stilla U (eds) (2006) ISPRS journal theme issue: "Airborne and spaceborne traffic monitoring". Int J Photogramm Remote Sens 61(3/4)

Hinz S, Meyer F, Eineder M, Bamler R (2007) Traffic monitoring with spaceborne SAR – theory, simulations, and experiments. Comput Vis Image Underst 106:231–244

Klemm R (ed.) (1998) Space-time adaptive processing. The Institute of Electrical Engineers, London

Livingstone C-E, Sikaneta I, Gierull C, Chiu S, Beaudoin A, Campbell J, Beaudoin J, Gong S, Knight T-A (2002) An airborne Synthetic Aperture Radar (SAR) experiment to support RADARSAT-2 Ground Moving Target Indication (GMTI). Can J Remote Sens 28(6):794–813

Meyer F, Hinz S, Laika A, Weihing D, Bamler R (2006) Performance analysis of the TerraSAR-X traffic monitoring concept. ISPRS J Photogramm Remote Sens 61(3–4):225–242

Müller R, Krauß T, Lehner M, Reinartz P (2007) Automatic production of a European orthoimage coverage within the GMES land fast track service using SPOT 4/5 and IRS-P6 LISS III data. Int Arch Photogramm Remote Sens Spat Info Sci 36(1/W51), on CD

Runge H, Laux C, Metzig R, Steinbrecher U (2006) Performance analysis of virtual multi-channel TS-X SAR-Modes. In: Proceedings of EUSAR'06, Germany

Sharma J, Gierull C, Collins M (2006) The influence of target acceleration on velocity estimation in dual-channel SAR-GMTI. IEEE Trans Geosci Remote Sens 44(1):134–147

Sikaneta I, Gierull C (2005) Two-channel SAR ground moving target indication for traffic monitoring in urban terrain. Int Arch Photogramm Remote Sens Spat Info Sci 61(3–4):95–101

Suchandt S, Eineder M, Müller R, Laika A, Hinz S, Meyer F, Palubinskas G (2006) Development of a GMTI processing system for the extraction of traffic information from TerraSAR-X data. In: Proceedings of EUSAR European Conference on Synthetic Aperture Radar

Weihing D, Hinz S, Meyer F, Suchandt S, Bamler R (2007) Detecting moving targets in dual-channel high resolution spaceborne SAR images with a compound detection scheme. In: Proceedings of International Geoscience and Remote Sensing Symposium (IGARSS'07), Barcelona, Spain, on CD

Chapter 5
Object Recognition from Polarimetric SAR Images

Ronny Hänsch and Olaf Hellwich

5.1 Introduction

In general, object recognition from images is concerned with separating a connected group of object pixels from background pixels and identifying or classifying the object. The indication of the image area covered by the object makes information which is implicitly given by the group of pixels, explicit by naming the object. The implicit information can be contained in the measurement values of the pixels or in the locations of the pixels relative to each other. While the former represent radiometric properties, the latter is of geometric nature describing the shape or topology of the object.

Addressing the specific topic of object recognition from Polarimetric Synthetic Aperture Radar (PolSAR) data this paper focuses on PolSAR aspects of object recognition. However, aspects related to general object recognition from images will be discussed briefly, where they meet PolSAR or remote sensing specific issues. In order to clarify the scope of the topic a short summary of important facets of the general problem of object recognition from imagery is appropriate here, though not specific to polarimetric SAR data.

The recognition of objects is based on knowledge about the object appearance in the image data. This is the case for human perception as well as for automatic recognition from imagery. This knowledge, commonly called object model, may be more or less complex for automatic image analysis, depending on the needs of the applied recognition method. Yet it cannot be left away, but is always present, either explicitly formulated, for example in the problem modeling or implicitly by the underlying assumptions of the used method – sometimes even without conscious intention of the user.

Object recognition is organized in several hierarchical layers of processing. The lowest one accesses the image pixels as input and the highest one delivers object

R. Hänsch (✉) and O. Hellwich
Technische Universität, Berlin Computer Vision and Remote Sensing, Franklinstr. 28/29, 10587 Berlin, Germany
e-mail: rhaensch@fpk.tu-berlin.de; hellwich@cs.tu-berlin.de

U. Soergel (ed.), *Radar Remote Sensing of Urban Areas*, Remote Sensing and Digital Image Processing 15, DOI 10.1007/978-90-481-3751-0_5,

instances as output. Human perception (Marr 1982; Hawkins 2004; Pizlo 2008) and automatic processing consist both of low-level feature extracting as well as of hypothesizing instances of knowledge-based concepts and their components, i.e., instances of the object models. Low-level feature extraction is data driven and generates output which is semantically more meaningful than the input. It is therefore the first step of the so-called bottom-up processing. Features may for instance be vectors containing radiometric parameters or parametric descriptions of spatial structures, such as edge segments. Bottom-up processing occurs on several levels of the processing hierarchy. Low-level features may be input to mid-level processing like grouping edge segments into connected components. An example of mid-level bottom-up-processing is the suggestion of a silhouette consisting of several edges. Predicting lower level object or object part instances on the basis of higher level assumptions is the inversion of bottom-up and therefore called top-down processing. It is knowledge driven and tries to find evidence for hypotheses in the data. Top-down processing steps usually follow preceding bottom-up steps giving reason to assume presence of an object. It generates more certainty with respect to a hypothesis for instance by searching missing parts, more complete connected components or additional proofs in spatial, or semantic context information. In elaborate object recognition methods bottom-up and top-down processing are mixed making the processing results more robust (see Borenstein and Ullman 2008, for example). For those hybrid approaches a sequence of hierarchical bottom-up results on several layers in combination with top-down processing yield to more certainty about the congruence of the real world and object models. Those conclusions were made by model knowledge about object relations and object characteristics like object appearance and object geometry. In this way specific knowledge about object instances is generated from general model knowledge.

Image analysis also tackles the problem of automatic object model generation by designing methods that find object parts, their appearance descriptions, and their spatial arrangement automatically. One example for optical imagery is proposed in Leibe et al. (2004) and is based on analysing sample imagery of objects using scale-invariant salient point extractors. Those learning based approaches are very important for analysing remote sensing imagery, for example polarimetric SAR data, as they ease the exchange of object types, which have to be recognized, as well as sensor types and image acquisition modes by automatically adjusting object models to new or changed conditions.

Remote sensing, as discussed here, is addressing geoinformation such as land use or topographic entities. In general those object categories are not strongly characterized by shape, in contrast to most other objects, usually to be recognized from images. Their outline often rather depends on spatial context such as topography and neighboring objects as well as on cultural context such as inheritance rules for farmland and local utilization customs. Therefore, remote sensing object recognition have to rely to a larger degree on radiometric properties than geometric features. In addition to the outline or other geometric attributes of an object, texture and color parameters are very important. Nevertheless, this does not mean that object recognition can rely on parameters observable within single pixels alone. Though this would be possible for tasks such as land use classification from low-resolution remote sensing imagery, object recognition from high-resolution remote sensing imagery requires the use of groups of pixels and also shape information – despite of the previous remarks. This is due to the relation of sensor resolution and pixel size and the way humans categorise their living environment semantically.

Though it may seem obvious that the sensor-specific aspects of object recognition are mainly related to radiometric issues rather than geometric ones, we nevertheless have to address geometric issues as well. This is due to the fact that the shape of the image of an object does not only depend on the object but also on sensor geometry. For instance in SAR image data we observe sensor-specific layover and shadow structures of three-dimensional objects and asterisk-shape processing artifacts around particularly bright targets outshining their neighborhood. In this paper we point out methods that are suitable to extract those structures enabling to recognize the corresponding objects in a better way.

The purpose of this chapter is to acquaint the reader with object recognition from polarimetric SAR data and to give an overview about this important part of SAR related research. Therefore, instead of explaining only a few state-of-the-art methods of object recognition in PolSAR data in detail, we rather try to provide information about advantages, limitations, existing or still needed methods, and prospects of future work.

We first explain the acquisition, representation, and interpretation of radiometric information of polarimetric SAR measurements in detail. After this general introduction to PolSAR we summarize object properties causing differences in the impulse response of the sensor, hence allowing differentiating between several objects. In addition, we address signal characteristics and models, which lead to algorithms for information extraction in SAR and PolSAR data. Besides general aspects of object recognition there are aspects that are specific to all approaches of object recognition from high-resolution remote sensing imagery. We shortly summarize those non-SAR-specific remote sensing issues. Furthermore, the specific requirements on models for object recognition in polarimetric SAR data will be discussed.

5.2 SAR Polarimetry

This section gives a short introduction to polarimetric SAR data and will briefly discuss acquisition, representation, basic features, and statistical models. Much more broader as well as more detailed information can be found in Lee and Pottier (2009) and Massonnet and Souyris (2008).

Synthetic Aperture Radar (SAR) measures the backscattered echo of an emitted microwave signal. Besides the known properties of the transmitted wave, amplitude and phase of the received signal depend strongly on geometric, radiometric, and physical characteristics of the illuminated ground. Electromagnetic waves, as those used by SAR, can be transmitted with a particular polarisation. While the electrical field component of a non-polarized transverse wave oscillates in all possible

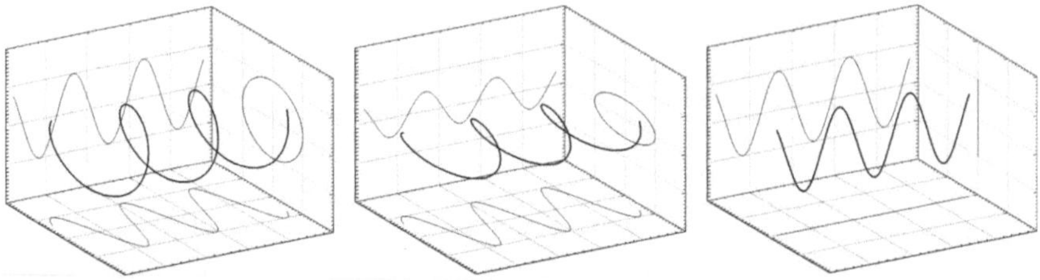

Fig. 5.1 From left to right: circular, elliptical, and linear (vertical) polarisation

Fig. 5.2 Single channel SAR (*left*) and PolSAR (*right*) image of Berlin Tiergarten (both TerraSAR-X)

directions perpendicular to the wave propagation, there are three different kinds of polarisations, i.e., possible restrictions of oscillation. These three polarisation types, namely circular, elliptical, and linear polarisation, are illustrated in Fig. 5.1.

The electrical field component of a linear polarised wave oscillates only in a single plane. This type of polarisation is the most common used in PolSAR, since it is the simplest one to emit from a technical point of view. However, a single polarisation is not sufficient to obtain fully polarimetric SAR data. That is why in remote sensing the transmit polarisation is switched between two orthogonal linear polarisations while co- and cross polarized signals are registered simultaneously. The most commonly used orientations are horizontal polarisation H and vertical polarisation V.

The advantage of a polarised signal is, that most targets show different behaviours regarding different polarisations. Furthermore, some scatterers change the polarisation of the incident wave due to material or geometrical properties. Because of this dependency, PolSAR signals contain more information about the scattering process, which can be exploited by all PolSAR image processing methods, like visualisation, segmentation, or object recognition.

Figure 5.2 shows an example explaining why polarisation is advantageous. The data is displayed in a false colour composite based on the polarimetric information.

The ability to visualize a colored representation of PolSAR data, where the colors indicate different scattering mechanism, makes visual interpretation easier.

PolSAR sensors have to transmit and receive in two orthogonal polarisations to obtain fully polarimetric SAR data. Since most sensors cannot work in more than one polarisation mode at the same time, the technical solutions always cause some loss in resolution and image size due to ambiguity rate and PRF constraints. Another answer to this problem is to waive one of the different polarisation combinations and to use, for example, the same mode for receiving as for transmitting, which results in dual-pol in contrast to quad-pol data.

The measurement of the backscattered signal of a resolution cell can be represented as complex scattering matrix **S**, which depends only on the geometrical and physical characteristics of the scattering process. Under the linear polarisation described above the scattering matrix is usually defined as:

$$\mathbf{S} = \begin{pmatrix} S_{HH} & S_{HV} \\ S_{VH} & S_{VV} \end{pmatrix} \tag{5.1}$$

where the lower indices of S_{TR} stand for transmit (T) and receive polarisation (R), respectively.

To enable a better understanding of the scattering matrix a lot of decompositions have been proposed. In general these decompositions are represented by a complete set $\boldsymbol{\Psi}$ of complex 2×2 basis matrices, which decompose the scattering matrix and are used to define a scattering vector **k**. The ith component of **k** is given by:

$$k_i = \frac{1}{2} tr(\mathbf{S} \cdot \boldsymbol{\Psi}_i), \tag{5.2}$$

where $\boldsymbol{\Psi}_i$ is an element of the set $\boldsymbol{\Psi}$ and $tr(\cdot)$ is the trace operator.
The most common decompositions are the lexicographic scattering vector $\mathbf{k}_L$ defined by

$$\mathbf{k}_L = (S_{HH}, S_{HV}, S_{VH}, S_{VV})^T, \tag{5.3}$$

which is obtained by using $\boldsymbol{\Psi}_L$ as set of basis matrices

$$\boldsymbol{\Psi}_L = \left\{ 2 \cdot \begin{pmatrix} 1 & 0 \\ 0 & 0 \end{pmatrix}, 2 \cdot \begin{pmatrix} 0 & 1 \\ 0 & 0 \end{pmatrix}, 2 \cdot \begin{pmatrix} 0 & 0 \\ 1 & 0 \end{pmatrix}, 2 \cdot \begin{pmatrix} 0 & 0 \\ 0 & 1 \end{pmatrix} \right\} \tag{5.4}$$

and the Pauli scattering vector $\mathbf{k}_P$ defined by

$$\mathbf{k}_P = \frac{1}{\sqrt{2}} \cdot (S_{HH} + S_{VV}, S_{HH} - S_{VV}, S_{HV} + S_{VH}, i(S_{HV} - S_{VH}))^T \tag{5.5}$$

where the Pauli matrices set $\boldsymbol{\Psi}_P$ is

$$\boldsymbol{\Psi}_P = \left\{ \sqrt{2} \cdot \begin{pmatrix} 1 & 0 \\ 0 & 1 \end{pmatrix}, \sqrt{2} \cdot \begin{pmatrix} 1 & 0 \\ 0 & -1 \end{pmatrix}, \sqrt{2} \cdot \begin{pmatrix} 0 & 1 \\ 1 & 0 \end{pmatrix}, \sqrt{2} \cdot \begin{pmatrix} 0 & -i \\ i & 0 \end{pmatrix} \right\} \tag{5.6}$$

While the lexicographic scattering vector is more related to the sensor measurements, the Pauli scattering vector enables a better interpretation of physical characteristics of the scattering process. Of course both are only two different representations of the same physical fact and there is a simple unitary transformation to convert each of them into the other.

A SAR system, where transmitting and receiving antenna are mounted on the same platform and are therefore nearly at the same place, is called monostatic SAR. In this case and under the basic assumption of reciprocal scatterers the cross-polar channels contain the same information:

$$S_{HV} = S_{VH} = S_{XX} \tag{5.7}$$

Because of this Reciprocity Theorem, which is valid for most natural targets, the above defined scattering vectors are simplified to:

$$\mathbf{k}_{L,3} = \left(S_{HH}, \sqrt{2}S_{XX}, S_{VV}\right)^T \tag{5.8}$$

and

$$\mathbf{k}_{P,3} = \frac{1}{\sqrt{2}} \cdot (S_{HH} + S_{VV}, S_{HH} - S_{VV}, 2 \cdot S_{XX})^T \tag{5.9}$$

The factor $\sqrt{2}$ in Eq. 5.8 is used to ensure the invariance regarding the vector norm.

Only scattering processes with one dominant scatterer per resolution cell can adequately be described by a single scattering matrix **S**. This deterministic scatterer changes the type of polarisation of the wave, but not the degree of polarisation. However, in most cases there is more than one scatterer per resolution cell, named partial scatterers, which change polarisation type and polarisation degree. This is no longer describable by a single scattering matrix and therefore needs second order statistics. That is the reason for representing PolSAR data by 3×3 covariance **C** or coherency matrices **T**, using lexicographic or Pauli scattering vectors, respectively:

$$\mathbf{C} = \langle \mathbf{k}_{L,3} \cdot \mathbf{k}_{L,3}{}^* \rangle \tag{5.10}$$

$$\mathbf{T} = \langle \mathbf{k}_{P,3} \cdot \mathbf{k}_{P,3}{}^* \rangle \tag{5.11}$$

where $(\cdot)^*$ means complex conjugation and $\langle \cdot \rangle$ is the expected value. These matrices are Hermitian, positive semidefinite, and contain all information about polarimetric scattering amplitudes, phase angles, and polarimetric correlations.

There are some, more or less, basic schemes to interpret the covariance or coherency matrices defined by Eqs. 5.10 and 5.11 (see Cloude and Pottier 1996, for an exhaustive survey). Since the coherency matrix **T** is closer related to physical properties of the scatterer, it is more often used. However, it should be stated, that both are similar and can be transformed into each other. An often applied approach to interpret **T** is based on an eigenvalue decomposition (Cloude and Pottier 1996):

$$\mathbf{T} = \mathbf{U} \cdot \mathbf{\Lambda} \cdot \mathbf{U}^* \tag{5.12}$$

where the columns of **U** contain the three orthonormal eigenvectors and the diagonal elements λ_{ii} of $\mathbf{\Lambda}$ are the eigenvalues λ_i of **T**, where $\lambda_1 \geq \lambda_2 \geq \lambda_3$. Due to the fact that **T** is a Hermitian and positive semidefinite complex 3×3 matrix, all three eigenvalues always exist and are non-negative. Based on this decomposition some basic features of PolSAR data, like entropy E or anisotropy A, can be calculated:

$$E = -\sum_i p_i \cdot \log_3 p_i \tag{5.13}$$

$$A = \frac{p_2 - p_3}{p_2 + p_3} \tag{5.14}$$

where $p_i = \lambda_i / \sum_j \lambda_j$ are pseudo-probabilities of the occurrence of a scattering process described by each eigenvector. Those simple features and an angle α describing the change of the wave and derived from the eigenvectors of **T** allow a coarse interpretation of the physical characteristics of the scattering process. The proposed classification scheme divides all possible combinations of E and α into nine groups and assign each of them a certain scattering process as illustrated in Fig. 5.3.

Different statistical models were utilized and evaluated to describe SAR data, in order to adopt best to clutter becoming highly non-Gaussian especially when dealing with high-resolution data or images of man-made objects. One possibility is modelling the amplitude of the complex signal as Rayleigh distributed under the assumption that real and imaginary part of the signal are Gaussian distributed and independent (Hagg 1998). Some other examples are based on physical ideas (using K- (Jakeman and Pusey 1976), Beta- (Lopes et al. 1990), or Weibull-distribution (Oliver 1993), Fisher laws (Tison et al. 2004)) or on mathematical considerations

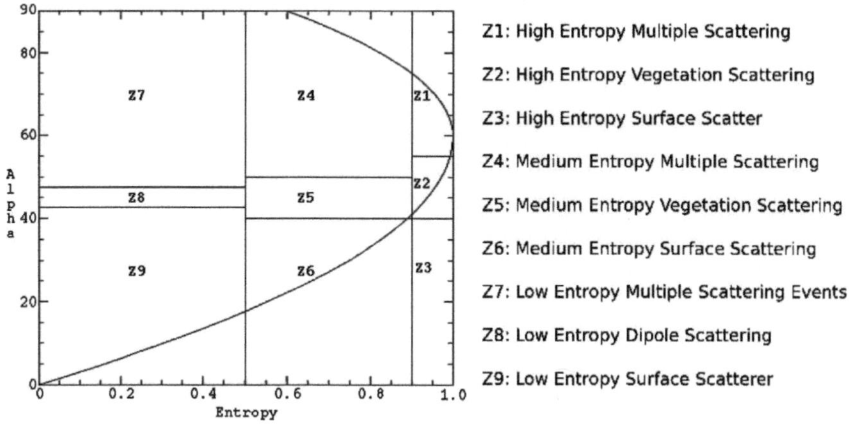

Fig. 5.3 Entropy-α classification thresholds based on Cloude and Pottier (1996)

(using Log-Normal- (Delignon et al. 1997) or Nakagami-Rice-distribution (Dana and Knepp 1986)). Each of those models has its own advantages, suppositions, and limitations.

For PolSAR data an often made basic assumption is that the backscattered signal of a distributed target, like an agricultural field, has a complex-Gaussian distribution with mean zero and variance σ. This is valid for all elements of the scattering vector, if there is a large amount of randomly distributed scatterers with similar properties in a large resolution cell, compared to the wavelength. Therefore, the whole vector can be assumed as complex-Gaussian distributed with zero mean and covariance matrix $\mathbf{\Sigma}$. That means, the whole distribution and therefore all properties of the illuminated resolution cell are governed by and can be described by the correct covariance matrix $\mathbf{\Sigma}$. This is another way to use the covariance or coherency matrix of Eqs. 5.10 and 5.11, respectively. According to those equations the covariance matrix can be estimated by averaging, which is mostly done locally due to the lack of multiple, registrated images:

$$\mathbf{C} = \frac{1}{n} \sum_i \mathbf{k}_i \cdot \mathbf{k}_i{}^H \tag{5.15}$$

where $(\cdot)^H$ means Hermitian transpose.

It is known (see Muirhead 2005, for more details), that the sum of squared Gaussian random variables with covariance matrix $\mathbf{\Sigma}$ is Wishart distributed with the probability density function p:

$$p_n(\mathbf{C}|\mathbf{\Sigma}) = \frac{n^{nq}|\mathbf{C}|^{n-q} \exp(-n \cdot tr(\mathbf{\Sigma}^{-1}\mathbf{C}))}{|\mathbf{\Sigma}|^n \cdot \pi^{q(q-1)/2} \prod_{k=1}^{q} \Gamma(n-k+1)}, \tag{5.16}$$

where q is the dimensionality of the scattering vector, n are the degrees of freedom, i.e., the number of independent data samples used for averaging and $\mathbf{\Sigma}$ is the true covariance matrix of the Gaussian distribution. The more data points are used for averaging, the more accurate is the estimation. However, too large regions are unlikely being located only in one homogeneous area. If the region used for local averaging covers more than one homogenous area, the data points belong to different distributions with different covariance matrices. In this case one basic assumption for using the Wishart distribution is violated. Especially in the vicinity of edges within the image, isotropic averaging will lead to non-Wishart distributed sample covariance matrices. Although it tends to fail in a lot of cases even on natural surfaces the Wishart distribution is a very common tool to model PolSAR data and was successfully used in many different algorithms for classification (Lee et al. 1999; Hänsch et al. 2008), segmentation (Hänsch and Hellwich 2008), or feature extraction (Schou et al. 2003; Jäger and Hellwich 2005).

5.3 Features and Operators

Several aspects of object recognition from PolSAR data are more related to general characteristics of SAR than to polarimetry. Although PolSAR is the main focus of this paper, they will be mentioned at the beginning of this section. A successful handling of those data makes a general understanding of those basic features indispensable.

One of the greatest difficulties when dealing with (Pol)SAR data arises from the coherent nature of the used microwave signal. In most cases there will be more than one scatterer per resolution cell. The coherent incident microwave is reflected by all those objects. Even if all scatter elements would have the same spectral properties, they have different distances to the sensor, which results in phase differences. Therefore, the received signal is a superposition of all those incoherent echoes, which interfere with each other. Because the interference can be either constructive or deconstructive, the phase of the received signal is purely random and the amplitude is distributed around a target specific mean value. This effect of random oscillations in the received signal intensity is called speckle effect, which is often characterised or even modeled as multiplicative noise. However, this denomination is incorrect, because noise is mostly associated with a random process. An image taken under identical circumstances will be the same, despite of changes due to noise. Hence, a SAR image taken under the same circumstances would have the same speckle. Therefore, speckle is only noise-like in terms of spatial randomness, but it is generated by a purely deterministic and not random process. Of course it is practically impossible to obtain two SAR images under identical circumstances, because of the steady change of the real world environment. Due to this fact it can be advantageous to imagine speckle as some kind of noise and to apply noise reduction techniques according to a specific model. However, one should keep in mind, that speckle is not like channel or thermal noise, one has to deal with in optical imagery.

Speckle results in a visual granularity in areas, which are expected to be homogenous (see Fig. 5.4). This granularity is one of the main reasons for the failure of standard image processing algorithms tailored for optical data.

There has been a lot of research on speckle reduction procedures ranging from simple spatial averaging to more sophisticated methods like anisotropic diffusion. Although speckle reduction techniques are often a helpful preprocessing step, many of them change the statistical characteristics of the measured signal, which has to be considered by subsequent steps.

Since speckle is produced by a deterministic process that is target specific, it contains useful information. There are some approaches, which take advantage of this information and use it for segmentation or recognition (Reigber et al. 2007a).

Two other SAR related effects are shadow and layover. The first one arises due to the side-looking acquisition geometry and stepwise height variations. It results in black areas within the SAR image, because there are regions at the ground, which could not be illuminated by the microwave signal due to occlusion. The shape of

Fig. 5.4 PolSAR image of agricultural area obtained by E-SAR over ailing

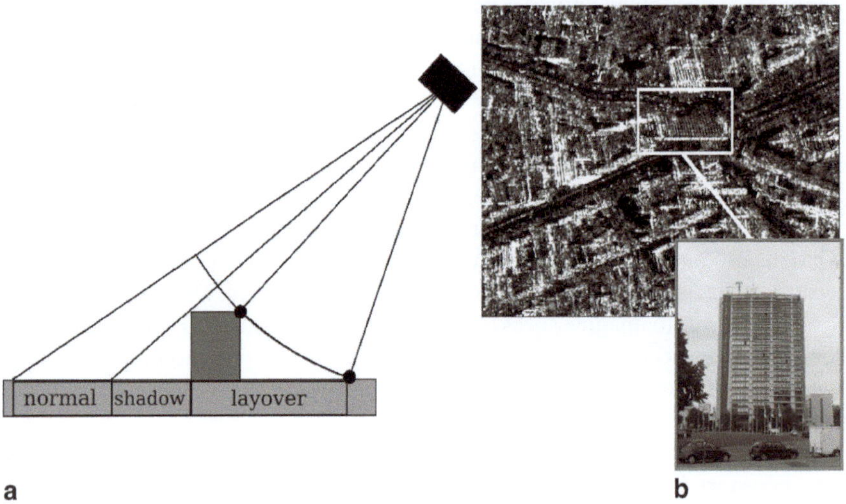

Fig. 5.5 Acquisition geometry of SAR (**a**), layover within TerraSAR-X image of Ernst-Reuter-Platz, Berlin (**b**)

this shadow is a function of sensor properties like altitude and incident angle and the geometric shape of terrain and objects. This feature is therefore highly variable, but also highly informative.

The second one emerges from the fact, that SAR measures the distance between sensor and ground by usage of an electromagnetic wave with a certain extension of the wave front in range direction. This results in ambiguities as there is more than one point with the same distance to the antenna as Fig. 5.5a illustrates. All points at the sphere will be projected into the same pixel. High objects, like buildings, will therefore be partially merged with objects right in front of them (see Fig. 5.5b). This adds further variability to the object characteristics. Different objects may belong to the same category, the ground in front of them usually does not. Nevertheless its properties will influence the features to some extent which are considered as describing the object category.

As stated above there exist different kinds of scatterers with different characteristics, for example distributed targets like agricultural fields and point targets like power poles, cars, or parts of buildings. The different properties of these diverse objects are at least partly measurable in the backscattered signal and can therefore be used in object recognition. However, they cause problems during the theoretical modeling of the data. Assumptions, which hold for one of them, do not hold for the other. Sample covariance matrices for example are Wishart distributed only for distributed targets. Furthermore, there exist different kinds of scattering mechanisms like volume scattering, surface scattering, or double bounces, which result in different changes of the polarisation of the received signal. Again, those varying properties are useful for recognition, because they add further information about the characteristics of a specific object, but have to be modeled adequately.

Another – more human related – problem is the different image geometry of SAR and optical sensors. While the first one measures a distance, the latter measures an angle. This leads to difficulties for manual interpretation of SAR images (as stated for example in Bamler and Eineder 2008) or during the manual definition of object models.

In general, images contain a lot of different information. This information can be contained in each pixels radiometric properties as well as in the relation with neighbouring pixels. In most cases only a minority of the available information is important, dependent on which task has to be performed. The great amount of information that is not meaningful in contrast to the small parts of information useful for solving the given problem, makes it more difficult to find a robust solution at all or in an acceptable amount of time. Feature extractors try to emphasize useful information and to suppress noise and irrelevant information. The extracted features are assumed to be less distortable by noise and more robust regarding the acquisition circumstances as individual pixels alone. Therefore, they provide a more meaningful description of the objects, which have to be investigated. The process of extracting features to use them for subsequent object recognition steps is called bottom-up, since the image pixels, as the most basic available information, are used to concentrate information on a higher level. The extracted features can be used by mid-level steps of object recognition or directly by classifiers, which answer the question, whether the features describe a wanted object. A lot of well-studied and good-performing feature operators for image analysis exist for close-range and even remote sensing optical imagery. However, those methods are in general not unmodifiedly applicable to SAR images, due to the different image statistics and acquisition geometries. In addition, even in optical images the exploitation of information distributed over the different radiometric channels is problematic. Similar difficulties arise in PolSAR data, where it is not always obvious how to combine the different polarisation channels. Most feature operators of optical data rely more or less on a Gaussian assumption and are not designed for multidimensional complex data. That is why they cannot be applied to PolSAR images. One approach to address the latter issue is to apply the specific method to each polarisation channel and combine the results afterwards using a fusion operator. However, that does not exploit the full polarimetric information. In addition, the fusion operator influences the results. Another possibility is to reduce the dimensionality of PolSAR data by combining the different channels into a single (maybe real valued) image. But that means a great loss of available information, too. Even methods, which can be modified to be applicable to PolSAR data, show in most cases only very suboptimal results, since they still assume other statistical properties, i.e., of optical imagery.

The probably most basic and useful feature operators for image interpretation are edge extractors or gradient operators. An edge is defined as an abrupt change between two regions within the image. The fact that human perception depends heavily on edges, is a strong cue that this information is very descriptive. Edge or gradient extraction is often used as preprocessing step for more sophisticated feature extractors, like interest operators. There exist a lot of gradient operators for optical data, for example Sobel- and DoG-operator. In Fig. 5.6b and c their application to

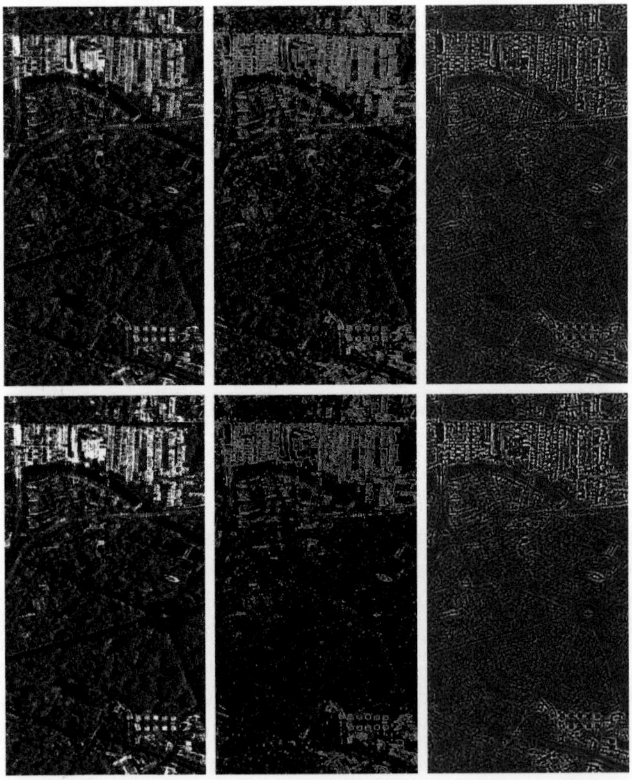

Fig. 5.6 From top-left to bottom-right: span image of Berlin (PolSAR, TerraSAR-X) (**a**), sobel (**b**), DoG (**c**), span image after speckle reduction (**d**), sobel after speckle reduction (**e**), DoG after speckle reduction (**f**)

a fully polarimetric SAR image is shown. Since both operators are not designed to work with multidimensional complex data, the span-image I_{span} (Fig. 5.6a) was calculated beforehand:

$$I_{span} = |S_{HH}|^2 + |S_{XX}|^2 + |S_{VV}|^2 \tag{5.17}$$

where $|z|$ is the amplitude of complex number z. As can be seen the most distinct edges were detected, while there are a lot of false positives due to the variations in intensity caused by speckle effect. Even after the application of a speckle reduction technique (Fig. 5.6d) the edge images (Fig. 5.6e and f) are not much better, i.e., do not contain less false positives. Speckle reduction may change image statistics and details can disappear, which could be vital for object recognition.

A good edge detector or gradient operator should indicate the position of an edge with high accuracy and have a low probability of finding an edge within a homogeneous region. Usually, operators designed for optical images fail to meet these two demands, because they are based on assumptions, that are not valid in PolSAR images. Figure 5.7a shows the result of an edge extractor developed especially for PolSAR data (Schou et al. 2003). Its basic idea is to compare two adjacent regions as illustrated by Fig. 5.7b. For each region the mean covariance matrix is calculated. An edge is detected, if the mean covariance matrices of the two regions are unlikely to be drawn from the same distribution. For that reason a likelihood-test-statistic based on Wishart distribution was utilized. The two covariance matrices $\mathbf{Z}_x$ and $\mathbf{Z}_y$ are assumed to be Wishart distributed:

$$\mathbf{Z}_x \sim W(n, \mathbf{\Sigma}_x) \tag{5.18}$$

$$\mathbf{Z}_y \sim W(m, \mathbf{\Sigma}_y) \tag{5.19}$$

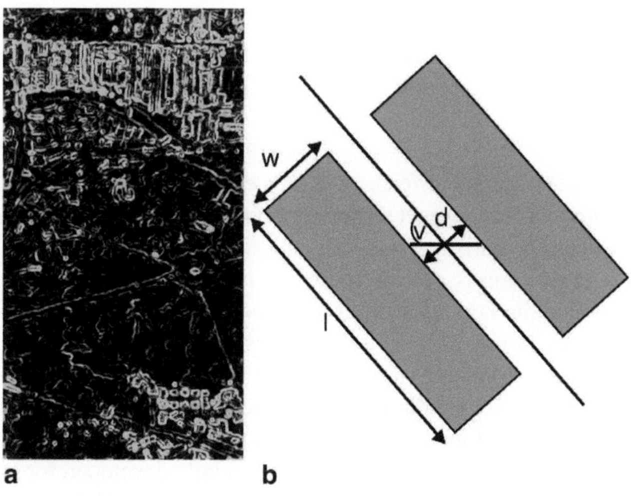

Fig. 5.7 PolSAR edge extraction (**a**), framework of CFAR edge detector (**b**)

Both matrices are considered to be equal, if the null hypothesis $H_0 : \mathbf{\Sigma}_x = \mathbf{\Sigma}_y$ is more likely to be true than the alternative hypothesis $H_1 : \mathbf{\Sigma}_x \neq \mathbf{\Sigma}_y$. The used likelihood-ratio test is defined by:

$$Q = \frac{(n+m)^{p(n+m)}}{n^{pn} m^{pm}} \cdot \frac{|\mathbf{Z}_x|^n |\mathbf{Z}_y|^m}{|\mathbf{Z}_x + \mathbf{Z}_y|^{n+m}} \tag{5.20}$$

As mentioned before, the Wishart distribution is defined over complex sample covariance matrices. To obtain these matrices from a single, fully polarimetric SAR image, spatial averaging has to be performed (Eq. 5.15). Of course it is unknown beforehand, where a homogeneous region ends. Therefore at the borders of regions pixel values will be averaged, which belong to different areas with different statistics, i.e., different true covariance matrices. These mixed covariance matrices cannot be assumed to follow the Wishart distribution, because one of its basic assumptions is violated. Since those problems occur especially in the neighborhood of edges and other abrupt changes within the image, the edge operator can lead only to suboptimal results. However, the operator is still quite useful, since it can be calculated relatively fast and provides better results compared to standard optical gradient operators.

Another possibility would be to make no assumptions about the true distribution of the data and to perform a non-parametric density estimation. However, two important problems make this solution impractical: firstly, non-parametric density estimations need usually a greater spatial support, which means, that fine details like few pixel wide lines will vanish. Secondly, such a density estimation would have to be performed in each pixel, which leads to very high computational load. This makes this approach clearly unfeasible in practical applications.

Another important feature is texture, the structured spatial repetition of signal pattern. Contemporary PolSAR sensors have achieved a resolution high enough to observe fine details of objects like buildings. Texture can therefore be a powerful feature to distinguish between different landuses and to recognize objects. An example of texture analysis for PolSAR data is given in De Grandi et al. (2004). It is based on a multi-scale wavelet decomposition and was used for image segmentation.

A lot of complex statistical features can be calculated more robustly, if the spatial support is known. The correct spatial support can be a homogeneous area, where all pixels have similar statistical and radiometrical properties. That is why it can be useful, if a segmentation is performed before subsequent processing steps. Unsupervised segmentation methods exploit low-level characteristics, like the measured data itself, to create homogeneous regions. These areas are sometimes called superpixels and are supposed to provide the correct spatial support, which is important for object recognition. Segmentation methods designed for optical data have similar problems like those mentioned above if applied to PolSAR data. However, there are some unsupervised segmentation algorithms especially developed for PolSAR data, which respect and exploit the specific statistics (Hänsch and Hellwich 2008).

A very important class of operators, extremely useful and often utilized by object recognition, are interest operators. Those operators define points or regions within the image, which are expected to be particularly informative due to geometrical or statistical properties. Common interest operators for optical images are Harris-, Foerstner-, and Kadir and Brady-operator (Harris and Stephens 1988; Förstner and Gülch 1987; Kadir and Brady 2001). Since all of them are based on the calculation of image gradients, which does not perform similarly well as in optical images, they cannot be applied to PolSAR data without modification. Until now, there are almost no such operators for PolSAR or SAR images. One of the very few examples was proposed in Jäger and Hellwich (2005) and is based on the work of Kadir and Brady. It detects salient regions within the image, like object corners or other pronounced object parts. It is invariant to scale, which obviously is a very important feature, because interesting areas are detected independently of their size. The saliency S is calculated by means of a circular image patch with radius s at location (x, y):

$$S(x, y, s) = H(x, y, s) \cdot G(x, y, s), \tag{5.21}$$

where $H(x, y, s)$ is the patch entropy and $G(x, y, s)$ describes changes in scale direction. Both of them are designed to fit the PolSAR data specific characteristics.

Despite those feature operators adopted from optical image analysis, there are other operators unique for (Pol)SAR data. Some basic low-level features can be derived by analysing the sample covariance matrix. Further examples of such features, besides those already given above, are interchannel phase differences and interchannel correlations. They measure the dependency of amplitude and phase on the polarisation.

More sophisticated features are obtained based on sublook analysis. The basic principle of SAR is to illuminate an object over a specific period, while the satellite or aircraft is passing by. During this time the object is seen from different squint angles. The multiple received echoes are measured and recorded in SAR raw data, which have to be processed afterwards. During this processing the multiple signals of the same target, which are distributed over a certain area in the raw image, are compressed in range and azimuth direction. Because the object was seen under different squint angles, the obtained SAR image can be decomposed into sub-apertures afterwards. Each of these subapertures correspond to a specific squint angle interval under which all objects in the newly calculated image are seen. Using the decomposed PolSAR image several features can be analysed. One example are coherent scatterers, caused by a deterministic point-like scattering process. These scatterers are less influenced by most scattering effects and allow a direct interpretation. In Schneider et al. (2006) two detection algorithms, based on sublook analysis, have been evaluated.

The first one uses sublook coherence γ defined by

$$\gamma = \frac{|\langle X_1 {X_2}^* \rangle|}{\sqrt{\langle X_1 {X_1}^* \rangle \langle X_2 {X_2}^* \rangle}}, \tag{5.22}$$

where X_i is the ith sublook image. The second one analyses the sublook entropy H:

$$H = -\sum_{i=1}^{N} p_i \log_N p_i, \tag{5.23}$$

where $p_i = \lambda_i / \sum_{j=1}^{N} \lambda_j$ and λ_i are the non-negative eigenvalues of the covariance matrix $\mathbf{C}$ of the N sublook images.

Another approach of subaperture analysis is the detection of anisotropic scattering processes. Normally isotropic backscattering is assumed, which means the received signal of an object is independent from the object alignment. This is only true for natural objects and even there exist exceptions, like quasiperiodic surfaces (for example rows of corn in agricultural areas). Due to the fact, that the polarisation characteristics of backscattered waves depend strongly on size, geometrical structure, and dielectric properties of the scatterer, man-made targets cannot be assumed to have isotropic backscattering. In fact, most of them show highly anisotropic scattering processes. For example double bounce, which is a common scattering type in urban areas, can only appear if an object edge is precisely parallel to the flight track. An analysis of the polarimetric characteristics under varying squint angles of subaperture images reveals objects with anisotropic backscattering. In Ferro-Famil et al. (2003) a likelihood ratio test has been used to determine whether the coherency matrices of a target in all sublook images are similar, in which case the object was supposed to exhibit isotropic backscattering.

5.4 Object Recognition in PolSAR Data

Pixelwise classification can be seen as some kind of predecessor in relation to object recognition. Objects are not defined as connected groups of pixels, which exhibit certain category-specific characteristics in their collectivity. But rather each pixel itself is assigned to a category dependent on its own properties and/or the properties of its neighbourhood. Especially unsupervised classification is an important step to a general image understanding, because it discovers structure within the data, which is hidden at the beginning, without the explicit usage of any high-level knowledge like object models. There are several of those methods, because most unsupervised clustering methods work without sophisticated feature extractors. Some of them are modified and adopted from optical data, others especially designed for SAR or PolSAR images. One of the first classification schemes was already mentioned above and is based on physical interpretation of features extracted from single covariance matrices. This approach was used by many other methods as basis for further steps. Another important classifier, which is widely considered as benchmark, was proposed in Lee et al. (1999) and is based on the Wishart distribution. Other examples are Hänsch et al. (2008), Lee et al. (1999), and Reigber et al. (2007b), all of them making use of statistical models and distance measures derived from them.

Of course such classification methods are only able to classify certain coarse distributed objects, which cause some more or less clear structures within the data space. That was sufficient for the applications of the last decades, because the resolution of PolSAR images was seldom high enough to recognize single objects, like buildings. However, contemporary PolSAR sensors are able to provide such resolution. New algorithms are now possible and necessary, which not only classify single pixels or image patches according to what they show, but which accurately find previously learned objects within those high-resolution images. There are a lot of different applications of such methods, ranging from equipment planning, natural risk prevention, and hazard management to defense.

Object recognition in close-range optical images often means either to find single specific objects or instances of an object class, which have very obvious visual features in common. An example of the first one is face recognition of previously known persons, an example of the latter face detection. In those cases object shape or object parts are very informative and an often used feature to detect and recognize objects in unseen images. In most of those cases the designed or learned object models have a clear and relatively simple structure. However, the object classes of object recognition in remote sensing are more variable as members of one class do not necessarily have obvious features in common. Their characteristics exhibit a great in-class variety. That is why it is more adequate to speak of object categories rather than of object classes. For example, in close-range data it is a valid assumption, that a house facade will have windows and doors, which have in most cases a very similar form and provide a strong correlation of corresponding features within the samples of one class. In remote sensing images the roof and the very skewed facade can be seen, which offer far less consistent visual features. Furthermore, object shape and object parts have a wide variation in remote sensing images. There often is no general shape, for example of roofs, forests, grassland, coast lines, etc. More important features are statistical properties of the signal within the object region. However, for some categories like streets, rivers, or agricultural fields, object shape is still a very useful and even essential information. Another difference to object recognition in close-range imagery is, that the task to recognize an individual object is unlikely in remote sensing. Here a more common problem is to search instances of a specific category. Therefore, object models are needed, which are able to capture both, the geometrical and radiometrical characteristics of an object category.

Due to the restricted incident angles of remote sensing sensors, pose variations seem to be rather unproblematic in comparison with close-range images. However, that is not true for SAR images, because a lot of backscattering mechanisms, like double bounce, are highly dependent on the specific positions of object structures, like balconies, with respect to the sensor. That is why the appearance even of an identical object can change significantly in different images due to different aspects and alignments during image acquisition. Furthermore, in close-range imagery there often exists an a priori knowledge about the object orientation. The roof of a house for example, is unlikely to be found at the bottom of the house. Since remote sensing images are obtained from air or space but in a side-looking manner, objects are always seen from atop, but all orientations are possible. Therefore, feature extraction operators as well as object models have to be rotation invariant.

Although SAR as active sensor is less influenced by weather conditions and independent from daylight, the spectral properties of objects can vary heavily within a category, because of physical differences, like nutrition or moisture of fields or grasslands.

Object models for object recognition in remote sensing with PolSAR data have to deal with those variations and relations, where the most problematic ones are:

- There exists a strong dependency on incident angle or object alignment for some object categories, like buildings, while other categories, for example grassland, totally lack this dependency.
- Object shape can be very informative for, e.g., agricultural fields or completely useless for categories like coast lines or forests.
- Due to the layover effect the ground in front of an object can influence the radiometric properties of the object itself.
- Usually there is a high in-class variability, due to physical circumstances, which are not class descriptive, but influence object instances.

Those facts make models necessary, which are general enough to cover all of those variations, but are not too general making recognition unstable or unfeasible in practical applications. Models, like Implicit Shape Model (ISM, see Leibe et al. 2004 for more details), which are very promising in close-range imagery, rely too strongly on object shape alone to be successfully transferable to remote sensing object recognition without modification.

In general, there are two possible ways to define an object model for object recognition: manual definition or automated learning from training images. The problems described above seem to make a manual definition of an object model advisable. For a lot of object categories an a priori knowledge about the object appearance exists, which can be incorporated in manually designed object models. It is for example known, that a street usually consists of two parallel lines with a relatively homogeneous area in between. However, this manual definition is only senseful, if the task is very specific, like extraction of road networks and/or if the objects are rather simple. Otherwise a manually designed object model wont be able to represent the complexity or variability of the object categories. Often a more general image understanding is requested, where the categories, which have to be learned, are not known before. In this case learning schemes are more promising, which do not depend on specific manually designed models, but derive them automatically from a given set of training images. Those learning schemes are based on the idea, that instances of the same category should possess similar properties, which appear consistently within the training images, while the background is unlikely to exhibit highly correlated features. These methods are more general and therefore applicable to more problems, without the need to develop and evaluate object models everytime when a new object category shall be learned. Furthermore, these methods are not biased by the human visual understanding, which is not used to the different perception geometry of SAR images. However, it should be considered, that the object model is implicitly given by the provided training set, which has to be chosen by human experts. The algorithms will consider features, which appear consistently in the training images, as part of the object or at least as informative for this object category. If the task is to recognize roads and all training images show roads in forests, one cannot expect, that roads in urban areas will be accurately recognized. In those cases the knowledge what is object and background has to be provided explicitly. The generation of the training set is therefore a crucial part. The object background should be variable enough to be recognized as background and the objects in the training images should vary to sample all possible object variations of the category densely enough, such that they can be recognized as corporate object properties. The generation of an appropriate training set is problematic for another reason, too. Obtaining PolSAR data or remote sensing images in general is very expensive. In most cases it is not possible to get a lot of images from different angles of view of a single object, as for example satellites follow a fixed orbit and the parameters available for image acquisition are limited. Furthermore, the definition of ground truth, which is important in many supervised (and for evaluation even in unsupervised) learning schemes, is even more difficult and expensive in remote sensing, than in close-range sensing. Despite the clear cut between different ways of defining object models for object recognition, it should be noted, that both require assumptions. The manual definition uses them very explicit and obviously automatic learning schemes depend on them implicitly, too. Not only the provided set of training images, but also feature extraction operators or statistical models, even the choice of a functional class of model frameworks influence the recognition result significantly.

The difficult image characteristics, the lack of appropriate feature extractors, the high in-class variety, and just recently available high-resolution PolSAR data are reasons, that there are very few successful methods, which address the problem of object recognition in PolSAR data. However, some work has been done for certain object categories. For example a lot of research was conducted for estimation of physical parameters of buildings, like building height. Also the detection of buildings in PolSAR images has been addressed in some recent publications, but is still a very active field of research (Quartulli and Datcu 2004; Xu and Jin 2007). The recognition of buildings is especially important, since it has various applications, for example identifying destroyed buildings after natural disasters, to plan and send best controlled humanitarian help as soon as possible. As SAR sensors have the advantage to be independent of daylight and nearly independent of weather conditions, they have a crucial role in those scenarios. Buildings cause very strong effects in PolSAR images due to the side-looking acquisition geometry of SAR and the stepwise height variations in urban areas. The layover and shadow effects are strong cues for building detection. Furthermore, buildings often have strong backscattering due to their dielectric properties, for example because of steel or metal in and on roofs and facades. If object edges are precisely parallel to the flight direction the microwave pulse can be reflected twice or even more times before received by the sensor, causing double bounce or trihedral reflections. Those scattering processes can be easily detected within the image, too. However, all those different effects make the PolSAR signal over man-made structures more complex. A lot of assumptions like the Reciprocity Theorem or Wishart distributed sample covariance matrices are not valid anymore in urban areas. Because of this, a lot of algorithms showing good performance at low resolution or at natural scenes, are not longer successfully applicable to high-resolution images of cities or villages. The statistical characteristics of PolSAR data in urban areas are still being investigated.

Despite of those difficulties, there are some approaches, which try to exploit building specific characteristics. One example is proposed in He et al. (2008) and exploits the a priori knowledge, that layover and shadow regions, which are caused by buildings, are very likely to be connected and of similar shape. A promising idea of this approach is, that it combines bottom-up and top-down methods. In a first step mean-shift segmentation (Comaniciu and Meer 2002) generates small homogenous patches. These regions, called superpixels, provide the correct spatial support for calculating more complex features, used in subsequent grouping steps. A few examples of these features are: mean of intensity, entropy, anisotropy, but also sublook coherence, texture and shape. Some of those attributes are characteristic for coherent scatterers, which appear often at man-made targets. The generated segments are classified into layover, shadow, or "other" regions in a Conditional Random Field (CRF), which was designed to account for the a priori knowledge that layover and shadow are often connected and exhibit a regular shape. An exemplary classification result is shown in Fig. 5.8.

Since this framework has been especially formulated for PolSAR data, it has to deal with all the problems mentioned above. Mean-shift for example, which is known to be a powerful segmentation method in optical images, is not designed to work with multidimensional complex data. That is why the log-span image was used during the segmentation phase instead of the polarimetric scattering vector. Furthermore, some assumptions about the distribution of pixel values had to be made, to make the usage of Euclidian distance and Gaussian-kernels reasonable. Nevertheless, the proposed framework shows promising results in terms of detection accuracy.

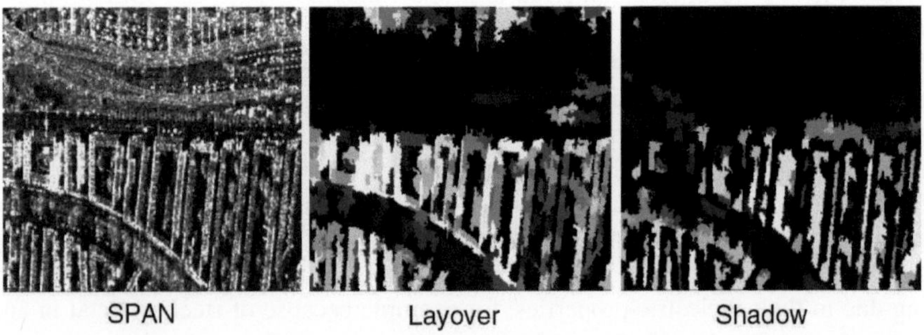

Fig. 5.8 *From left to right*: span image of PolSAR image of Copenhagen obtained by EMISAR, detected layover regions, detected shadow regions

5.5 Concluding Remarks

A lot of research on polarimetric SAR data has been done in recent years. Different methods, originally designed for SAR or even optical image processing, have been adopted to meet the PolSAR specific requirements and their applicability has been evaluated. Furthermore, new ideas, models, and algorithms have been developed especially for PolSAR data. Several possible interpretations of PolSAR measurements have been proposed, some based on physical ideas, others on mathematical concepts. All those considerations and developments lead to an initiating progress in object recognition for polarimetric SAR imagery.

However, due to the specific properties of SAR and PolSAR data most basic image analysing techniques, like gradient operators, for example Sobel-operator, which perform well in optical data, yield to very bad results, if applied to (Pol)SAR images. Operators which exploit the PolSAR data specific structure are needed to significantly improve the results of all subsequent steps. To obtain recognition results, which are competitive with those obtained with optical data, the first step has to be to define PolSAR specific feature extraction methods. This still is and has to be an active field of research.

Although different statistical models have been utilized to meet the challenges of SAR and PolSAR data most of them do neither perform well in high-resolution imagery nor in urban scenes. However, both gain increasing importance in contemporary image understanding in remote sensing. Therefore, new models and algorithms are necessary, which are successfully applicable to those kinds of data. The described problems within the different levels of object recognition explain the slow progress of object recognition in PolSAR images.

Despite the mentioned difficulties, using PolSAR imagery as information source is highly advantageous. In addition to the well known SAR related positive properties like independence from daylight, etc., it provides a lot of features, which are not contained in any other remote sensing imagery. Those characteristics can be used to effectively distinguish between object regions and background in localisation tasks and to classify the detected object instances. To achieve this goal it is absolutely necessary to finally leave the realm of only pixel based classification of for instance land uses and to continue on research about recognition of more complex objects.

New modern satellites like TerraSAR-X and Radarsat-2 – to mention only two of them – make high-resolution PolSAR data available in a sufficiently large amount to support the scientific community.

First results of object recognition in PolSAR data are promising and sanctify expectations that results will be obtained within the next years which are competitive to those of object recognition from optical data. Furthermore, future work will include fusion of PolSAR images with other kinds of data. With good prospects are fusion of SAR and optical imagery or the usage of polarimetric interferometric SAR (PolInSAR) data. The former adds radiometric information not contained in SAR images, while the latter augments the polarimetric characteristics with topography related information.

Summing up all mentioned facts about advantages and limitations, features and methods, solved and unsolved problems, one can easily catch the increasing importance of PolSAR data and object recognition from those images.

Acknowledgements The authors would like to thank the German Aerospace Center (DLR) for providing E-SAR and TerraSAR-X data. Furthermore, this work was supported by grant of DFG HE 2459/11.

References

Bamler R, Eineder M (2008) The pyramids of Gizeh seen by TerraSAR-X – a prime example for unexpected scattering mechanisms in SAR. IEEE Geosci Remote Sens Lett 5(3):468–470

Borenstein E, Ullman S (2008) Combined top-down/bottom-up segmentation. IEEE Trans Pattern Anal Mach Intell 30(12):2109–2125

Cloude S-R, Pottier E (1996) An entropy based classification scheme for land applications of polarimetric SAR. IEEE Trans Geosci Remote Sens 35(1):68–78

Cloude S-R, Pottier E (1996) A review of target decomposition theorems in radar polarimetry. IEEE Trans Geosci Remote Sens 34(2):498–518

Comaniciu D, Meer P (2002) Mean shift: a robust approach toward feature space analysis. IEEE Trans Pattern Anal Mach Learn 24(5):603–619

Dana R, Knepp D (1986) The impact of strong scintillation on space based radar design II: noncoherent detection. IEEE Trans Aerosp Electron Syst AES-22:34–46

De Grandi G et al (2004) A wavelet multiresolution technique for polarimetric texture analysis and segmentation of SAR images. In Proceedings of the IEEE International Geoscience Remote Sensing Symposium, IGARSS'04, vol 1, pp 710–713

Delignon Y et al (1997) Statistical modeling of ocean SAR images. IEEE Proc Radar Son Nav 144(6):348–354

Ferro-Famil L et al (2003) Scene characterization using sub-aperture polarimetric interferometric SAR data. In: Proceedings of the IEEE International Geoscience and Remote Sensing Symposium 2003, IGARSS'03, vol 2, pp 702–704

Förstner W, Gülch E (1987) A fast operator for detection and precise location of distinct points, corners and centers of circular features. In: Proceedings of the ISPRS intercommission workshop on fast processing of photogrammetric data, Interlaken, Switzerland, pp 281–305

Hagg W (1998) Merkmalbasierte Klassifikation von SAR-Satellitenbilddaten. Dissertation, University Karlsruhe, Fortschritt-Berichte VDI, Reihe 10, no. 568, VDI Verlag, Düsseldorf

Harris C, Stephens M (1988) A combined corner and edge detector. In: Proceedings of the 4th Alvey vision conference, Manchester, England. The British Machine Vision Association and Society for Pattern Recognition (BMVA), see http://www.bmva.org/bmvc, pp 17–151

Hawkins J (2004) On intelligence. Times Book, ISBN-10 0805074562

Hänsch R, Hellwich O (2008) Weighted pyramid linking for segmentation of fully-polarimetric SAR data. In: Proceedings of ISPRS 2008 – International archives of photogrammetry and remote sensing, vol XXXVII/B7a, Beijing, China, pp 95–100

Hänsch R et al (2008) Clustering by deterministic annealing and Wishart based distance measures for fully-polarimetric SAR-data. In: Proceedings of EUSAR 2008, vol 3, Friedrichshafen, Germany, pp 419–422

He W et al (2008) Building extraction from polarimetric SAR data using mean shift and conditional random fields. In: Proceedings of EUSAR 2008, vol 3, Friedrichshafen, Germany, pp 439–442

Jakeman E, Pusey N (1976) A model for non-Rayleigh sea echo. IEEE Trans Anten Propag AP-24:806–814

Jäger M, Hellwich O (2005) Saliency and salient region detection in sar polarimetry. In: Proceedings of IGARSS'05, vol 4, Seoul, Korea, pp 2791–2794

Kadir T, Brady M (2001) Scale, saliency and image description. Int J Comput Vis 45(2):83–105

Lee JS, Pottier E (2009) Polarimetric radar imaging: from basics to applications. CRC press, ISBN-10 142005497X

Lee JS et al (1999) Unsupervised classification using polarimetric decomposition and the complex wishart classifier. IEEE Trans Geosci Remote Sens 37(5):2249–2258

Leibe B et al (2004) Combined object categorization and segmentation with an implicit shape model. In: ECCV04 workshop on statistical learning in computer vision, Prague, pp 17–32

Lopes A et al (1990) Statistical distribution and texture in multilook and complex SAR images. In Proceedings IGARSS, Washington, pp 20–24

Marr D (1982) Vision: a computational investigation into the human representation and processing of visual information. W. H. Freeman and Co., ISBN 0-7167-1284-9

Massonnet D, Souyris J-C (2008) Imaging with synthetic aperture radar. EPFL Press, ISBN 0849382394

Muirhead RJ (2005) Aspects of multivariate statistical theory. Wiley, ISBN-10: 0471094420

Oliver C (1993) Optimum texture estimators for SAR clutter. J Phys D Appl Phys 26:1824–1835

Pizlo Z (2008) 3D shape: its unique place in visual perception. ISBN-10 0-262-16251-2

Quartulli M, Datcu M (2004) Stochastic geometrical modeling for built-up area understanding from a single SAR intensity image with meter resolution. IEEE Trans Geosci Remote Sens 42(9):1996–2003

Reigber A et al (2007a) Detection and classification of urban structures based on high-resolution SAR imagery. In: Urban remote sensing joint event, pp 1–6

Reigber A et al (2007b) Polarimetric fuzzy k-means classification with consideration of spatial context. In: Proceedings of POLINSAR07, Frascati, Italy

Schneider RZ et al (2006) Polarimetric and interferometric characterization of coherent scatterers in urban areas. IEEE Trans Geosci Remote Sens 44(4):971–984

Schou J et al (2003) CFAR edge detector for polarimetric SAR images. IEEE Trans Geosci Remote Sens 41(1):20–32

Tison C et al (2004) A new statistical model for markovian classification of urban areas in high-resolution SAR images. IEEE Trans Geosci Remote Sens 42(10):2046–2057

Xu F, Jin Y-Q (2007) Automatic reconstruction of building objects from multiaspect meter-resolution SAR images. IEEE Trans Geosci Remote Sens 45(7):2336–2353

Chapter 6
Fusion of Optical and SAR Images

Florence Tupin

6.1 Introduction

There are nowadays many kinds of remote sensing sensors: optical sensors (by this we essentially mean the panchromatic sensors), multi-spectral sensors, hyper-spectral sensors, SAR (Synthetic Aperture Radar) sensors, LIDAR, etc. They have all their own specifications and are adapted to different applications, like land-use, urban planning, ground movement monitoring, Digital Elevation Model computation, etc. But why using jointly SAR and optical sensors? There are two main reasons: first, they hopefully provide complementary information; secondly, SAR data only may be available in some crisis situations, but previously acquired optical data may help their interpretation.

The first point needs clarification. For human interpreters, optical images are usually really easier to interpret (see Figs. 6.1 and 6.2). Nevertheless, SAR data bring lots of information which are not available in optical data. For instance, the localization of urban areas is more easily seen on the SAR image (first row of Fig. 6.1). Beyond that, further information can be extracted if different combinations of polarization are used (Cloude and Pottier 1997). SAR is highly sensitive to geometrical configurations and can highlight objects appearing with a low contrast on the optical data, like flooded areas (Calabresi 1996) or man-made objects in urban areas. Besides, polarimetric data have a high capability to discriminate phenological stages of plants like rice (Aschbacher et al. 1996). However, the speckle phenomenon strongly affects such signals, leading to imprecise object borders, which calls for a combination with optical data. The characteristics of optical and SAR data will be detailed and compared in the following section.

The second point is related to the all weather – all time data acquisition capability of SAR sensors. Although many problems can more easily be solved with optical data, the availability of such images is not guaranteed. Indeed, they can be strongly affected by atmospheric conditions and in many rainy or humid areas, useful optical

F. Tupin (✉)
Institut TELECOM, TELECOM ParisTech, CNRS LTCI, 46 rue Barrault, 75 013 Paris, France
e-mail: florence.tupin@telecom-paristech.fr

U. Soergel (ed.), *Radar Remote Sensing of Urban Areas*, Remote Sensing and Digital Image Processing 15, DOI 10.1007/978-90-481-3751-0_6,

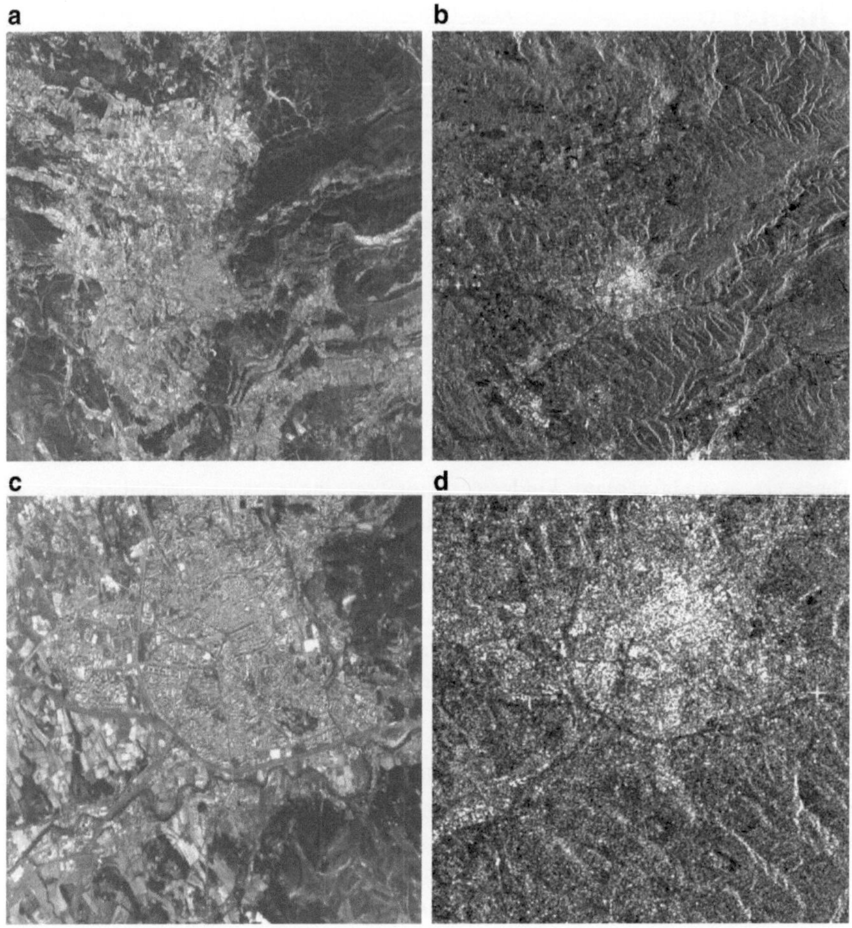

Fig. 6.1 Coarse resolution. Example of optical (SPOT, images **a** and **c**) and SAR (ERS-1, images **b** and **d**) data of the city of Aix-en-Provence (France). Resolution is approximately 10 m for both sensors. *First row:* the whole image, second row a zoom on the city and the road network

images are not always available due to the cloud cover. However, in emergency situations like natural disasters, e.g., earth-quake, tsunami, etc., fast data access is a crucial point (Wang et al. 2005). In such cases, additional information from optical data can drastically advance SAR data processing, even if it is acquired at different dates and with different resolutions. Indeed, object boundaries and area delimitations are usually stable in the landscape and can be introduced in the SAR processing.

Nevertheless, optical and SAR fusion is not an easy task. The first fusion step is registration. Due to the different appearance of objects in SAR and optical imagery, adapted methods have been developed. This problem is studied in Section 6.3. In the section thereafter (Section 6.4), some recent methods for joint classification of

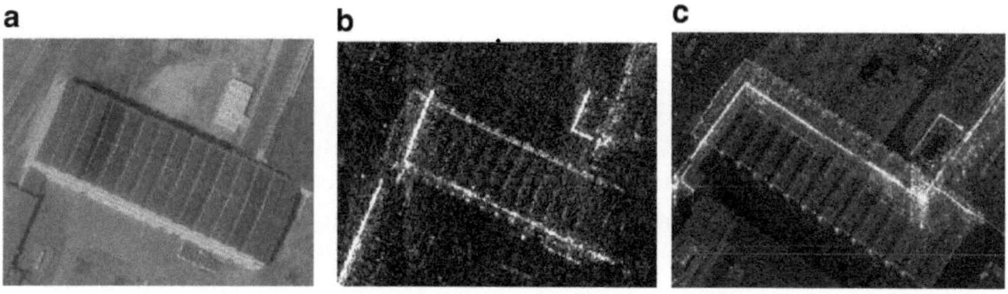

Fig. 6.2 Very high-resolution (VHR) images. Example of optical (©IGN, *on the left*) and SAR (RAMSES ©ONERA, S-band *in the middle* and X-band *on the right*) images of a building. Resolution is below 1 m. The speckle noise present on the SAR images strongly affects the pixel radiometries, and the geometrical distortions lead to a difficult interpretation of the building

optical and SAR data are presented. Section 6.5 deals with the introduction of optical information in the SAR processing. It is not exactly "fusion" in the classical sense of the word, since both data are not considered at the same level. Two applications are described: the detection of buildings using SAR and optical images and 3D reconstruction in urban areas with high-resolution data. For this last application, two different approaches based on a Markovian framework for 3D reconstruction are described.

6.2 Comparison of Optical and SAR Sensors

SAR and optical sensors differ by essentially three points:

- Optical sensors are passive, using the sun illumination of the scene, whereas SAR sensors are active, having their own source of electro-magnetic waves; therefore, optical sensors are sensitive to the cloud cover while SAR sensors are able to acquire data independently of the weather and during the night.
- Both sensors are sensitive to very different features; SAR backscattering strongly depends on the roughness of the object with respect to the wavelength, the electromagnetic properties, the humidity, etc., whereas the optical signal is influenced by the reflectance properties.
- The "noise" is very different (additive for optical images and multiplicative for SAR images) leading to different models for the radiometric distributions.
- The geometrical distortions caused by the acquisition systems are different, and the distance sampling of SAR sensors appears disturbing to human interpreters at first.

Such differences are fully developed when dealing with high-resolution (HR) or VHR images (Fig. 6.2).

6.2.1 Statistics

Most of the optical images present some noise which can be well modeled by an additive white Gaussian noise of zero mean. It is not at all the case for SAR signal. The interferences of the different waves reflected inside the resolution cell lead to the so-called "speckle" phenomenon strongly disturbing the SAR signal. It can be modeled as a multiplicative noise (Goodman 1976) following a Gamma distribution for intensity images and a Nakagami one for amplitude data. The Nakagami distribution has the following form (Fig. 6.3):

$$p_A(u|L,\mu) = \frac{2}{\mu}\frac{\sqrt{L}}{\Gamma(L)}\left(\frac{\sqrt{L}u}{\mu}\right)^{2L-1}\left(e^{-\frac{\sqrt{L}u}{\mu}}\right)^2, u \geq 0 \tag{6.1}$$

with $\mu = \sqrt{R}$ where R is proportional to the backscattering coefficient of the imaged pixel, and L is the number of looks, i.e number of averaged samples to reduce the speckle effect. In case of textured areas like urban or vegetated ones, Fisher distributions are appropriate models (Tison et al. 2004). The shapes of such distributions with three parameters are illustrated in Fig. 6.3.

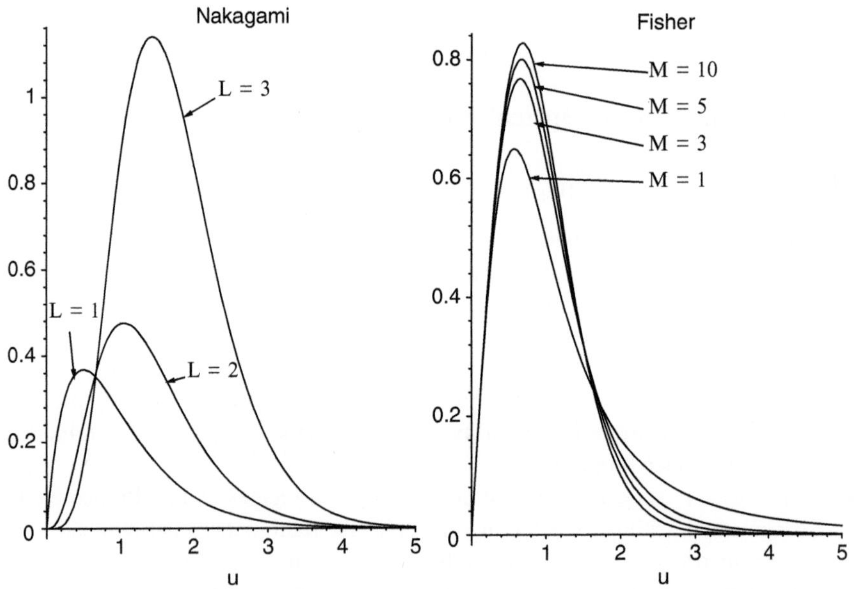

Fig. 6.3 Distribution of radiometric amplitudes in SAR images: probability density function $p_A(u|L,\mu)$ versus u. *On the left*, the Nakagami distribution and *on the right* the Fisher distribution. Both of them have "heavy tails" (Tison et al. 2004)

6.2.2 Geometrical Distortions

The principle of the SAR acquisition system is that the object position in the image depends on the range measurement. The scene is "distance sampled", which means that two points at the same distance from the sensor will be imaged in the same pixel. Besides, the higher an object, the closer to the sensor it is mapped in the image (see Figs. 6.2 and 6.4).

The distance sampling leads to two effects. The first one is the layover effect. It corresponds to areas where different signals of different ground objects are mixed since they are located at the same distance. The second one is the appearance of shadow areas, where no information is available due to the presence of obstacle on the electromagnetic wave path.

Of course there are also shadows in the optical data, depending on the object elevation and on the sun position. For building detection, the fact that the shadows do not correspond in optical and SAR data hampers algorithms based on pixel level fusion.

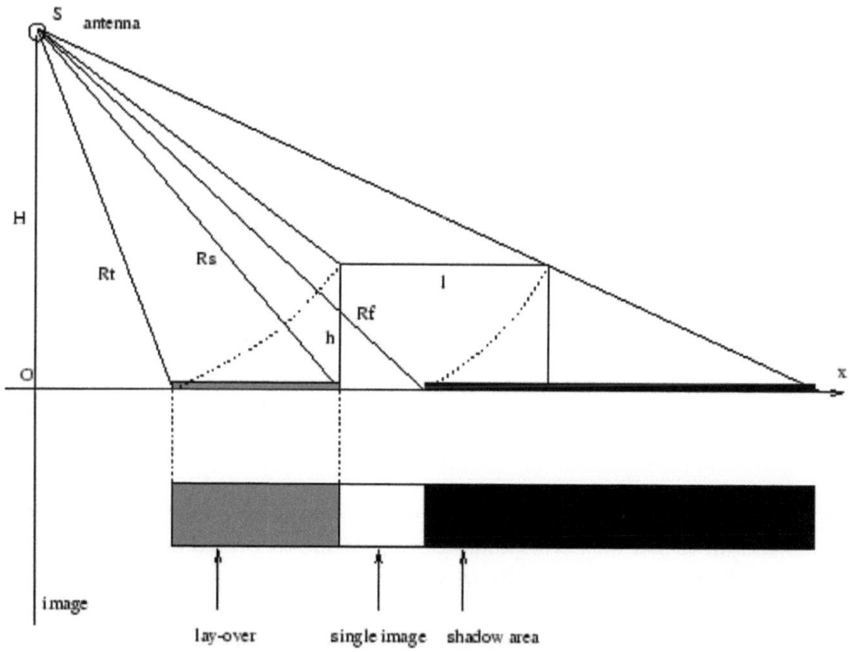

Fig. 6.4 Geometrical distortions due to distance sampling. The overlay part corresponds to mixed signals from ground, roof and facade of the building, whereas in the shadow area, no information is available

6.3 SAR and Optical Data Registration

The preliminary step before fusion usually is registration, allowing to obtain the data in the same ground geometry. Two main situations can be distinguished: in the first one, the sensor parameters are well known and the projection equations can be used; in the second one, they are not available and polynomial deformations are usually computed.

6.3.1 *Knowledge of the Sensor Parameters*

In this section we recall the geometrical equations of image formation for SAR and optical sensors. It has to be mentioned that the new products delivered by space agencies are more and more geo-coded. This fact enables direct fusion of the data with the drawback of strong dependence on the accuracy of the used Digital Terrain Model. In addition, interpolation functions can lead to artefacts.

In order to project points from optical to SAR data and inversely, some transformation functions are used. They are based on the computation of the 3D coordinates of the point and on the knowledge of the sensor acquisition system parameters.

The principle of the SAR system is based on the emission of electromagnetic waves which are then backscattered by ground objects. For a given time t of acquisition, the imaged points lie in the intersection of a sphere of range $R = ct$ and a cone related to the pointing direction of the antenna (see Fig. 6.5). More precisely, let us denote by S the sensor position, by $\boldsymbol{V}$ the speed of the sensor, and by θ_D the Doppler angle which is related to the Doppler frequency f_D and the speed by $\cos(\theta_D) = \frac{\lambda f_D}{2|V|}$, the SAR equations are then given by:

$$SM^2 = R^2 \tag{6.2}$$

$$R\sin(\theta_D)V = \boldsymbol{SM}.\boldsymbol{V} \tag{6.3}$$

Knowing the line i and column j of a pixel and making a height hypothesis h, the 3D coordinates of the corresponding point M are recovered using the previous equations. R is given by the column number j, the resolution step δR, and the Nadir range R_o, by $R = j \times \delta R + R_o$. Thus the 3D point M is the intersection of a sphere with radius R, the Doppler cone of angle θ_D and a plane with altitude h. The coordinates are given as solutions of a system with three equations and two unknowns, since the height must be given.

Inversely, knowing the 3D point M allows to recover the (i, j) pixel image coordinates, by computing the sensor position for the corresponding Doppler angle (which provides the line number) and then deducing the sensor – point distance, which permits to define the column number, since $j = \frac{R-R_O}{\delta R}$.

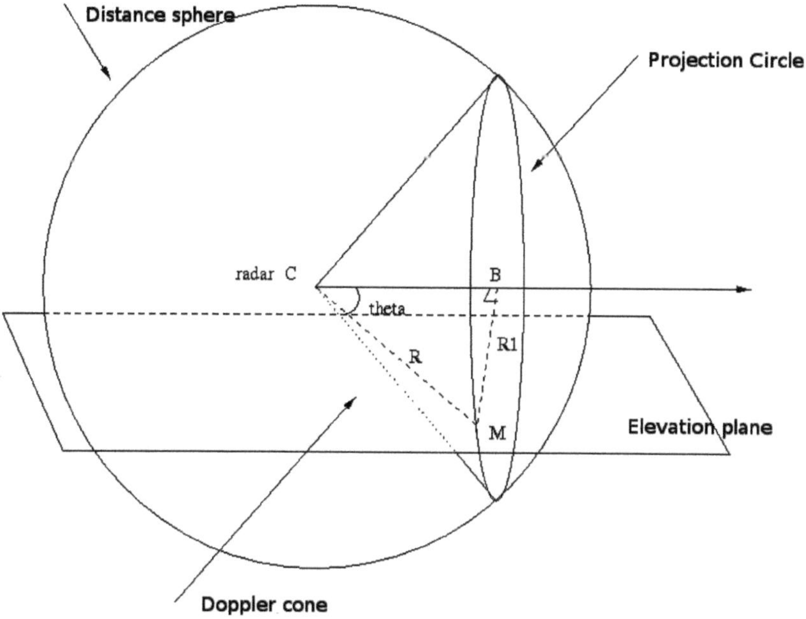

Fig. 6.5 Representation of the distance sphere and the Doppler cone in SAR imagery. If an elevation hypothesis is available, using the corresponding plane, the position of the 3D point M can be computed

The geometrical model for optical image acquisition in case of a pine-hole camera is completely different and is based on the optical center. Each point of the image is obtained from the intersection of the image plan and the line joining the 3D point M and the optical center C. The collinear equations between the image coordinates (x_m, y_m) and the 3D point M (X_M, Y_M, Z_M) are given by:

$$x_m = \frac{a_{11}X_M + a_{12}Y_M + a_{13}Z_M + a_{14}}{a_{31}X_M + a_{32}Y_M + a_{33}Z_M + a_{34}} \qquad (6.4)$$
$$y_m = \frac{a_{21}X_M + a_{22}Y_M + a_{23}Z_M + a_{24}}{a_{31}X_M + a_{32}Y_M + a_{33}Z_M + a_{34}}$$

where the a_{ij} represent parameters of both the interior orientation and the exterior orientation of the sensor. Once again, a height hypothesis is necessary to obtain M from an image point (x_m, y_m). Figure 6.6 illustrates the two different acquisition systems. A point of the SAR image is projected to the optical image for different heights. Since the point is on the same circle for the different elevations, it is always imaged as the same point in the SAR data. But its position is changing in the optical image.

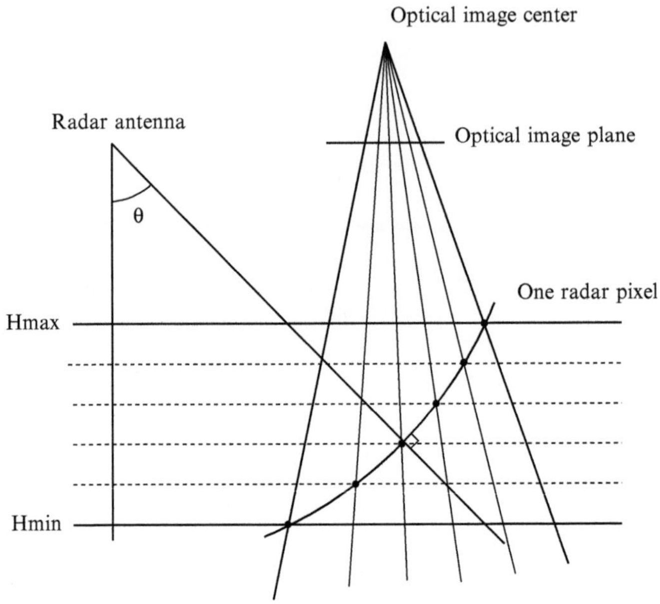

Fig. 6.6 Illustration of the two different sensor acquisition geometries. A point of the SAR image is projected to the optical image for different heights. Since the point is on the same circle for the different elevations, it is always imaged as the same point in the SAR data. But its position is changing in the optical image

6.3.2 Automatic Registration

The previous equations can only be used with a good knowledge of the sensor parameters. Many works have been dedicated to automatic registering of SAR and optical data with polynomial approaches (Dare and Dowman 2000; Moigne et al. 2003; Hong and Schowengerdt 2005). Most of them proceed in two steps: first some similarity measure between the two sensors is defined to obtain a set of matching points; then some optimization algorithm is used to compute the best parameters of the transformation.

The definition of similarity measures is not an easy task since as we have seen in Section 6.2 the appearance of objects is very different for the two sensors. Two main approaches have been developed:

- Feature-based approaches which rely on the extraction of edges or lines in both sensors (Dare and Dowman 2000; Inglada and Adragna 2001; Lehureau et al. 2008).
- Signal-based approaches which rely on the computation of a radiometric similarity measure on local windows.

Concerning the feature-based approaches, the main problem is that the shapes of the features are not always similar for both data. For instance for VHR images, the corner between the wall and the ground of a building usually appears as a very bright line in the SAR data (see for instance Fig. 6.2). However, it corresponds to an edge in the optical image. Therefore, different detectors have to be used.

Concerning the radiometric similarity measures, different studies have been dedicated to the problem. In Inglada and Giros (2004) and Shabou et al. (2007), some of them are analyzed and compared. One of the best criterion is the mutual entropy between the two signals.

6.3.3 A Framework for SAR and Optical Data Registration in Case of HR Urban Images

In Lehureau et al. (2008) a complete framework has been proposed to do automatic registration between HR optical and SAR data. The different steps of the proposed method are the following. First, a rigid registration is applied, which is computed using Fourier–Mellin invariant. Nevertheless, the deformations between optical and SAR images are not only translation, rotation and scale. An improvement of the first estimations through the use of a polynomial transformation is thus done.

As said previously, due to the radiometric differences, it is not easy to register the data using directly the pixel intensity. In this work, edges of the optical images and lines of the SAR images are extracted. First a coarse registration is looked for and the assumption is made that the transformations are rigid; which means only translation, rotation and scaling. The similarity measure used is the correlation. In order to optimize the computation time, the frequency domain is used in a multi-scale way.

The features that are to be matched must be some elements present in both images, that can be points, regions, and edges for example. In this work, the matching is actually based on matching corresponding lines (SAR) and edges (optical). For the optical image, the Canny edge detector gives the contour of roads and buildings. The detector of Tupin et al. (1998) extracts lines of the SAR images, that often match with building edges. These lines often correspond to ground-wall double reflexion. Figure 6.8 shows the extracted features.

6.3.3.1 Rigid Deformation Computation and Fourier–Mellin Invariant

The registration method uses Fourier–Mellin invariant as described in Reddy and Chatterji (1996). It is an extension of the phase correlation technique. This frequency-based approach is used to estimate the translation between two images.

Let f_1 and f_2 be two images differing only by a translation, and F_1 and F_2 their corresponding Fourier Transforms:

$$f_2(x, y) = f_1(x - \delta x, y - \delta y) \tag{6.5}$$

$$F_2(u, v) = e^{-j2\pi(u\delta x + v\delta y)} F_1(u, v) \tag{6.6}$$

$$\frac{F_1(u, v) F_2'^{*}(u, v)}{|F_1(u, v) F_2'(u, v)|} = e^{j2\pi(u\delta x + v\delta y)} \tag{6.7}$$

By taking inverse Fourier Transform, an impulse is obtained corresponding to the translation $(\delta x, \delta y)$.

The Fourier–Mellin invariant extends the phase correlation to rotation and scaling, by using a log-polar transform. Let g_1 and g_2 be two images differing by a rotation of θ_0 and a scale of α, and G_1, G_2 be their corresponding Fourier Transforms:

$$g_2(x, y) = g_1(\alpha(x \cos\theta_0 + y \sin\theta_0), \alpha(-x \sin\theta_0 + y \cos\theta_0)) \tag{6.8}$$

According to the Fourier transform properties, a rotation becomes a rotation of the same angle in the frequency domain and a scaling becomes an inverse scaling.

$$G_2(u, v) = \frac{1}{|\alpha|} G_1\left(\frac{u}{\alpha} \cos\theta_0 + \frac{v}{\alpha} \sin\theta_0, \frac{-u}{\alpha} \sin\theta_0 + \frac{v}{\alpha} \cos\theta_0\right) \tag{6.9}$$

By converting in log-polar coordinates, rotation and scaling become translations:

$$G_2(\log\rho, \theta) = \frac{1}{|a|} G_1(\log\rho - \log\alpha, \theta - \theta_0) \tag{6.10}$$

Yet, this method is highly sensitive to the features that are to be matched. In order to increase robustness, a coarse-to-fine strategy is employed in which a multi-scale pyramid is constructed. Three levels of the pyramid are built, corresponding to three resolutions.

On the first one, the dark lines, usually corresponding to the roads of the SAR image are extracted, the research space of the parameters is limited to [–90°;90°] and the scaling between [0.95;1.05]. This supposes to have a knowledge of the approximate resolution and the orientation of the images.

On the other levels, bright lines are extracted corresponding to the building corner reflectors. The registration is initialized with the previous result and the research space is restricted to [–10°;10°] and [0.95;1.05].

In order to accurately determine the translation parameters, Fourier–Mellin invariant is not fully sufficient. Indeed, as explained previously, the features taken are not exactly the same in both images. Once the rotation and scaling have been estimated, accurate determination of the translation parameters based on pixel intensity and mutual information becomes possible. An exhaustive search on the center of the optical image is made to determine its location in the SAR image. The differences in the coordinates give the parameters of the global translation.

6.3.3.2 Polynomial Deformation

In the case of SAR and optical images, the assumption of a rigid deformation between both data is not fully verified. A parallax effect appears in metric resolution imagery that cannot be corrected merely with a rigid transformation. In order to improve the registration, a polynomial deformation is looked for. To define the coefficients of the deformation, some pairs of associated points in both images are searched.

Points of interest are extracted from the optical image, using the Harris corner detector (Harris and Stephens 1988). This is a popular point detector that measures the local changes of the signal in different directions. Interest points are extracted like corners or intersections. Among all the points, just few of them are kept. In each section of a grid of size 5×5, a point is selected, then those in the border are rejected. Finally, a set of interest points distributed over the entire image is found. The use of Harris detector ensures that the points are not in homogenous area, but in fact the point selection phase has not a big importance. Indeed, large windows are used around each point to find the corresponding point in the optical image.

Once the points are selected in the optical image, the location of the corresponding points in the SAR image is searched. For this purpose, a similarity measure is needed. Among all the criteria that can be used for multisensor image registration, the mutual information (MI) is selected.

The MI is a measure of statistical dependency between two data sets. For two random variables X and Y, it is given by:

$$MI(X,Y) = H(Y) - H(Y|X) \tag{6.11}$$
$$= H(X) + H(Y) - H(X,Y) \tag{6.12}$$

where $H(X) = -E_X(\log(P(X)))$ represents the entropy of the variable X, $P(X)$ is the probability distribution of X and E_X the expectation. This registration method is based on the maximization of MI and works directly with image intensities.

The MI is applied on the full intensity of optical image and on the SAR image quantified in 10 gray levels. This quantification step is used to fasten the computation time and reduce the speckle influence. Because a rigid transformation has already been applied, it is assumed that for each point, its corresponding point in the SAR image is around the same place. An exhaustive search of the MI maximum on a neighborhood of 60 pixels around the optical point location is done to find it. Since a large window size is used to compute MI, the influence of elevated structures is limited.

A final registration is performed by estimating the best deformation fitting the couples of associated points. The model used is a second order polynomial transfora b

Fig. 6.7 Original images: on the *left* the optical SAR image ©CNES and on the *right* the original SAR image ©ONERA (Office National d'Etudes et de Recherches Arospatiales)

mation. In a preliminary step, the couples of points are filtered with respect to their similarity value. The final model is then estimated via a least square method.

6.3.3.3 Results

Some results of the proposed algorithm for the original images of Fig. 6.7 are presented in the following figures. Figure 6.8 shows the used primitives, lines of the SAR images and edges of the optical data, superimposed after the Fourier–Mellin rigid transform, and after the polynomial registration result (see also Fig. 6.9). The evaluation of the results has been made using points taken manually in both data. An error of 30 pixels has been found after the rigid registration. This result was improved to 11 pixels with the polynomial registration, which in this case corresponds to approximately 5 m.

6.4 Fusion of SAR and Optical Data for Classification

6.4.1 State of the Art of Optical/SAR Fusion Methods

Since the beginning of SAR imagery, there have been works on the problem of fusion with other sensors (Brown et al. 1996). Some of them deal with the extraction of some specific objects, like oil tanks (Wang et al. 2004), buildings (Tupin and Roux 2003) or bridges (Soergel et al. 2008). SAR imagery is the essential data source for defining regions of interest or the initialization of object search. Many

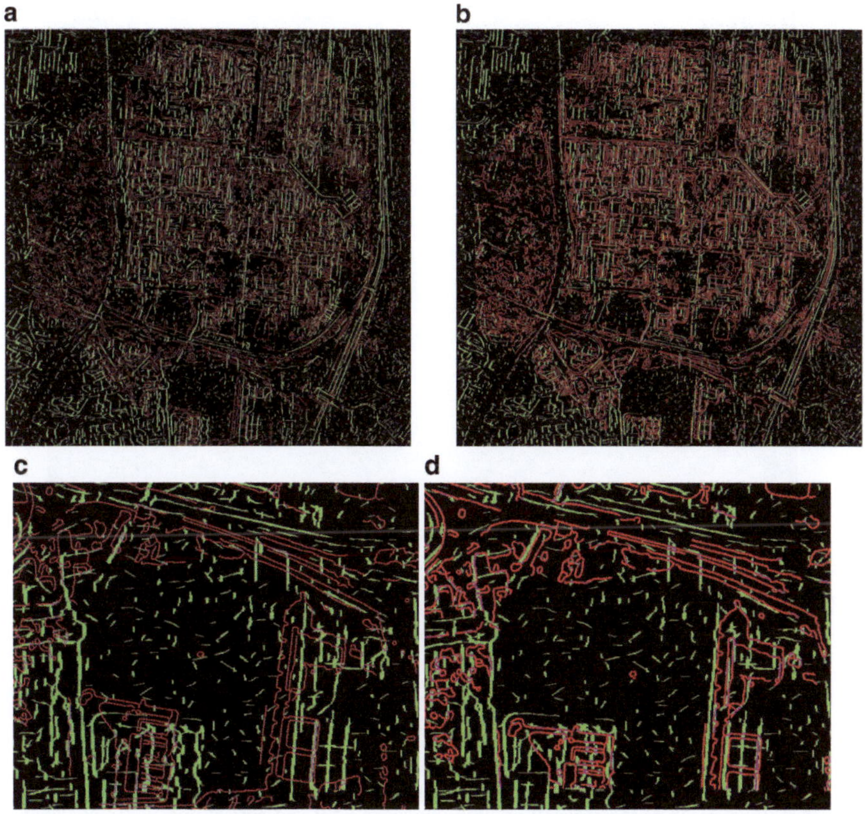

Fig. 6.8 Results of the proposed method: (**a**) and (**c**) Fourier–Mellin invariant result, and (**b**) and (**d**) after polynomial transformation. *Green lines* correspond to the optical extracted features after registration and red lines to the SAR features (from Lehureau et al. 2008)

different approaches that merge complementary information from SAR and optical data have been investigated (Chen and Ho 2008). Different kinds of data can be used with SAR sensors: multi-temporal series, polarimetric data, multi-frequencies data, interferometric (phase and coherence) images, depending on the application framework.

One family of methods is given by Maximum Likelihood based approaches and extensions where the signals from the different sensors are concatenated in one vector. In this case, the main difficulty relies on establishing a good model for the multisource data distribution. In Lombardo et al. (2003) a multivariate lognormal distribution seems to be an appropriate candidate, but multivariate Gaussian distributions have also been used. More sophisticated methods introducing contextual knowledge inside Markovian framework have been developed (Solberg et al. 1996). Other works are based on the evidential theory of Dempster and Shafer to consider union of classes and represent both imprecision and uncertainty (Hegarat-Mascle

Fig. 6.9 Final result of the registration with interlaced SAR and optical images (from Lehureau et al. 2008). The optical image is registered to the SAR ground range image using the polynomial transformation

et al. 2002a; Bendjebbour et al. 2002). This is specially useful when taking into account the "cloud" class in the optical images (Hegarat-Mascle et al. 2002b). Unsupervised approaches based on Isodata classification have also been proposed (Hill et al. 2005 for agricultural types classification with polarimetric multi-band SAR).

Another family is given by neural networks which have been widely used for remote sensing applications (Serpico and Roli 1995). The 2007 data fusion contest on urban mapping using coarse SAR and optical data has been won using such a method with pre- and post-processing steps (Pacifici et al. 2008). SVM approaches are also widely used for such fusion (Camps-Valls et al. 2008) at the pixel level.

Instead of working at the pixel level, different methods have been developed to combine the sensors at the decision level. The idea is to use an ensemble of classifiers and then merge them to improve the classification performances. Examples of such approaches can be found in Briem et al. (2002), Waske and Benediktsson (2008) and Waske and der Linden (2008).

It is not really easy to draw general conclusions concerning the performances of such methods, since the used data are usually different, as well as the applicative framework. In the following section (Section 6.4.2) we will focus on 3D reconstruction using SAR and optical data.

6.4.2 *A Framework for Building Detection Based on the Fusion of Optical and SAR Features*

In this section we describe an approach for the detection of building outlines in semi-urban areas using both SAR features and optical data (Tupin and Roux 2003). The proposed method is divided into two main steps: first, extraction of partial potential building footprints on the SAR image, and then shape detection on the optical one using the previously extracted primitives. Two methods of shape detection have been developed, the simplest one finding the "best" rectangular shape, and the second one searching for a more complicated shape in case of failure of the first one. Note that both sources are not used at the same level: the SAR image only focuses a region of interest in the optical image and provides orientation information about the potential building, whereas the building shape is searched in the optical image.

Using the detector proposed in Tupin et al. (1998), bright lines are extracted. The SAR primitives are then projected in optical geometry using the geometrical equations an height hypothesis corresponding to the ground height (here a flat ground of 8m is supposed). Only the extremities of the segment are projected and a straight line approximation is made. This is not exact but since the lines are quite short, this approximation gives acceptable results. In the following, a SAR primitive is a projected segment representing the side of a potential building. The aim is then to associate to each SAR primitive a building shape with a confidence level, allowing the suppression of the false alarms of the previous step. The detection difficulty is related to many parameters: shape complexity of the building, contrast between the building and the background, presence of structures on the roof.

6.4.2.1 Method Principle

Two approaches have been developed (Tupin and Roux 2003) for the shape detection step. The first one is faster but provides only rectangular shapes and the second one is slower but is able to detect more complicated shapes.

Both of them are applied on a set of segments extracted from the optical image by the following steps:

- Application of the Canny–Deriche edge detector
- Thinning of the edges
- Polygonal approximation of the segments to obtain a vectorial representation

A filtering of the optical segments is also applied based on proximity and direction criteria:

- First, for each SAR primitive, an interest area is computed using the sensor viewing direction.
- Secondly, only the segments which are parallel or perpendicular to the SAR primitive are selected, with an angular tolerance.

Both the set of filtered segments and the Canny–Deriche response image will be used in the following.

6.4.2.2 Best Rectangular Shape Detection

First, the building side in the optical image is detected using the SAR primitive, and then an exhaustive box search is done using only optical segments.

The building side is defined as the parallel optical segment $\boldsymbol{s_o}$ which is the closest to the SAR primitive and with the higher mean of the edge detector responses. Since the extremities of the segment, denoted by M_o^1 and M_o^2, may be not exactly positioned, a new detection is applied along the previously detected segment $\boldsymbol{s_o}$. Three candidate extremities are kept for each extremity. To do so, a search area around M_o^i is defined (Fig. 6.10) and each point M in this area is attributed a score depending on the edge detector responses along a small segment $\boldsymbol{s_o^p}(M)$ perpendicular to $\boldsymbol{s_o}$. The three points with the best scores are kept for each M_o^i. They are denoted by $M_o^i(p)$, with $1 \leq p \leq 3$.

The rectangular box detection is then applied for each possible pair of extremities $(M_o^1(p), M_o^2(q))$, with $1 \leq p \leq 3$, $1 \leq q \leq 3$. For each pair, a rectangular box of variable width w is defined and an associated score is computed. For each side k of the box, the mean of edge detector responses is computed $\mu(k)$. Then the score of

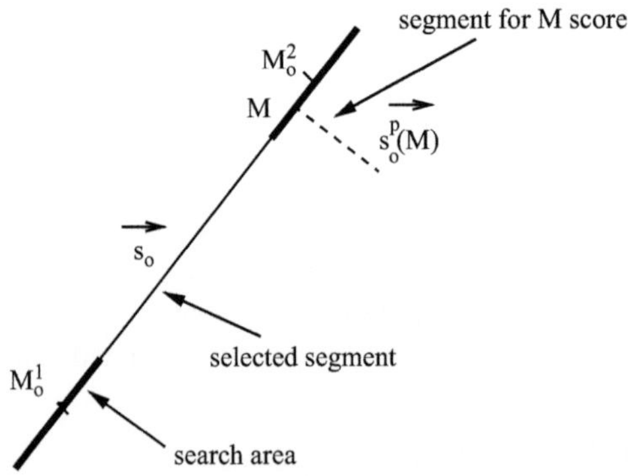

Fig. 6.10 Detection of candidates around each detected extremity M_o^i. Around each M_o^i a search area is defined (*bold segment*). In this area, for each tested point M, the segment $s_o^p(M)$ perpendicular to the original segment is considered, and the mean of the edge responses along it is computed defining the score of M. The three best points are selected and denoted by $M_o^i(p)$, with $1 \leq p \leq 3$

the box $S(M_o^1(p), M_o^2(q), w)$ is defined by:

$$S(M_o^1(p), M_o^2(q), w) = \min_k \mu(k) \tag{6.13}$$

This fusion method, based on the minimum response, gives a weak score to boxes which have a side that does not correspond to an edge. For each extremity pair $(M_o^1(p), M_o^2(q))$, the width w giving the best score is selected. The final box among all the possible pairs is then given by the best score.

This method gives quite good results for rectangular buildings and for good SAR primitives (well positioned in the optical image and with a satisfying size).

6.4.2.3 Complex Shape Detection

In the case of more complicated shapes, a different approach should be used. It is based on the detection of specific features, specially on corners, to define a building as a set of joined corners.

First of all, a set of candidate corners is detected using the filtered optical segments. For each segment, two corners are detected. As in the previous section, a search area is defined around each extremity and the corner with the best score is selected. A corner is defined as two intersecting segments, the score of a segment is defined as the mean of the edge detector responses as previously, and the corner score as the minimum score along the two segments. The corners are filtered and only the corners with a score above a threshold are selected. The threshold has been manually set.

Secondly, a starting segment $\mathbf{s}_o$ is detected in the same way as before. Starting from this segment a search area is defined as previously but with a much bigger size since the building shape can be quite complicated. In this case the SAR primitive is often only a small part of the building.

Starting from $\mathbf{s}_o$ and its corners, a path joining a set of corners is searched. To do so, a search tree is built starting from a corner. Let us denote by $(M_i, \mathbf{s}_i, \mathbf{t}_i)$ a corner i ($\mathbf{s}_i$ and $\mathbf{t}_i$ are the two small segments defining the corner). The set of prolonging segments of corner i is then detected. A corner j is said to potentially prolong the corner i if the following conditions are fulfilled:

- The projection of M_j on the line $(M_i, \mathbf{t}_i)$ is close to M_i.
- $\mathbf{s}_j$ or $\mathbf{t}_j$ is parallel and with an opposite direction compared to $\mathbf{s}_i$ – we will denote by u_j the concerned vector in the following.
- Denoting $M_i' = M_i + \mathbf{s}_i$ and $M_j' = M_j + \mathbf{u}_j$, then $\mathbf{M_iM_i'}.\mathbf{M_jM_j'} < 0$.

In the search tree, all the corner candidates are sons of i, and the tree is iteratively built. A branch stops when a maximum number of levels is reached or when the reached node corresponds to the root. In the last case, a path joining the corners has been detected. All the possible paths in the search tree are computed and a score is attributed. Once again, the path score corresponds to the score minimum of the segments joining the corners. The best path gives the searched building shape.

6.4.2.4 Results

Some results of this approach are presented in Fig. 6.11 for the two described methods. The following comments can be made on this approach:

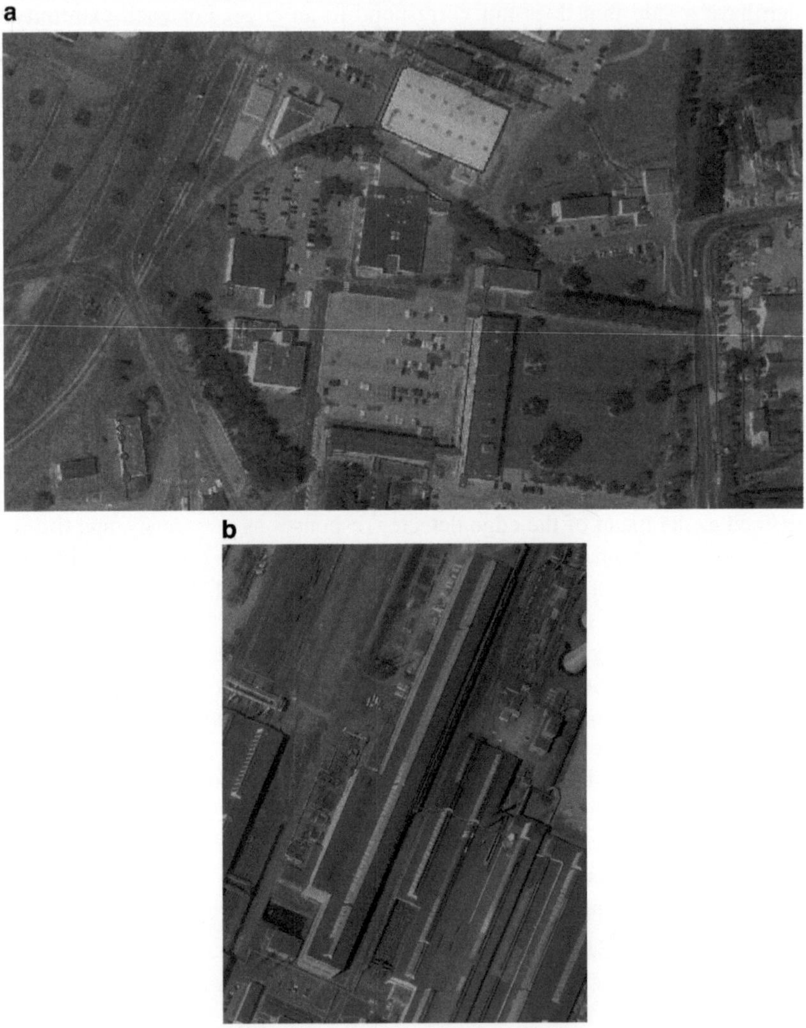

Fig. 6.11 Example of results of the proposed method. (**a**) Results of the best rectangular box detection. The group of three circles correspond to the candidate extremities which have been detected. The SAR primitive and the best box are also shown. (**b**) Example of building detection using the corner search tree (the SAR primitive is also shown). Figures from Tupin and Roux (2003)

- The detection of big buildings is difficult for many reasons. First, the SAR primitives are disconnected, and correspond to a small part of the building. Besides, the method based on the corner search tree has the following limitations:
  - The limited depth of the tree due to combinatorial explosion
  - The weak contrast of some building corners which are therefore not detected
  - The limited size of the search area, although quite large
  - The presence of roof structures which lead to a partial detection
- The detection of middle and small buildings is rather satisfying since they often have a simple shape. Both methods give similar results except in the case of more complex shapes, but the rectangular box method is also less restrictive on the extremity detection. In both cases, the only criteria which are taken into account are the edge detector responses without verification of the region homogeneity. For both methods the surrounding edges can lead to a wrong candidate.

6.5 Joint Use of SAR Interferometry and Optical Data for 3D Reconstruction

SAR and optical data can be jointly exploited to derive 3D information. Indeed, using associated points and geometrical equations, it is possible to recover the point elevation (in Toutin and Gray (2000) with manual interaction and using satellite images, in Tupin and Roux (2004) or Junjie et al. (2006) with VHR images). In this part, we are interested in a different subject, dealing with 3D SAR information like interferometric or radargrammetric data and an optical image of the same area. We have proposed a methodology based on a Markovian framework to merge both information. In such a situation, the optical data mainly provides the shapes of the building footprints whereas the SAR images bring their elevation. Let us note that the sensor parameters are supposed to be well known, and the optical data is acquired with an almost vertical viewing direction.

6.5.1 Methodology

The main idea of the proposed approach is to feed an over-segmentation of the optical image with 3D SAR features. Then the height of each region is computed using the SAR information and contextual knowledge expressed in a Markovian framework.

The first step is the extraction of 3D SAR information. It can be provided either by interferometric phases of points, or, as in this example, by matching of points in two SAR images (stereo-vision principle called radargrammetry). In Tupin and Roux (2005), a feature based approach is proposed. First, point-like and linear features are extracted in the two SAR images and matched afterwards. An associated height h_t is computed for each primitive t having a good matching score, defining a set S^{SAR}.

Starting from a set of regions computed on the optical data and denoted by S, a graph is defined. Each region corresponds to a node of the graph and the relationship between two regions is given by their adjacency, defining a set E of edges. The graph G is then $G = (S, E)$. For each region $s \in S$, $\mathcal{R}_s^{opt}$ is the corresponding part of the optical image. To each region s is associated a set of SAR primitives P_s such that their projection (or the projection of the middle point for segments) on the optical image belongs to $\mathcal{R}_s^{opt}$: $P_s = \{t \in S^{SAR} / I^{opt}(t, h_t) \in \mathcal{R}_s^{opt}\}$ with $I^{opt}(t, h_t)$ the image of the SAR primitive t projected in the optical image using the height information h_t. For segment projection, the two end-points are projected and then linked, which is not perfectly exact but is a good approximation.

One of our main assumptions is that in urban areas the height surface is composed of planar patches. Because of the lack of information in our radargrammetric context, a model of flat patches, instead of planar or quadratic surfaces (Maitre and Luo 1992), has been used. But in the case of interferometric applications for instance, more complicated models could be easily introduced in the proposed framework. The problem of height reconstruction is modeled as the recovery of a height field H defined on the graph G, given a realization y of the random observation field $Y = (Y_s)_{s \in S}$. The observation y_s is given by the set of heights of P_s: $y_s = \{h_t, t \in P_s\}$. To clearly distinguish between the height field and the observation, we denote by $y_s(t)$ the height associated to $t \in P_s$, and therefore $y_s = \{y_s(t), t \in P_s\}$. To introduce contextual knowledge, H is supposed to be a Markov random field for the neighborhood defined by region adjacency. Although Markov random fields in image processing are mostly used on the pixel graph (Geman and Geman 1984), they have also proved to be powerful models for feature based graph, like region adjacency graph (Modestino and Zhang J 1992), characteristic point graph (Rellier et al. 2000) or segment graph (Tupin et al. 1998). The searched realization $\hat{h}$ of H is defined to maximize the posterior probability $P(H|Y)$. Using the Bayes rule:

$$P(H|Y) = \frac{P(Y|H)P(H)}{P(Y)} \tag{6.14}$$

If some conditional independence assumptions are made: the observation for a region only depends on the true height of this region, the probability $P(Y|H)$ becomes:

$$P(Y|H) = \Pi_s P(Y_s|H_s) = \exp\left(-\sum_s -\log(P(Y_s|H_s))\right) = \exp\left(-\mathcal{U}(y|h)\right) \tag{6.15}$$

This assumption is quite reasonable and does not imply the independence of the regions. As far as the prior $P(H)$ is concerned, we propose to use a Markovian model. Indeed, a local knowledge around a region is usually sufficient to predict its height. Therefore, H is supposed to be a Markov random field for the neighborhood defined by the adjacency relationship. This means that $P(H)$ is a Gibbs distribution and is written:

$$P(H) \propto \exp -\mathcal{U}(h) = \exp\left(-\sum_{c \in C} V_c(h_s, s \in c)\right) \tag{6.16}$$

with C the set of cliques of the graph. Using both results for $P(Y|H)$ and $P(H)$, the posterior field is also Markovian (Geman and Geman 1984). $\hat{h}$ minimizes an energy $\mathcal{U}(h, y) = \mathcal{U}(y|h) + \mathcal{U}(h)$ composed of two terms: a likelihood term $\mathcal{U}(y|h)$ and a prior term of regularization $\mathcal{U}(h)$.

Since the $(\mathcal{R}_s^{opt})_{s \in S}$ form a partition of the optical image, each SAR primitive belongs to a unique optical region (in the case of segments, the middle point is considered). But many primitives can belong to the same region, and possibly with different heights. Due to the conditional independence assumption of the observations, the likelihood term is written $\mathcal{U}(y|h) = \sum_s U_s(y_s, h_s)$. Another assumption is made about the independence of the SAR primitives conditionally to the region height h_s, which implies: $U_s(y_s, h_s) = \sum_{t \in P_s} u_s(y_s(t), h_s)$. Without real knowledge about the distribution of the SAR height primitive conditionally to h_s, a Gaussian distribution could be used, which leads to a quadratic energy. To take into account possible outliers in the height hypotheses, a truncated quadratic expression is chosen:

$$U_s(y_s, h_s) = \sum_{t \in P_s} \min\left[(h_s - y_s(t))^2, c\right] \tag{6.17}$$

This energy is zero if no SAR primitive belongs to the optical region.

The searched for solution is constituted of objects (buildings) on a rather smooth ground. Besides, inside a building, the different parts should have a rather similar height. This knowledge is introduced in the definition of the clique potential of the graph. Only order two cliques are considered (the other clique potentials are set to zero). Two constraints are introduced in the potential definition. The first one is that the height field is naturally discontinuous. Although the height is regular inside a building or part of it, there are strong discontinuities between buildings and ground. Due to the height discontinuities, an implicit edge process is introduced. Different functions preserving discontinuities could have been used but once again a truncated quadratic function has been used.

The second constraint is related to the radiometry of the optical image. We would like to take into account the fact that a contrasted edge between two regions often implies a height discontinuity. Therefore, a weighting coefficient γ_{st} is associated to the graph edges st. This coefficient tends to 0 when the interaction between the two adjacent regions should be suppressed and 1 otherwise. The following prior energy is eventually used: $\mathcal{U}(h) = \beta \sum_{(s,t)} \gamma_{st} \min[(h_s - h_t)^2, k]$.

This energy favors configurations where adjacent regions have close heights, except if γ_{st} is small which means the presence of an edge between the two regions. If the two heights are different, the penalty is limited to k, thus preserving the

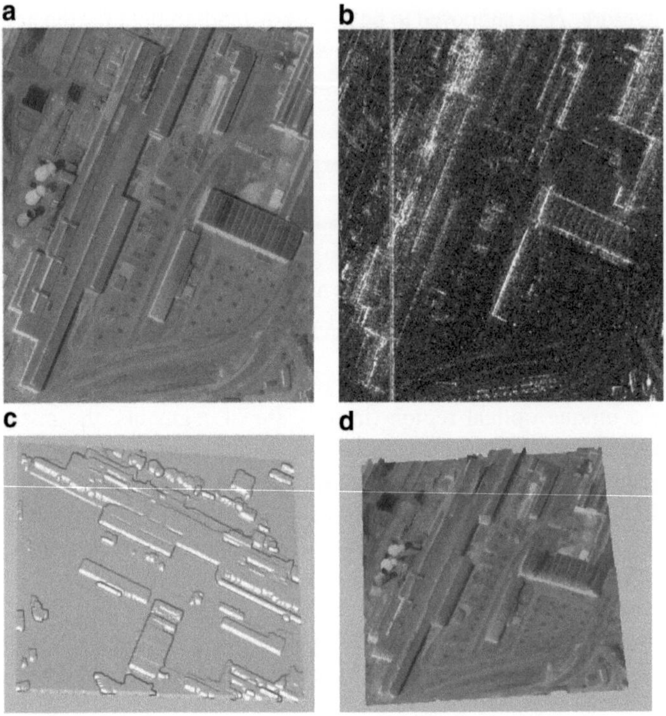

Fig. 6.12 (**a**) original optical image copyright IGN and (**b**) original SAR image copyright DGA on the top. On the bottom, perspective views of the result (radargrammetric framework) (**c**) without and (**d**) with superimposition of the optical image. Figures form Tupin and Roux (2005)

discontinuities naturally present in the image. The global energy is optimized using an Iterated Conditional Mode algorithm (ICM) (Besag 1986) with an initialization done by minimizing the likelihood term for each region.

Figure 6.12 shows some results obtained using the proposed methodology.

6.5.2 Extension to the Pixel Level

In some recent work (Denis et al. 2009b), we have investigated a different approach working at pixel level and more adapted to the interferometric case. This time the height field is defined on the pixel graph and the regularization term is based on the minimization of the Total Variation. The idea is to introduce the discontinuities which are present on the optical image to weight the regularization potential. By this way, the shapes of the objects on the optical image are introduced. Besides a new fast approximate optimization algorithm (Denis et al. 2009a) is used.

The whole approach is described by the following steps. The height map in the world coordinates is obtained by projection of the points from the radar image (steps 1–2). The cloud of points is then triangulated (step 3). A valued graph is then built with nodes corresponding to each of the points in the cloud and values set using the SAR amplitude, height and the optical information (step 5). To ease the introduction of optical information, the optical image is regularized (smoothed) prior to graph construction (step 4). Once the graph is built, a regularized height mesh is computed by defining a Markov field over the graph (step 6).

The first step is done by projecting the SAR points using the elevation given by the interferometric phase and using the equation of Section 6.2.1. Before projecting the points from radar geometry to world coordinates, shadows are detected (step 1) to prevent from projecting points with unknown (i.e., random) height. This detection is made using the Markovian classification described in Tison et al. (2004). The projection of this cloud on a horizontal plane is then triangulated with Delaunay algorithm to obtain a height mesh (step 3). The height of each node of the obtained graph can then be regularized. Although the graph is not as dense as the optical image pixels, it is denser than the Region Adjacency Graph used previously.

As in the previous subsection, the height field is regularized. The joint information of amplitude and interferometric data is used together with the optical data. Let us denote by a_s the amplitude of pixel s. Under the classical model of Goodman, the amplitude a_s follows a Nakagami distribution depending on the square root of the reflectivity $\hat{a}_s$. And the interferometric phase ϕ_s follows a Gaussian distribution with mean $\hat{\phi}_s$ leading to a quadratic energy. With these assumptions the energy to minimize is the following, where the two first terms correspond to the likelihood term and the third one to the regularization term:

$$E(\hat{a},\hat{\phi}|a,\phi) = \frac{1}{\beta_a}\sum_s M\left[\frac{a_s^2}{\hat{a}_s^2} + 2\log \hat{a}_s\right] \tag{6.18}$$

$$+\frac{\gamma}{\beta_\phi}\sum_s \frac{(\phi_s - \hat{\phi}_s)^2}{\hat{\sigma}_{\phi_s}^2} + \sum_{(s,t)} V_{(s,t)}(\hat{a}_s, \hat{a}_t, \hat{\phi}_s, \hat{\phi}_t). \tag{6.19}$$

β_a and β_ϕ are some weightings of the likelihood terms introduced in order to balance the data fidelity and regularization terms. The standard deviation $\hat{\sigma}_{\phi_s}^2$ at site s is approximated by the Cramer–Rao bound $\hat{\sigma}_{\phi_s}^2 = \frac{1-\rho_s^2}{2L\rho_s^2}$ (with L the number of average samples and ρ_s the coherence of site s). For low coherence areas (shadows or smooth surfaces, denoted $Shadows$ in the following), this Gaussian approximation is less relevant and a uniform distribution model is preferred: $p(\phi_s|\hat{\phi}_s) = \frac{1}{2\pi}$.

Concerning the regularization model for $V_{(s,t)}(\hat{a}_s, \hat{a}_t, \hat{\phi}_s, \hat{\phi}_t)$, we propose to introduce the optical image gradient as a prior (in this case the optical image can be seen as an external field). Besides, the proposed method aims at preserving *simultaneously* phase and amplitude discontinuities. Indeed, the phase and amplitude information are hopefully linked since they reflect the same scene. Amplitude discontinuities are thus usually located at the same place as phase discontinuities and conversely. We propose in this approach to perform the joint regularization of phase and amplitude. To combine the discontinuities a disjunctive max operator is chosen. This will keep the discontinuities of both data. The joint prior model with optical information is eventually defined by (prior term):

$$E(\hat{a}, \hat{\phi}) = \sum_{(s,t)} G_{opt}(s,t) \max(|\hat{a}_s - \hat{a}_t|, \gamma|\hat{\phi}_s - \hat{\phi}_t|), \tag{6.20}$$

with γ a parameter that can be set to 1, and otherwise accounts for the relative importance given to the discontinuities of the phase ($\gamma > 1$) or of the amplitude ($\gamma < 1$). $G_{opt}(s,t)$ is defined by:

$$G_{opt}(s,t) = \max(0, 1 - k|o_s - o_t|) \tag{6.21}$$

with o_s and o_t the gray values in the optical image for sites s and t. When the optical image is constant between sites s and t, $G_{opt}(s,t) = 1$ and the classical regularization is used. When the gradient $|o_s - o_t|$ is high (corresponding to an edge), $G_{opt}(s,t)$ is low thus reducing the regularization of amplitude and phase.

In Denis et al. (2009a), an efficient optimization algorithm for this kind of energy has been proposed.

Figure 6.13a shows a height mesh with the regularized optical image used as texture. The mesh is too noisy to be usable. We performed a joint amplitude/phase

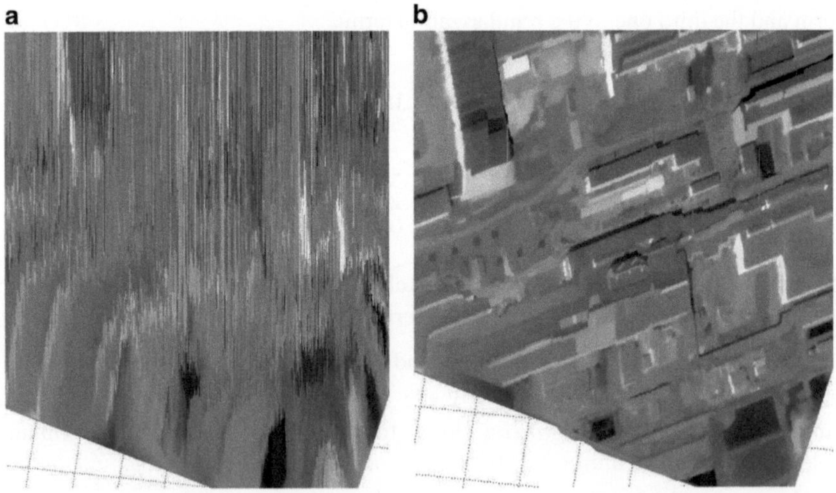

Fig. 6.13 Perspective views of the result; (**a**) Original elevation directly derived from the interferometric phase and projected in optical geometry; this figure is very noisy due to the noise of the interferometric phase, specially in shadow areas. (**b**) Elevation after the regularization approach. Figure from Denis et al. (2009b)

regularization using the gradient of the optical image as a weight that eases the apparition of edges at the location of the optical image contours. The obtained mesh is displayed on Fig. 6.13b. The surface is a lot smoother with sharp transitions located at the optical image edges. Buildings are clearly above the ground level (be aware that the shadows of the optical image create a fake 3D impression).

This approach requires a very good registration of the SAR and optical data, implying knowledge of all acquisition parameters which is not always possible depending on the source of images. The optical image should be taken with normal incidence to match the radar data. The image displayed on Fig. 6.13 was taken with a slight angle that displaces the edges and/or doubles them. For the method to work well, the edges of structures must be visible in both optical and InSAR images. A more robust approach would require a higher level analysis with, e.g., significant edge detection and building detection.

6.6 Conclusion

In spite of the improvement of sensor resolution, fusion of SAR and optical data remains a difficult problem. There is nowadays an increased interest to the subject with the recent launch of sensors of a new generation like TerraSAR-X, CosmoSkyMed, Pleiades. Although low level tools can help the interpretation process, to take the best of both sensors, high-level methods have to be developed working at the object level, especially in urban areas. Indeed, the interactions of the scattering mechanisms and the geometrical distortions require a full understanding of the local structures. Approaches based on hypothesis testing and fed by SAR signal simulation tools could bring interesting answers.

Acknowledgment The authors are indebted to ONERA Office National d'Etudes et de Recherches Arospatiales and to DGA Dlgation Gnrale pour l'Armement for providing the data. They also thank CNES for providing data and financial support in the framework of the scientific proposal R-S06/OT04-010.

References

Aschbacher J, Pongsrihadulchai A, Karnchanasutham S, Rodprom C, Paudyal D, Toan TL (1996) ERS SAR data for rice crop mapping and monitoring. Second ERS application workshop, London, UK, pp 21–24

Bendjebbour A, Delignon Y, Fouque L, Samson V, Pieczynski W (2002) Multisensor image segmentation using Dempster–Shafer fusion in Markov fields context. IEEE Trans Geo Remote Sens 40(10):2291–2299

Besag J (1986) On the statistical analysis of dirty pictures. J R Statist Soc B 48(3):259–302

Briem G, Benediktsson J, Sveinsson J (2002) Multiple classifiers applied to multisource remote sensing data. IEEE Trans Geosci Remote Sens 40(10):2291–2299

Brown R et al (1996) Complementary use of ERS-SAR and optical data for land cover mapping in Johor, Malaysia, year = 1996, Second ERS application workshop, London, UK, pp 31–35

Calabresi G (1996) The use of ERS data for flood monitoring: an overall assessment. Second ERS application workshop, London, UK, pp 237–241

Camps Valls G, Gomez-Chova L, Munoz-Mari J, Rojo-Alvarez J, Martinez-Ramon M, Serpico M, and Roli F (2008) Kernel-based framework for multitemporal and multisource remote sensing data classification and change detection. IEEE Trans Geosci Remote Sens 46(6):1822–1835

Chen C, Ho P (2008) Statistical pattern recognition in remote sensing. Pattern Recogn 41(9):2731–2741

Cloude SR, Pottier E (1997) An entropy based classification scheme for land applications of polarimetric SAR. IEEE Trans Geosci Remote Sens 35(1):68–78

Dare P, Dowman I (2000) Automatic registration of SAR and SPOT imagery based on multiple feature extraction and matching, IGARSS'00, pp 24–28

Denis L, Tupin F, Darbon J, Sigelle M (2009a) SAR image regularization with fast approximate discrete minimization. IEEE Trans. on Image Processing 18(7):1588–1600. http://www.tsi.enst.fr/%7Etupin/PUB/2007C002.pdf, 2009

Denis L, Tupin F, Darbon J, Sigelle M (2009b) Joint regularization of phase and amplitude of InSAR data: application to 3D reconstruction,Geoscience and Remote Sensing, 47(11):3774-3785, http://www.tsi.enst.fr/%7Etupin/PUB/article-2009-9303.pdf, 2009

Geman S, Geman D (1984) Stochastic relaxation, Gibbs distribution, and the Bayesian restoration of images. IEEE Trans Pattern Anal Machine Intel PAMI-6(6):721–741

Goodman J (1976) Some fundamental properties of speckle. J Opt Soc Am 66(11):1145–1150

Harris C, Stephens M (1988) A combined corner and edge detector. In: Proceedings of the 4th Alvey vision conference, Manchester, pp 147–151

Hegarat-Mascle SL, Bloch I, Vidal-Madjar D (2002a) Application of Dempster–Shafer evidence theory to unsupervised classification in multisource remote sensing. IEEE Trans Geosci Remote Sens 35(4):1018–1030

Hegarat-Mascle SL, Bloch I, Vidal-Madjar D (2002b) Introduction of neighborhood information in evidence theory and application to data fusion of radar and optical images with partial cloud cover. Pattern Recogn 40(10):1811–1823

Hill M, Ticehurst C, Lee J-S, Grunes M, Donald G, Henry D (2005) Integration of optical and radar classifications for mapping pasture type in Western Australia. IEEE Trans Geosci Remote Sens 43:1665–1681

Hong TD, Schowengerdt RA (2005) A robust technique for precise registration of radar and optical satellite images, Photogram Eng Remote Sens 71(5):585–594

Inglada J, Adragna F (2001) Automatic multi-sensor image registration by edge matching using genetic algorithms, IGARSS'01, pp 113–116

Inglada J, Giros A (2004) On the possibility of automatic multisensor image registration. IEEE Trans Geosci Remote Sens 42(10):2104–2120

Junjie Z, Chibiao D, Hongjian Y, Minghong X (2006) 3D reconstruction of buildings based on high-resolution SAR and optical images, IGARSS'06

Lehureau G, Tupin F, Tison C, Oller G, Petit D (2008) Registration of metric resolution SAR and optical images in urban areas. In: EUSAR 08, june 2008

Lombardo P, Oliver C, Pellizeri T, Meloni M (2003) A new maximum-likelihood joint segmentation technique for multitemporal SAR and multiband optical images. IEEE Trans Geosci Remote Sens 41(11):2500–2518

Maitre H, Luo W (1992) Using models to improve stereo reconstruction. IEEE Trans Pattern Anal Machine Intel, pp. 269–277

Modestino JW, Zhang J (1992) A Markov random field model-based approach to image interpretation. IEEE Trans Pattern Anal Machine Intel 14(6):606–615

Moigne JL, Morisette J, Cole-Rhodes A, Netanyahu N, Eastman R, Stone H (2003) Earth science imagery registration, IGARSS'03, pp 161–163

Pacifici F, Frate FD, Emery W, Gamba P, Chanussot J (2008) Urban mapping using coarse SAR and optical data: outcome of the 2007 GRSS data fusion contest. IEEE Geosci Remote Sens Lett 5:331–335

Reddy BS, Chatterji BN (1996) A FFT-based technique for translation, rotation and scale-invariant image registration. IEEE Trans Image Proces 5(8):12661271

Rellier G, Descombes X, Zerubia J (2000) Deformation of a cartographic road network on a SPOT satellite image. Int Conf Image Proces 2:736–739

Serpico S, Roli F (1995) Classification of multisensor remote-sensing images by structured neural networks. IEEE Trans Geosci Remote Sens 33(3):562–578

Shabou A, Tupin F, Chaabane F (2007) Similarity measures between SAR and optical images, IGARSS'07, 4858–4861, 2007

Soergel U, Cadario E, Thiele A, Thoennessen U (2008) Building recognition from multi-aspect high-resolution InSAR data in urban areas. IEEE J Selected Topics Appl Earth Observ Remote Sens 1(2):147–153

Solberg A, Taxt T, Jain A (1996) A Markov random field model for classification of multisource satellite imagery. IEEE Trans Geosci Remote Sens 34(1):100 – 113

Tison C, Nicolas J, Tupin F, Maître H (2004) A new statistical model of urban areas in high-resolution SAR images for Markovian segmentation. IEEE Trans Geosci Remote Sens 42(10):2046–2057

Toutin T, Gray L (2000) State of the art of elevation extraction from satellite SAR data. ISPRS J Photogram Remote Sens 55:13–33

Tupin F, Roux M (2003) Detection of building outlines based on the fusion of SAR and optical features. ISPRS J Photogram Remote Sens 58(1-2):71–82

Tupin F, Roux M (2004) 3D information extraction by structural matching of SAR and optical features. In: ISPRS'2004, Istanbul, Turquey, 2004

Tupin F, Roux M (2005) Markov random field on region adjacency graphs for the fusion of SAR and optical data in radargrammetric applications. IEEE Trans Geosci Remote Sens 43(8):1920–1928

Tupin F, Maître H, Mangin J-F, Nicolas J-M, Pechersky E (1998) Detection of linear features in SAR images: application to road network extraction. IEEE Trans Geosci Remote Sens 36(2):434–453

Wang Y, Tang M, Tan T, Tai X (2004) Detection of circular oil tanks based on the fusion of SAR and optical images, Third international conference on image and graphics, Hong Kong, China

Wang X, Wang G, Guan Y, Chen Q, Gao L (2005) Small satellite constellation for disaster monitoring in China, IGARSS'05, 2005

Waske B, Benediktsson J (2008) Fusion of support vector machines for classification of multisensor data. IEEE Trans Geosci Remote Sens 45(12):3858–3866

Waske B, der Linden SV (2008) Classifying multilevel imagery from SAR and optical sensors by decision fusion. IEEE Trans Geosci Remote Sens 46(5):1457 – 1466

Chapter 7
Estimation of Urban DSM from Mono-aspect InSAR Images

Céline Tison and Florence Tupin

7.1 Introduction

The extraction of 3D city models is a major issue for many applications, such as protection of the environment or urban planning for example. Thanks to the metric resolution of new SAR images, interferometry can now address this issue. The evaluation of the potential of interferometry over urban areas is a subject of main interest concerning the new high-resolution SAR satellites like TerraSAR-X, SAR Lupe, CosmoSkymed. For instance, TerraSAR-X spotlight interferograms provides very accurate height estimation over buildings (Eineder et al. 2009).

This chapter reviews methods to estimate DSM (Digital Surface Model) from mono-aspect InSAR (Interferometric SAR) images. Emphasis is put on one method based on a Markovian model in order to illustrate the kinds of results which can be obtained with such data. In order to fully assess the potential of interferometry, we focus on the use of one single interferometric pair per scene. The following chapter presents multi-aspect interferometry.

An interferogram is the phase difference of two SAR images which are acquired over the same scene with slightly different incidence angles. Under certain coherence constraints, this phase difference (the interferometric phase) is linked to scene topography. The readers would find details on interferometry principles in Massonnet and Rabaute (1993), Madsen et al. (1993), Rosen et al. (2000) and Massonnet and Souyris (2008). The interferometric phase ϕ and the corresponding coherence ρ are, respectively, the phase and the magnitude of the normalized complex hermitian

C. Tison (✉)
CNES, DCT/SI/AR, 18 avenue Edouard Belin, 31 400 Toulouse, France
e-mail: celine.tison@cnes.fr

F. Tupin
Institut TELECOM, TELECOM ParisTech, CNRS LTCI, 46 rue Barrault, 75 013 Paris, France
e-mail: florence.tupin@telecom-paristech.fr

U. Soergel (ed.), *Radar Remote Sensing of Urban Areas*, Remote Sensing and Digital Image Processing 15, DOI 10.1007/978-90-481-3751-0_7,

product of two initial SAR images (s_1 and s_2). In order to reduce noise, an averaging over a $L \times L$ window is added:

$$\rho e^{j\phi} = \frac{\sum_{i=1}^{L^2} s_1(i) s_2^*(i)}{\sqrt{\sum_{i=1}^{L^2} |s_1(i)|^2 \sum_{i=1}^{L^2} |s_2(i)|^2}} \tag{7.1}$$

ϕ has two contributions: the orbital phase ϕ_{orb}, linked to the geometrical variations of the line-of-sight vector along the swath and the topographical phase ϕ_{topo}, linked to the DSM. By Taylor expanding to first order, the height h of every pixel is proportional to ϕ_{topo} and depends on the wavelength λ, the sensor-target distance R, the perpendicular baseline $B_\perp$ and the incidence angle θ:

$$h = \frac{\lambda}{2\pi p} \frac{R \sin\theta}{B_\perp} \phi_{topo} \tag{7.2}$$

with p equal to 2 for the mono-static case and to 1 for the bistatic case. ϕ_{orb} is only geometry dependent and can be easily removed from ϕ (Rosen et al.2000). Therefore, in the following, the interferometric phase should be understood as the topographic phase (the orbital phase was removed previously). The height is derived from this phase.

Although Eq. (7.2) looks simple, its direct inversion does not lead to an accurate DSM. In many cases, the knowledge of the phase modulo 2π which requires a phase unwrapping step, is the main reason that prevents direct inversion. The height corresponding to a phase equal to 2π is called the ambiguity altitude. Generally this ambiguity altitude is much higher than the heights of buildings, which prevents phase unwrapping over urban areas. Therefore, phase unwrapping is not addressed when processing urban scenes. Users have to carefully choose the baseline so that the ambiguity height is higher than the highest building.

For high-resolution images of urban areas, the difficulties arise from geometrical distortions (layover, shadow), multiple reflections, scene geometry complexity and noise. As a consequence, high level algorithms are required to overcome these problems and to have a good understanding of the scene. In Section 7.2, a review of existing methods is proposed. All these methods are object oriented. Height filtering and edge preservation require specific processing for the different objects of the scene (e.g., a building with a roof should not be filtered the same way as vegetation). Then, Section 7.3 details the requirements on data quality to achieve accurate DSM estimation. Finally an original method, based on Markovian fusion, is proposed in Section 7.4 and evaluated on real data. The challenge is to get both an accurate height and an accurate shape description of each object in the scene.

7.2 Review of Existing Methods for Urban DSM Estimation

Four processing families for DSM estimation from InSAR images can be found in the literature:

- Shape-from-shadow methods: building footprints and heights are estimated from shadows detected in amplitude images.
- Stochastic geometry: the 3D shapes and position of buildings are optimized through energy criteria.
- Approximation by planar surfaces: filtering of interferograms to detect planar surfaces.
- Filtering of interferograms and 3D reconstruction using a classification.

These methods are all object orientated because they tend to process each building individually after its detection. Table 7.1 summarizes the different methods, their advantages and their drawbacks. The approach outlined in the fourth row of Table 7.1 can advantageously combine the other methods to get a joint classification and DSM. More details of the mentioned methods in the table are provided in the following paragraphs.

Note that all these methods had been published some years ago. Recent works use mostly multi-aspect interferograms as explained in the following chapter or they are based on manual analysis (Brenner and Roessing 2008; Eineder et al. 2009).

Table 7.1 Summary of existing works on urban DSM estimation with SAR interferometry

Methods	References	Advantages	Limits
Shape-from-shadow	Bolter et al.: Bolter and Pinz (1998); Bolter and Leberl (2000); Bolter (2000) Cellier et al.: Cellier (2006, 2007)	– Estimation of a precise building footprint, – Good detection rate	– Requirements of at least two (ideally four) images acquired on orthogonal tracks, – Failure if buildings are too close (shadow coverages)
Approximation of roofs by planar surfaces	Gamba and Houshmand: Houshmand and Gamba (2001); Gamba and Houshmand (1999, 2000); Gamba et al. (2000)	– Model of ridged roof – Precise description of buildings	– Limited to high and large buildings only – Failure on small buildings – Requires an accurate identification of connected roof parts
Stochastic geometry	Quartulli et al.: Quartulli and Dactu (2001)	– Precise model of buildings – Insensitive to noise at local scale	– Long computation time – Limitation to some building shapes
3D estimation based on prior segmentation	Soergel et al.: Soergel et al. (2000a,b, 2003) Tison et al.: Tison et al. (2007) Petit: Petit (2004)	– No a priori building model – Usable on various kinds of cities – Large choices of algorithms	– Over-segmentation on some buildings – Merging of some buildings into a unique one – Mandatory post-processing

7.2.1 Shape from Shadow

In SAR images, shadow size s is linked to object height h and incidence angle θ: $s = \frac{h}{\cos\theta}$. As a consequence, shadows provide valuable information on object height but also on object shape. Actually the edges of the shadow match one of the edges of the object. For instance, for rectangular buildings, the closest shadow edge to the near range is one of the four edges of the building. If shadows are detected in four SAR images whose tracks are either perpendicular or opposite, they describe all the edges of buildings. Then the building height can be estimated from the shadow length (see equation above) or the building height can be estimated from an interferogram. In this last case, the shadows help only to detect the building footprints.

In Bolter and Pinz (1998), Bolter and Leberl (2000) and Bolter (2000), building footprints are estimated from shadows in two or four SAR images. In these works, the height is estimated from interferograms over the footprint extracted by shadow analysis, whereas in Cellier (2006, 2007) heights are derived from shadows and compared to interferometry. Bolter et al. have shown that the estimation error on height is lower when using the interferograms (± 1.56 m) instead of using the shadows (± 1.86 m). However, the footprints are better estimated when using the shadows ($\pm 27.80\,\mathrm{m}^2$ error on surface) rather than interferometric analysis ($\pm 109.6\,\mathrm{m}^2$).

Basically, this approach combines interferometric analysis to estimate heights and the shape-from-shadow method to get building footprints. The main problem is the need of at least two images with perpendicular tracks of the same area. In addition, the method fails in dense urban areas where layovers and shadows occlude part of the buildings. Shape-from-shadow cannot be used alone for efficient estimation of a DSM in urban areas; it has to be combined with interferometry. A related method to estimate building heights exists taking advantage of the layover part of the signal (Tupin 2003) to estimate building height.

7.2.2 Approximation of Roofs by Planar Surfaces

In Gamba and Houshmand (1999) and Houshmand and Gamba (2001), interferograms are processed as set of 3D points in order to fit planar surfaces. The main steps of the algorithm are Gamba and Houshmand (2000) and Gamba et al. (2000):

- Image segmentation into areas of similar level: each level corresponds here to an averaged height.
- Search of seeds representing planar surfaces: seeds are defined as the intersection of three or two level segments, whose lengths are greater than a defined threshold.
- Iterative region growing to get a planar surface from the seeds.
- Approximation by horizontal planes which minimize a quadratic error criterion.

Different thresholds have to be set; they have strong impact on the final results as, if badly chosen, they can lead to over or under segmentations. To restrict this effect, pyramidal approach has also been suggested. The height accuracy obtained for large buildings is ± 2.5 m. The algorithm has been tested on AIRSAR images in C band with a range resolution of 3.75 m.

This method provides accurate results on large and isolated buildings. Image resolutions have a strong impact on the kind of area that can be processed with this method.

7.2.3 Stochastic Geometry

Stochastic geometry for DSM extraction has been first proposed for optical images (Ortner et al. 2003) with successful results. An adaptation to SAR interferometric images has been developed in Quartulli and Dactu (2001) and Quartulli and Dactu (2003a); Quartulli and Datcu (2003b). Stochastic geometry optimizes model parameters taking into account amplitude, coherence and interferometric phase. Buildings are modelled as parallelepipeds with a gabled roof. A probabilistic model is used to optimize the model parameters like the slope of the roof, its length, its width and the position of the parallelepiped buildings.

In order to reduce computation time, the building shape model is restricted to a unique model, which limits the representation of this approach. Nonetheless, this method is very promising as it is completely object oriented. In addition, it allows for the integration of contextual relationships between the object of the scene. The main limit is the computing time, which should be greatly reduced in the next years.

7.2.4 Height Estimation Based on Prior Segmentation

Many DSM estimation methods are based on a first step which aims at computing a segmentation or a classification of the scene (Soergel et al. 2003; Petit 2004; Tison et al. 2007). A very advanced processing chain is proposed in Soergel et al. (2003, 2000a,b). An extension to multi-aspect images is induced in these works. The basic idea is to segment the images to get building footprints, then to determine an averaged height value for each roof and finally to gather elementary shapes to get more complex roofs. Four main steps are proposed:

- Filtering and segmentation: intensity images are filtered to remove speckle; features, like bright lines, are detected.
- Detection: the interferometric heights are used to determine the ground altitude; parts above the ground are matched with the previously extracted features to estimate rectangles representing buildings.
- Reconstruction: rectangle shapes are improved with contextual information (such as road orientations and orthogonality between walls) to correct their orientations

and dimensions; three roof types are modelled (flat roofs, gabled roofs and domes); in case multi-aspect interferograms are available, merging is made at this step to avoid layover and shadows.
- Iterative improvement: iterative gathering of rectangles are authorized if two rectangles are adjacent without big statistical differences; comparison with initial images are made.

This method has been compared to ground truth provided by LIDAR data showing good accuracy of the results. The images that have been used, are DO-SAR X-band images (resolution 1.2×1.2 m). In Tison et al. (2007), similar scheme has been adopted. However, it is restricted to mono-aspect interferometry and the focus is on the fusion strategy. This algorithm is discussed extensively in Section 7.4.

In Petit (2004), a classification is also used from the very beginning of the process. Fuzzy classification helps to retrieve shadows, roads, grass, trees, urban structures, bright and very bright pixels. First validation on real data led to accurate results.

7.3 Image Quality Requirements for Accurate DSM Estimation

Figure 7.1 presents three kinds of interferometric data of semi-dense urban areas, acquired by airborne and spaceborne sensors. Ground resolution is around 50 cm for airborne data and around 1 m for spaceborne data. The TerraSAR-X images are repeat pass interferometric images, which leads to lower coherence values. Single pass interferometry guarantees that no temporal changes occur on the scene (mostly on the vegetated areas) and that the atmospheric conditions are the same; the coherence is then higher. In addition, airborne images benefit from a higher signal to noise ratio.

The AES interferogram has been computed over a very difficult area for DSM reconstruction: the urban density is very high leading to many shadows and layovers. In such areas, the coherence is low.

7.3.1 Spatial Resolution

Spatial resolution is of course the main limiting factor for accurate estimation of DSM on urban areas. Interferogram computation requires spatial averaging to reduce noise. Thus, the final resolution of the interferogram will be, at least, two or three times lower than the initial sensor resolution. In this paper, we consider that small buildings are detached houses or buildings with only one or two floors. Large buildings have more than two floors and a footprint greater than 900 m^2.

For instance, TerraSAR-X data, with 1 m ground resolution, enable to identify large buildings. Confusion will occur in very dense urban areas where buildings

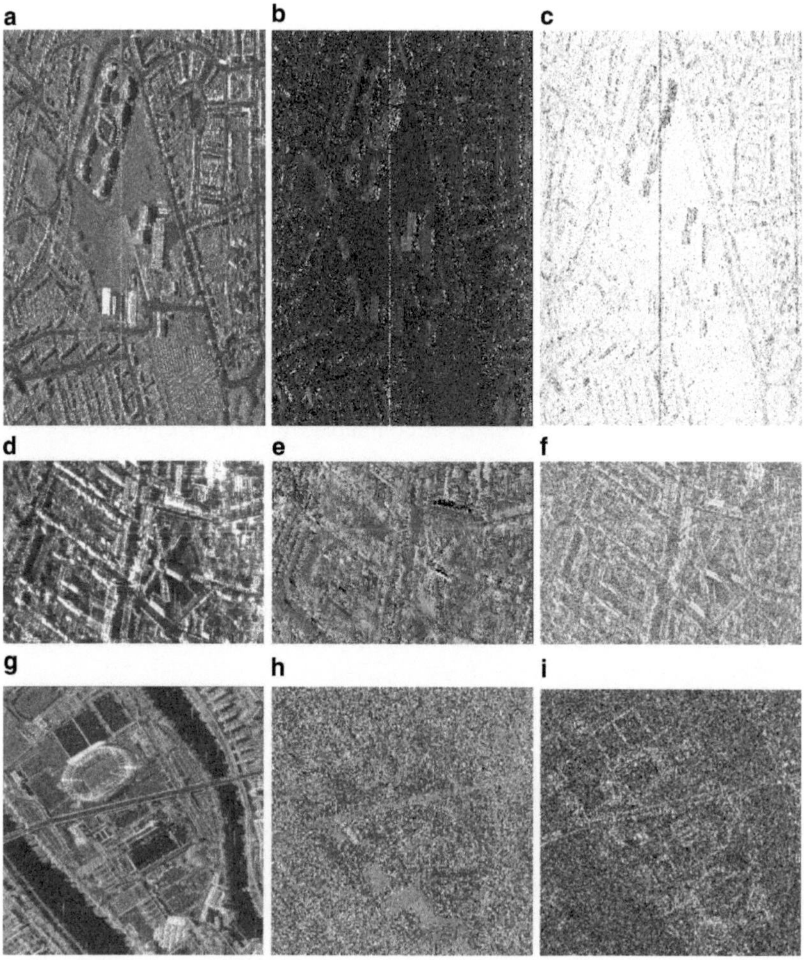

Fig. 7.1 Examples of interferograms of urban areas acquired with different sensors: *first line*, RAMSES airborne sensor, *second line*, AES-1 airborne sensor and *third line*, TerraSAR-X satellite sensor. TerraSAR-X images have been acquired in Spotlight mode (1 m ground resolution) in repeat pass. Airborne images are submetric images. For each scene, amplitude, interferometric phase and coherence over a small district are presented

are smaller. Visually, 1 m ground resolution appears to be a resolution limit for DSM estimation on urban areas. Thus, Spotlight mode is mandatory if using spaceborne data.

The spatial resolution is not the only parameter that can guarantee to detect the building footprint or not. The incidence angle has also to be taken into account. For low incidence angle, the size of layovers is large and for high incidence angle, the size of shadows is large. So, the choice of incidence angle has to be carefully made to reach the best compromise between shadows and layovers. For semi-dense urban areas, where buildings are far the one to another, it is better to avoid layovers. For dense urban areas, shadows hide some buildings: layovers may be preferable to get the right number of buildings. But in any case, it is really hard to delineate precisely the building footprints.

7.3.2 Radiometric Resolution

All the same, spatial resolution is not the only crucial factor. Radiometric resolution has to be taken into account to derive the altimetric accuracy. If accuracy is too low, averaging window has to be bigger, which decrease the final spatial resolution. Hence, spatial resolution and altimetric accuracy are also linked.

Altimetric accuracy is a function of ambiguity altitude and signal to noise ratio (SNR). The ambiguity altitude h_{amb} is computed from Eq. (7.2) with $\Phi_{topo} = 2\pi$:

$$h_{amb} = \frac{\lambda R \sin\theta}{p B_\perp} \tag{7.3}$$

The height accuracy δh depends on the phase standard deviation $\hat{\sigma}_\phi$:

$$\delta h = \frac{h_{amb}}{2\pi} \hat{\sigma}_\phi \tag{7.4}$$

Firstly, as can be seen in the two above equations, an important parameter is the radar wavelength λ. The height accuracy is proportional λ. As a consequence, X-band images allow for better accuracy than L-band images. In addition, small wavelengths are more suitable to image man-made structures, where the details are quite small.

Secondly, as a first approximation, $\hat{\sigma}_\phi$ is a function of SNR and the number of looks L:

$$\hat{\sigma}_\phi = \frac{\sqrt{1-\rho^2}}{\sqrt{2L}\rho} \quad \text{and} \quad \rho = \frac{SNR}{1+SNR} \tag{7.5}$$

Too noisy images lead to poor height quality. For instance, in Fig. 7.1, the SNR on ground is very low for AES-1 and TerraSAR-X. For the latter, signal noise may come from a lack of optimization during interferometric processing. Further work is needed to better select the common frequency bandwidth between both images. Noisy interferograms prevents accurate DSM estimation, especially on ground. Reliable ground reference will be difficult to get.

In the case of RAMSES images, SNR is very high even on ground. The interferogram is easier to analyze because the information on ground remain reliable. When interferogram is noisy, the need of a classification becomes obvious.

Finally, note that the altimetric accuracy has a direct impact on geo-referencing because DSM is needed to project slant range geometry on ground geometry. An error δh in the height estimation implies a projection error of $\delta X = \frac{\delta h}{\tan\theta}$. This error has to be added to the location error coming from sensor specification.

7.4 DSM Estimation Based on a Markovian Framework

In this section, a method for DSM retrieval from mono-aspect interferometry is presented. From the state of the art, it appears that two main strategies can be chosen when working with a single interferometric pair: either stochastic geometry or reconstruction based on prior segmentation. The latter has been selected as it leads to less constraints on building models (Tison et al. 2007; Soergel et al. 2000b).

7.4.1 *Available Data*

The available dataset are single pass interferometric SAR images acquired by RAMSES (ONERA[1] SAR sensor) over Dunkerque (North of France). The X-band sensor was operated at sub-metric resolution. The baseline is about 0.7 m, which leads to an average ambiguity altitude of 180 m. This ambiguity altitude is much higher than the elevation variations of the scene.

Unfortunately the theoretical SNR was not available, thus $\hat{\sigma}_\phi$ has been estimated on a planar surface. It is about 0.1 radians, which leads to a height accuracy of about 2–3 m (Eq. 7.5). This value is too high for a good DSM retrieval of small houses but good results can be expected for large buildings.

An IGN BD Topo©[2] is available for the area: this database gives building footprints (1 m resolution) and average height of building edges (1 m accuracy). Unfortunately, the lack of knowledge of SAR sensor parameters prevents us from registering the SAR data on the BD Topo© precisely. Therefore, a manual comparison is performed between the estimated DSM and the BD Topo©. The BD Topo© has been completed by ground truth campaign.

Figures 7.1a–c and 7.7 represent the available images over Bayard district. The focus is on the Bayard College which is in the middle of the images (three large buildings). All the processing steps are performed on slant range images. Only the refining step requires a projection on ground; this projection is included in the processing.

7.4.2 *Global Strategy*

As explained in the introduction, SAR images over urban areas are very complex. Due to SAR acquisition geometry, building signatures are highly complex: layover (mixing ground, wall and roof backscatterings), roof, shadow and bright lines, associated to ground-wall corner. Part of the interferometric information is corrupted in

[1] ONERA = Office National d'Etudes et de Recherches Aérospatiales.

[2] Dataset of the French geographical institute.

the layovers and shadows. A main issue is to identify the corrupted pixels to estimate the building height on the reliable areas only.

In order to ease the analysis, a classification into regions of interest is performed. Three main classes have been defined: ground, vegetation and buildings. DTM (Digital Terrain Model), i.e., ground altitudes, should be very smooth; only large scale changes are meaningful. A DSM of buildings should at least provide average roof heights and, at best, simplified roof model. The objective is to get a DSM with well identified building footprints. In vegetated areas, DSM can vary a lot as in real world.

Moreover, classification in this approach is also linked to the height: for instance, roads are lower than rows of trees located next to them. The global idea is to merge several features to get, at the same time, a classification and a DSM. Mimicking a fusion method developed for SAR image interpretation (Tupin et al. 1999), joint classification and height maps are computed from low level features extracted from amplitude, coherence and interferogram images.

Figure 7.2 summarizes the method which consists of three main steps: feature detection, merging and improvement. First, input images are processed to get six feature images: the filtered interferogram, a first classification, a corner reflector

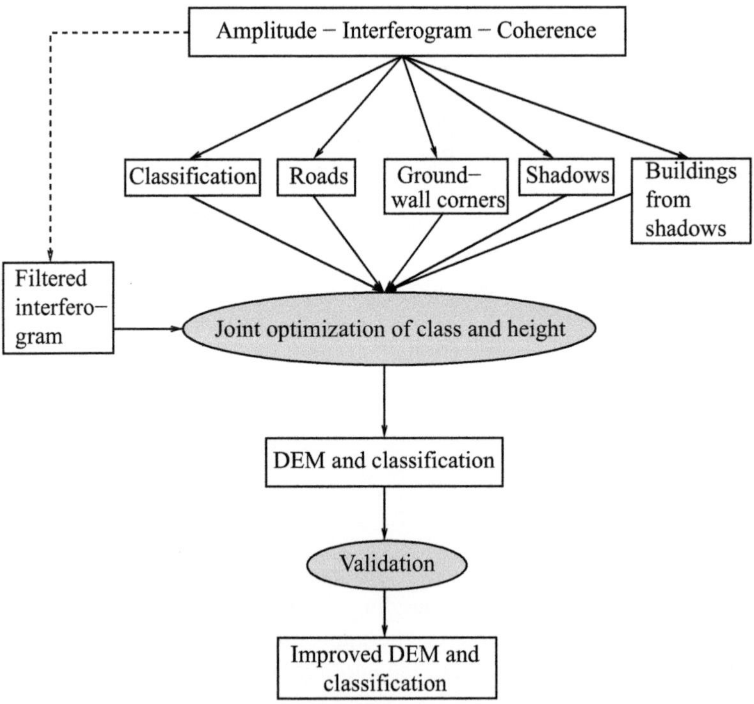

Fig. 7.2 General scheme for joint height and class estimation. The three main processing steps are: (1) the extraction of feature images, (2) the joint optimization of class and height from these features, (3) the validation and improvement of the estimations

map, a road map, a shadow map and a building-from-shadow map. The SLC (Single Look Complex) resolution is kept when processing the amplitude image to get accurate detection results. Six looks images (in slant range) are used when processing interferometric data.

Second, the previously extracted features are merged for joint classification and height retrieval. Height and class values are described by probability functions in a Markovian field. Optimization is made on the energy of this Markovian field.

Third, as in Soergel et al. (2003), last step is an improvement step where shadows and layover areas are computed from the estimated DSM. Comparisons are made with the estimated classification and corrections are performed.

The main contributions of this method are to use only one interferometric pair, to have no constraint on building shape and to retrieve jointly height and class. Note that the proposed features (number and meaning) are not limited and can be changed without modifying the global processing scheme. This process is very flexible and can be adapted easily to any other SAR images.

7.4.3 First Level Features

The input data are the amplitude of the SAR image, the interferogram and the corresponding coherence. These three images are processed to get improved or higher level information. Six algorithms are proposed for this purpose (each algorithm refers to one mathematical operator). They are not claimed to be the most efficient ones to extract urban landscapes. Users may implement their own information extraction algorithms with no consequence on the fusion scheme. Therefore, we deliberately do not detail the algorithms at this stage.

Most of the algorithms were developed especially for this study and have been published; the others are well known methods, which are helpful to solve part of the problem. The readers can refer to the references for more details.

The six operators which have been used in this work can be divided in three groups:

- Classification operator
  A first classification, based on amplitude statistics, is computed (Tison et al. 2004a). The statistical model is a Fisher distribution; this model is dedicated to high-resolution SAR data over urban areas. The results are improved with the addition of coherence and interferogric phase (Tison et al. 2007). The output is a classified image with seven classes (ground, dark vegetation, light vegetation, dark roof, medium roof, light roof/corner reflector and shadow).
- Filtering operator
  the interferogram is filtered to remove global noise with an edge preserving Markovian filtering (Geman and Reynolds 1992); it is a low-level operator which improves the information. The output is a filtered interferogram.

- Structure extraction operators
  specific operators dedicated to the extraction of the main objects which structure the urban landscape (roads, Lisini et al. 2004; corner reflectors, Tison et al. 2007; shadows and isolated buildings extracted from shadow, Tison et al. 2004b), have been developed. The outputs are binary images (1 for the object sought after, 0 elsewhere).

Therefore, six new inputs (i.e., the filtered interferogram, the classification, the road map, the corner reflector map, the shadow map and the building from shadow map) are now available from the three initial images. This new information is partly complementary and partly redundant. For instance, the corner reflectors are detected both with the dedicated operator and the classification. Generally speaking, the redundancy comes from very different approaches: the first one is local (classification) and the other one is structural (operators), accounting for the shape. This redundancy leads to a better identification of these important structures.

7.4.4 *Fusion Method: Joint Optimization of Class and Height*

In this step, the scene is divided into six classes: ground G, low vegetation (grass) Gr, high vegetation (trees) T, building B, wall-ground corner CR and shadows S. The height is regularized taking into account the classes (and inversely). The aim of this fusion is to use simultaneously all the available information derived previously and to add contextual relationships between regions. Contextual relationships take into account both height and class. The optimization is performed on a region graph instead of on pixels to keep region-based analysis.

7.4.4.1 Definition of the Region Graph

Once feature extractions are performed, an edge detector (Sobel algorithm) is applied individually to each result. This latter is a label or a binary map leading, in any case, to trivial edge detection. In addition, edge detection is also applied to the filtered interferogram to get regions with constant altitudes.

Thus, for each feature, regions with homogeneous values are defined. At this stage, a region map is defined for each feature. Then union of all these region maps is made to associate a single feature vector to each region (use of the $\cup$ operator). As a result, the final region map contains smaller regions than the initial feature region maps. A watershed is applied to assure that each region is closed. A partition of the images is computed (Fig. 7.3) and a Region Adjacency Graph (RAG) can be defined (Modestino and Zhang 1992). A feature vector $d^k = [d_1^k, d_2^k, ..., d_n^k]$ (n being the number of features) is associated to each region. The unique value d_i^k corresponds to the ith feature of region k.

Fig. 7.3 Partition (*white lines*) obtained by intersecting all the feature maps. The partition is superimposed on the RAMSES amplitude image over the Bayard College

The filtered interferogram is not considered as one of the n features even if the interferogram has been used to define RAG. Actually, the filtered height map is not binary and can thus be processed in a different way. For each region, the height $\bar{h}$ is taken equal to the mode of the histogram.[3]

7.4.4.2 Fusion Model: Maximum A Posteriori Model

In the following, bold characters are used for vectors. When possible, capitals are used for random variables and normal size characters for samples.

Two fields are defined on the RAG: the height field H and the label field L. The height values are quantized in order to get discrete values from 0 to ambiguity altitude h_{amb} with a 1 m step. There is a small oversampling of the height regarding the expected accuracy. H_s, the random variable associated to node s takes its value in $\mathbb{Z} \cap [0, h_{amb}]$ and L_s takes its value in the finite set of urban objects: {Ground (G), Grass (Gr), Tree (T), Building (B), Corner Reflector (CR), Shadow (S)}. These classes have been chosen to model all the main objects of cities as they appear in SAR images.

[3] The mode is the value that occurs the most frequently in a dataset or a probability distribution.

The six outputs of Section 7.4.3 define two fields $\bar{H}$ and $\mathbf{D}$, that are used as inputs of this merging step. $\bar{H}$ is the filtered interferogram and $\mathbf{D}$ is the observation field given by the classification and the structure extractions.

A value $\bar{h}_s$ of $\bar{H}$ for a region s is defined as the mean height of the filtered interferogram over this region. A value $\mathbf{d}_s = (d_s^i)_{1 \leq i \leq n}$ of $\mathbf{D}$ for a region s is defined as a vector containing the classification result and the object extraction results. This vector contains labels for the classification operator (here six classes are used) and binary values for the other operators (i.e., corner reflector, road, shadow, building estimated from shadows). They are still binary or "pure" classes because of the over-segmentation induced by the RAG definition.

The aim is subsequently to find the configuration of the joint field (L, H) which maximizes the conditional probability $P(L, H|\mathbf{D}, \bar{H})$. It is the best solution using a Maximum A Posteriori (MAP) criterion. With the Bayes equation:

$$P(L, H|\mathbf{D}, \bar{H}) = \frac{P(\mathbf{D}, \bar{H}|L, H)P(L, H)}{P(\mathbf{D}, \bar{H})} \tag{7.6}$$

and the product rule to estimate the joint probability $P(L, H)$ is:

$$P(L, H) = P(L|H)P(H) \tag{7.7}$$

Finally, using Eq. (7.7), the joint probability $P(L, H|\mathbf{D}, \bar{H})$ conditional to $(\mathbf{D}, \bar{H})$ is equal to:

$$P(L, H|\mathbf{D}, \bar{H}) = \frac{P(\mathbf{D}, \bar{H}|L, H)P(L|H)P(H)}{P(\mathbf{D}, \bar{H})} \tag{7.8}$$

Instead of supposing L and H independent, $P(L|H)$ is kept to constrain the class field by the height field. It usually allows one to take into account simple considerations on real architecture such as:

- Roads are lower than adjacent buildings.
- Grass and road are approximately at the same height.
- Shadows are close to high objects, i.e., building and tree.
- Corner reflector are lower than adjacent buildings.
- Corner reflector are close to buildings.

This link between H and L is the main originality and advantage of this approach.

Knowing the configurations $\mathbf{d}$ and $\bar{h}$, the denominator $P(\mathbf{D}, \bar{H})$ is a constant $\frac{1}{k}$ and thus, is not implied in the optimization of (L, H). Therefore, by simplifying Eq. (7.8), the final probability to be optimized is:

$$P(L, H|\mathbf{D}, \bar{H}) = kP(\mathbf{D}, \bar{H}|L, H)P(L|H)P(H) \tag{7.9}$$

with k a constant. Terms of Eq. (7.9) are defined in the following section.

Energy Terms

Assuming that both fields H and $L|H$ (field L conditionally dependent on field H) are Markovian, their probabilities are Gibbs fields. Adding the hypothesis of region to region independence, conditionally dependent on L and H, the likelihood term $P(\mathbf{D}, \bar{H}|L, H)$ is also a Gibbs field.

Hence, $P(\mathbf{D}, \bar{H}|L, H) = \prod_s P(\mathbf{D}_s, \bar{H}_s|L, H)$ and assuming that the observation of regions does not depend on the other regions, $P(\mathbf{D}, \bar{H}|L, H) = \prod_s P(\mathbf{D}_s, \bar{H}_s|L_s, H_s)$. As a consequence, the energy is defined with a clique singleton. The posterior field is thus Markovian and the MAP optimization of the joint field (L, H) is equivalent to the search for the configuration that minimizes its energy.

For each region s, the conditional local energy U is defined as a function of the class l_s and the height h_s conditional to the observed parameters of its neighbourhood V_s: $U(l_s, h_s|\mathbf{d}_s, \bar{h}_s, l_t, h_{t\,t\in V_s})$. These observed parameters are: the detector values $\mathbf{d}_s$, the observed height $\bar{h}_s$, the configuration of the fields L and H of its neighbourhood V_s. In the following, the neighbourhood V_s is defined by all the adjacent regions of a region s under consideration.

The energy is made up of two terms: the likelihood term U_{data} (coming from $P(\mathbf{D}, \bar{H}|L, H)$) corresponding to the influence of the observations, and the different contributions of the regularization term U_{reg}(coming from $P(L|H)P(H)$) corresponding to the prior knowledge that is introduced on the scene. They are weighted by a regularization coefficient β and by the surface area A_s of the region via a function α. The choice of the weights (β and α) is empirical. The results do not change drastically with small (i.e., 10%) variations of β and α.

Taking into account the decomposition of the energy term into two energies (U_{reg} and U_{data}) and the weighting by the weight β of the regularization term and by the surface function α, the following energy form is proposed:

$$U(l_s, h_s|\mathbf{d}_s, \bar{h}_s, l_t, h_{t\,t\in V_s}) = (1-\beta)\left(\sum_{t\in V_s} A_t A_s\right)\alpha(A_s)U_{data}(\mathbf{d}_s, \bar{h}_s|l_s, h_s) + \beta \sum_{t\in V_s} A_t A_s U_{reg}(l_s, h_s, l_t, h_t) \quad (7.10)$$

A_t is the surface area of the neighbour region t of region s. α is a linear function of A_s. If A_s is large then the influence of the neighbourhood is reduced ($\forall x, 1 \leq \alpha(x) \leq 2$). In addition, the different contributions of the regularization term are weighted by the surface product $A_t A_s$ in order to give more credit to the largest regions. The factor $(\sum_{t\in V_s} A_t A_s)$ is a normalization factor.

Likelihood Term The likelihood term describes the probability $P(\mathbf{D}, \bar{H}|L, H)$. $\mathbf{D}$ and $\bar{H}$ are conditionally independent thus $P(\mathbf{D}, \bar{H}|L, H) = P(\mathbf{D}|L, H) \times P(\bar{H}|L, H)$. Moreover, $\mathbf{D}$ is independent from H and $\bar{H}$, the observed height, is independent from L. The dependence between class and height is between H and L and not $\bar{H}$ and L.

Finally, $P(\mathbf{D}, \bar{H}|L, H) = P(\mathbf{D}|L) \times P(\bar{H}|H)$. Therefore, the likelihood term is considered equal to:

$$U_{data}(\mathbf{d}_s, \bar{h}_s | l_s, h_s) = \sum_{i=1}^{n} U_D(d_s^i | l_s) + (h_s - \bar{h}_s)^2 \tag{7.11}$$

The likelihood term of the height is quadratic because of the Gaussian assumption over the interferometric phase probability (Rosen et al. 2000). There is no analytical expression of the probability density function of $P(d_s^i|l_s)$; it is thus determined empirically.

The values of $U_D(d_s^i|l_s)$ are determined by the user, based on his a priori knowledge of the detector qualities. The d_s^i values are part of finite sets (almost binary sets) because detectors' outcomes are binary maps or classification. So, the number of $U_D(d_s^i|l_s)$ values to be defined is not too high. Actually d_s^1 is the classification operator result and has six possible values. The other four feature maps (the corner reflector map d_s^2, the road map d_s^3, the "building from shadow" map d_s^4 and the shadow map d_s^5) are binary map values. Hence, the users have to define 96 values (see Table 7.2). Nevertheless, for binary maps, most of the values are equal, because only one class is detected (the other ones are processed equally), which restricts the number of values to approximately fifty. An example of the chosen values is given in Table 7.2. To simplify the user choices, only eight values can be chosen: 0.0, 0.5, 0.8, 1.0, 3.0 and −3.0, −2.0, −10.0. Intermediate values do not have any impact on the results. The height map is robust towards changes of values whereas the classification is more sensitive to small changes (from 0.8 to 0.5 for instance). Some confusion may arise between buildings and trees for such parameter changes.

Moreover, these values are defined once over the entire dataset, and are not modified regarding the particularities of the different parts of the global scene

Regularization Term The contextual term, relating to $P(L|H)P(H)$, introduces two constraints and is written in Eq. (7.12). The first term, γ, comes from $P(L|H)$ and imposes constraints on two adjacent classes l_s and l_t depending on their heights. For instance, two adjacent regions with two different heights cannot belong to the same road class. A set of such simple rules is built up and introduced in the energy term.

The second term, ψ, comes from $P(H)$ and introduces contextual knowledge on the reconstructed height field. Since there are many discontinuities in urban areas, the regularization should both preserve edges and smooth planar regions (ground, flat roof).

$$U_{reg}(l_s, h_s, l_t, h_t) = \gamma_{(h_s,h_t)}(l_s, l_t) + \psi(h_s - h_t) \tag{7.12}$$

For the class conditionally dependent on the heights, a membership of the class is evaluated based on the relative height difference between two neighbours. Three cases have been distinguished: $h_s \approx h_t$, $h_s < h_t$ and $h_s > h_t$ and an adjacency matrix is built for each case. In order to preserve symmetry, the matrix of the last case is equal to the transposed matrix of the second case.

Table 7.2 $U_D(d_s^i|l_s)$ values for every class and every detector. The lines correspond to the different values that each element d_s^i of d_s has, whereas the column corresponds to the different classes considered for l_s. Each value in the table is thus $U(d_s^i|l_s)$ given the value of d_s^i and the value of l_s. The minimum energy value is 0.0 (meaning "it is the good detector value for this class") and the maximum energy value is 1.0 (meaning "this detector value is not possible for this class"). There are three intermediate values: 0.3, 0.5 and 0.8. Yet, if some detectors bring obviously strong information, we underline their energy by using ±2, ±3 or −10 regarding the confidence level. In this way, corner reflector and shadow detectors are associated to low energy because these detectors contribute trustworthy information which cannot be contested. The merging is robust regarding small variations of energy values.
CR = corner reflectors, R = roads, BS = buildings from shadows, B = building, S = shadow. The classification values d_s^1 mean: 0 = ground, 1 = vegetation, 2 = dark roof, 3 = mean roof, 4 = light roof, 5 = shadow.
The classes are: Ground (G), Grass (Gr), Tree (T), Building (B), Corner Reflector (CR), Shadow (S)

d_s^i \ l_s		G	Gr	T	B	CR	S
Classification	$d_s^1=0$	0.0	1.0	1.0	1.0	1.0	1.0
	$d_s^1=1$	1.0	0.0	0.8	1.0	1.0	1.0
	$d_s^1=2$	1.0	0.5	0.0	0.0	1.0	1.0
	$d_s^1=3$	1.0	1.0	0.5	0.0	1.0	1.0
	$d_s^1=4$	1.0	1.0	1.0	0.0	0.0	1.0
	$d_s^1=5$	1.0	1.0	1.0	1.0	1.0	−3.0
CR	$d_s^2=0$	1.0	1.0	1.0	1.0	3.0	1.0
	$d_s^2=1$	1.0	1.0	1.0	1.0	−2.0	1.0
R	$d_s^3=0$	1.0	1.0	1.0	1.0	1.0	1.0
	$d_s^3=1$	−10.0	1.0	1.0	1.0	1.0	1.0
BS	$d_s^4=0$	0.0	0.0	0.3	0.5	0.0	0.0
	$d_s^4=1$	1.0	1.0	0.3	0.0	0.3	1.0
S	$d_s^5=0$	1.0	1.0	1.0	1.0	1.0	3.0
	$d_s^5=1$	1.0	1.0	1.0	1.0	1.0	−2.0

$h_s \approx h_t$:

$$\gamma_{(h_s,h_t)}(l_s,l_t) = 0 \text{ if } (l_s,l_t) \in \{B, CR, S\}^2 \quad (7.13)$$

$$\gamma_{(h_s,h_t)}(l_s,l_t) = \delta(l_s,l_t) \text{ else} \quad (7.14)$$

δ is the Kronecker symbol.

In this case, the two adjacent regions have similar height and they should belong to the same object. Yet, if the region is a shadow or a corner reflector, the height may be noisy and could be close, in average, to that of the building.

$h_s < h_t$:

$$\gamma_{(h_s,h_t)}(l_s,l_t) = c(l_s,l_t) \quad (7.15)$$

$h_s > h_t$:

$$\gamma_{(h_s,h_t)}(l_s,l_t) = c(l_t,l_s) \quad (7.16)$$

These last two cases relate the relationship between classes with respect to their heights based on architectural rules. The user has to define the values $c(l_s,l_t)$ regarding real urban structure. But there is a unique set of values for an entire dataset. An example of the chosen values is given in Table 7.3.

Table 7.3 $c(l_s, l_k)$ values, i.e., $\gamma_{(h_s,h_k)}(l_s, l_k)$ values if $h_s < h_k$. The symmetric matrix gives the values of $\gamma_{(h_s,h_k)}(l_s, l_k)$ when $h_s > h_k$. Four values are used from 0.0 to 2.0. 0.0 means that it is highly probable to have class l_s close to class l_k, whereas 2.0 means the exact contrary (it is almost impossible).
The classes are: Ground (G), Grass (Gr), Tree (T), Building (B), Corner Reflector (CR), Shadow (S)

l_s \ l_k	G	Gr	T	B	CR	S
G	1.0	2.0	0.5	0.5	2.0	1.0
Gr	2.0	1.0	0.5	0.5	2.0	1.0
T	2.0	2.0	0.0	1.0	2.0	1.0
B	1.0	1.0	1.0	0.0	0.0	0.0
CR	2.0	2.0	2.0	0.0	0.0	1.0
S	1.0	1.0	1.0	0.0	1.0	0.0

For the height, the regularization is calculated with an edge preserving function (Geman and Reynolds 1992):

$$\psi(h_s, h_t) = \frac{(h_s - h_t)^2}{1 + (h_s - h_t)^2} \tag{7.17}$$

This function is a good compromise in order to keep sharp edges while smoothing planar surfaces.

7.4.4.3 Optimization Algorithm

Due to computational constraints, the optimization is processed with an ICM (Iterative Conditional Modes) algorithm (Besag 1986). The classification initialization is computed from the detector inputs. The maximum likelihood value is assigned to the initial class value, i.e., for each region, the initial class l_s is the one which minimizes $\sum_{i=1}^{n} U_D(d_s^i|l_s)$. The initialization of the height map is the filtered interferogram. This initialization is close to the expected results, which allows for an efficient optimization through the ICM method.

The algorithm is run with specific values: the regularization coefficient β is given a value of 0.4 and the α function is equal to $\alpha(A) = \frac{A - \min(A_s)}{\max(A_s) - \min(A_s)} + 1$. $\min(A_s)$ and $\max(A_s)$ are, respectively, the minimum and the maximum region surfaces of the RAG. The energy terms defined by the user are presented in Tables 7.2 and 7.3. These values are used for the entire dataset; they are not adapted to each extract separately. For a given dataset, the user has thus to define these values only once.

7.4.4.4 Results

The fusion has been performed in 8-connexity with $\beta = 0.3$ and $\alpha_{max} = 2.0$. In Figs. 7.4 and 7.5, results are illustrated for the Bayard College area. Some conflicts happen between high vegetation and building classes. For instance, some part of the

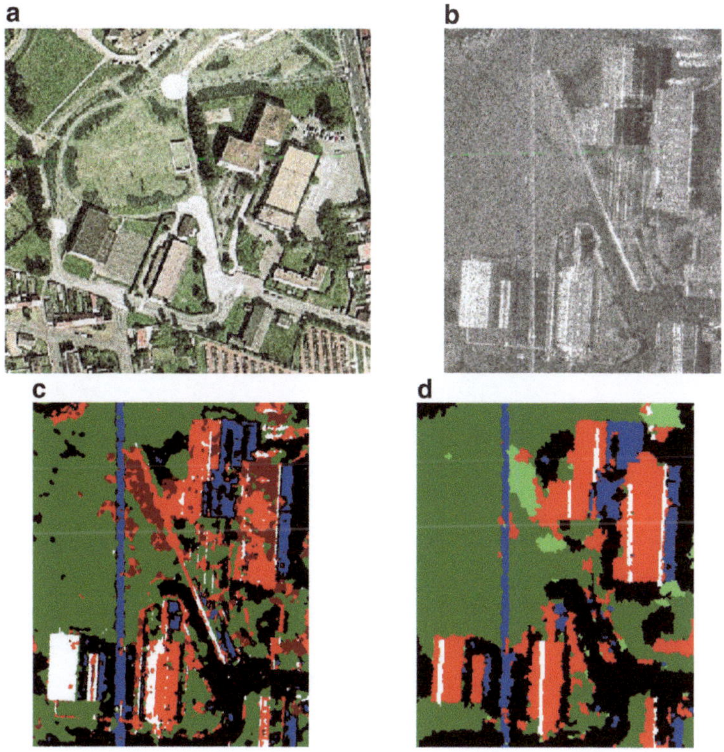

Fig. 7.4 Bayard College area. The College consists of the three top right buildings. The bottom left building is a gymnasium, the bottom centre building is a swimming pool and the bottom right building is a church: (**a**) is the IGN optic image, (**b**) is the amplitude image, (**c**) is the classification obtained at the first processing step and (**d**) is the classification obtained by the fusion scheme. This last classification is clearly less noisy with accurate results for most of parts of the scene. Colour coding: *black* = streets, *dark green* = grass, *light green* = trees, *red* = buildings, *white* = corner reflector, *blue* = shadow

poplar alley is classified as building. Part of the church roof is classified as road. The error comes from the road detector to which great confident is given in the merging process. But in the DSM, the roof appears clearly above the ground. Nevertheless, roads are well detected and the global classification is correct. The DSM is smooth (compared to the initial altimetric accuracy) over low vegetation and buildings. On roads, coherence is quite low, leading to a noisy DSM.

7.4.5 Improvement Method

The final step will correct some errors in the classification and the DSM by checking the coherency between them. In this part, two region adjacency graphs are considered: the one defined for the merging step (based on regions) and a new one

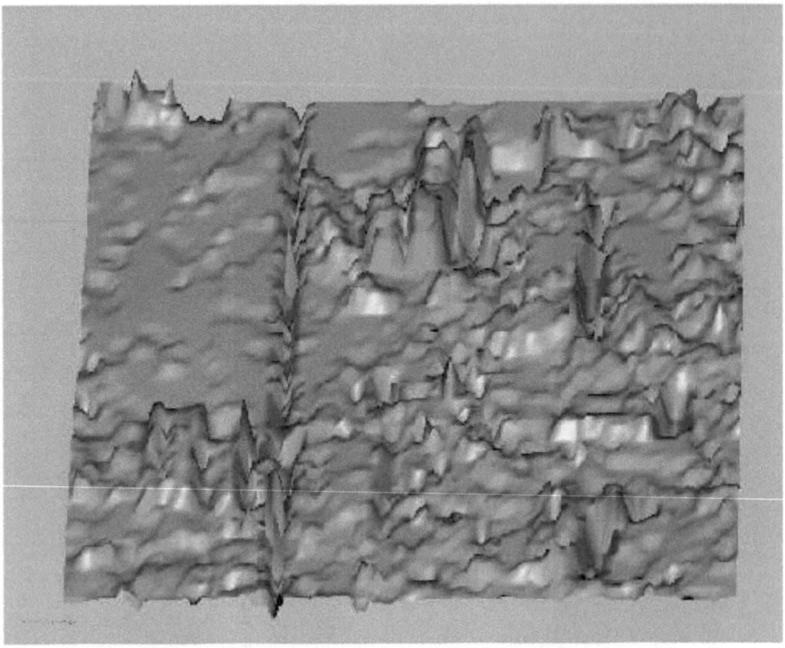

Fig. 7.5 3D view of the DSM computed for the Bayard College

constructed from the final classification l. The regions of the same class, in the first graph, are merged to obtain the complete object, leading to an object adjacency graph.

The corrections are performed for each object. When an object is flagged as misclassified, it is split in regions again (according to the previous graph) in order to correct only the misclassified parts of the objects.

The correction steps include:

- Rough projection of the estimated DSM on ground geometry.
- Computation of the "layover and shadow map" from the DSM in ground geometry (ray tracing technique).
- Comparison of the estimated classification with the previous map l, detection of problems (for instance, layover parts that lay on ground class or layover parts that do not start next to a building).
- Correction of errors: for each flagged object, the partition of regions is reconsidered and the region not compliant with the layover and shadow maps is corrected. For layover, several cases are possible: if layovers appear on ground regions, the regions are corrected as trees or buildings depending on their size; for buildings that do not start with a layover section, the regions in front of the layover are changed into grass. The height is not modified at this stage.

Thanks to this step, some building edges are corrected and missing corner reflectors are added. The effects of the improvement step on the classification

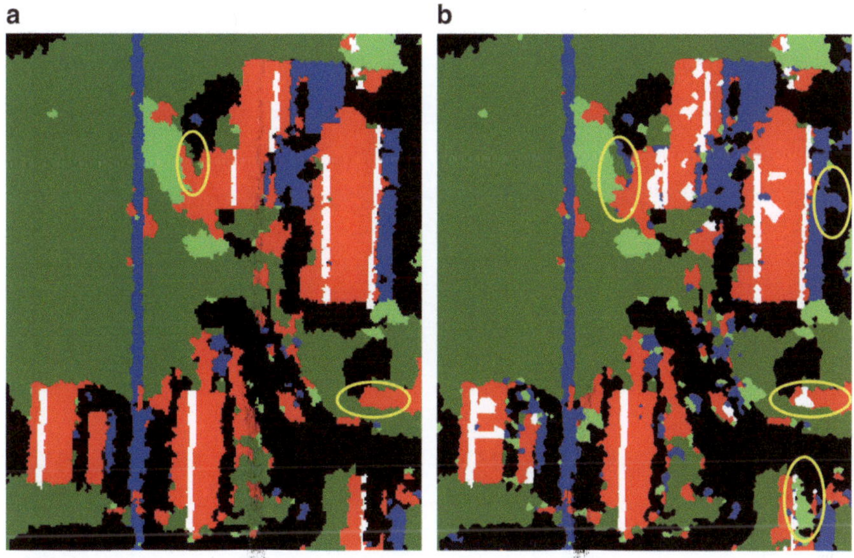

Fig. 7.6 Illustration of classification correction step (**b**). Initial classification to be corrected is plotted in (**a**). Interesting parts are *circle* in *yellow*

are illustrated on Fig. 7.6. The comparison of layover start and building edges allows the edges to be relocated. In some cases, the building edges are badly positioned due to small objects close to the edges. They are discarded through layover comparison.

In the very last step, the heights of vegetation regions are re-evaluated: it is nonsense to have a mean value for a region of trees. Thus the heights of the filtered interferogram are kept in each pixel (instead of a value per region). Actually tree regions do not have a single height and the preservation of the height variations over these regions enables us to stay closer to reality.

7.4.6 Evaluation

The final results obtained for the Bayard district are presented on Fig. 7.7.

A manual comparison between ground truth and estimated DSM has been conducted for nineteen buildings of the Bayard area. They have been picked out to describe a large variety of buildings (small and large ones, regular and irregular shapes). The mean height of the estimated building height is compared to the mean height of the BD Topo©(ground truth). For each building, the estimated height is compared to the expected height. The rms error is around 2.5 m, which is a very good result in view of the altimetric accuracy (2–3 m).

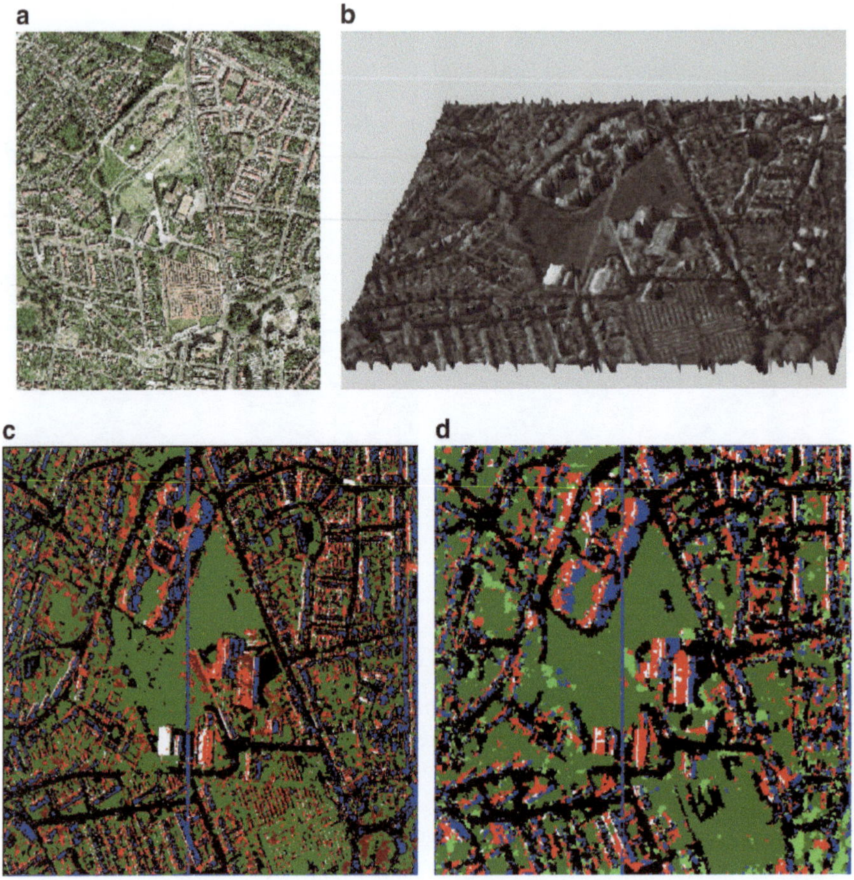

Fig. 7.7 Results of the Bayard district: (**a**) optical image (IGN), (**b**) 3D view of the DSM with SAR amplitude image as texture, (**c**) classification used as input, (**d**) final classification. (*black* = streets, *dark green* = grass, *light green* = trees, *red* = buildings, *white* = corner reflector, *blue* = shadow)

Firstly, altimetric and spatial image resolutions have a very strong impact on the quality of the result. They cannot be ignored for result analysis. From these results, the spatial resolution has to be better than 50 cm and the altimetric accuracy better than 1 m to preserve all the structures for a very accurate reconstruction of dense urban areas (containing partly small houses). When these conditions are not met, one should expect poor quality results for the smallest objects, which can be observed in our dataset. This conclusion is not linked with the reconstruction method.

Secondly, a typical confusion is observed in all scenes: buildings and trees are not always well differentiated. They both have similar statistical properties and can only be differentiated based on their geometry. In fact, building shape is expected to be very regular (linear or circular edges, right angles, etc.) compared with vegetation areas (at least, in cities). A solution may be the inclusion of geometrical constraints to discriminate buildings from vegetation. Stochastic geometry is a possible investigation field to add a geometrical constraint after the merging step.

This problem appears mostly for the industrial areas where there are no trees. Actually, some buildings have similar heights and statistical properties like trees (e.g., because of chimneys) and confusions occur. In this case, the user may add an extra-information in the algorithm (for instance suppression of the tree class) to reach a better result. This has been successfully tested. This example proves that an expert will get better results than a novice, or a fully automatic approach. Actually, the complexity of the algorithm and the data requires expertise. The user has to fix some parameters for the merging step level (energy, weighting values). Nevertheless, once the parameters have been assigned for a given dataset, the entire dataset can be processed with these values. Yet locally some extra information may be required, e.g., a better selection of the class.

However, the method remains very flexible: users can change detection algorithms or energy terms to improve the final results without altering the processing chain architecture. For instance, the detection of shadows is not optimum so far and better detection will certainly improve the final result.

7.5 Conclusion

SAR interferometry provides an efficient tool for DSM estimation over urban areas, for special applications, e.g., after natural hazards or military purposes. SAR image resolution has to be around 1 m to efficiently detect the buildings. The main advantage of interferometry, compared to SAR radargrammetry, is to provide a dense height map. Yet, the inversion from this height map to an accurate DSM with identified urban objects (such as building) is definitely not straightforward because of the radar geometry, the image noise and the scene complexity. Efficient estimation requires some properties on the images: the spatial resolution should obviously be much lower to the size of the buildings to be reconstructed, the interferometric coherence should be high and the signal to noise ratio has to be high to guarantee a good altimetric accuracy.

Nevertheless, even high quality images will not lead directly to a precise DSM. High level processings are required to obtain to an accurate DSM. This chapter has reviewed the four main algorithm families which are proposed in the literature to estimate 3D models from mono-aspect interferometry. They are based on shape-from-shadow, modelling of roofs by planar surfaces, stochastic analysis and analysis based on prior classification.

A special focus is made on one of this method (classification based) to detail the different processing steps and the associated results. This method is based on a Markovian merging framework. The method has been evaluated on real RAMSES images with accurate results.

Finally, we have shown that mono-aspect interferometry can provide valuable information on height and building shape. Of course, merging with multi-aspect data or multi-sensor data (such as optical images) should improve the results. However, for some geographical areas, the datasets are poor and knowing that with only one high-resolution interferometric pair accurate result can be derived,is important information.

Acknowledgment The authors are indebted to ONERA and to DGA[4] for providing the data. They also thank DLR for providing interferometric images in the framework of the scientific proposal MTH224.

References

Besag J (1986) On the statistical analysis of dirty pictures. J Roy Stat Soc 48:259–302
Bolter R (2000) Reconstruction of man-made objects from high-resolution SAR images. In: IEEE aerospace conference, vol 3, pp 287–292
Bolter R, Pinz A (1998) 3D exploitation of SAR images. In: MAVERIC European Workshop
Bolter R, Leberl F (2000) Phenomenology-based and interferometry-guided building reconstruction from multiple SAR images. In: EUSAR 2000, pp 687–690
Brenner A, Roessing L (2008) Radar imaging of urban areas by means of very high-resolution SAR and interferometric SAR. IEEE Trans Geosci Remote Sens 46(10):2971–2982
Cellier F (2006) Estimation of urban DEM from mono-aspect InSAR images. In: IGARSS'06
Cellier F (2007) Reconstruction 3D de bâtiments en interférométrie RSO haute résolution: approche par gestion d'hypothèses. PhD dissertation, Tlcom ParisTech
Eineder M, Adam N, Bamler R, Yague-Martinez N, Breit H (2009) Spaceborne SAR interferometry with TerraSAR-X. IEEE Trans Geosci Remote Sens 47(5):1524–1535
Gamba P, Houshmand B (1999) Three dimensional urban characterization by IFSAR measurements. In: IGARSS'99, I. . International, Ed., vol 5, pp 2401–2403
Gamba P, Houshmand B (2000) Digital surface models and building extraction: a comparison of IFSAR and LIDAR data. IEEE Trans Geosci Remote Sens 38(4):1959–1968
Gamba P, Houshmand B, Saccani M (2000) Detection and extraction of buildings from interferometric SAR data. IEEE Trans Geosci Remote Sens 38(1):611–617
Geman D, Reynolds G (1992) Constrained restoration and the recovery of discontinuities. IEEE Trans Pattern Anal Mach Intel 14(3):367–383
Houshmand B, Gamba P (2001) Interpretation of InSAR mapping for geometrical structures. In: IEEE/ISPRS joint workshop on remote sensing and data fusion over urban areas, Nov. 2001, rome
Lisini G, Tison C, Cherifi D, Tupin F, Gamba P (2004) Improving road network extrcation in high resolution SAR images by data fusion. In: CEOS, Ulm, Germany
Madsen S, Zebker H, Martin J (1993) Topographic mapping using radar interferometry: processing techniques. IEEE Trans Geosci Remote Sens 31(1):246–256
Massonnet D, Rabaute T (1993) Radar interferometry: limits and potentials. IEEE Trans Geosci Remote Sens 31:445–464
Massonnet D, Souyris J-C (2008) Imaging with synthetic aperture radar. EPFL Press, 2008, ch. SAR interferometry: towards the ultimate ranging accuracy

[4] DGA=Délégation Générale pour l'Armement.

Modestino JW, Zhang J (1992) A Markov random field model-based approach to image interpretation. IEEE Trans Pattern Anal Mach Intel 14(6):606–615
Ortner M, Descombes X, Zerubia J (2003) Building extraction from digital elevation model. In: ICASSP'03
Petit D (2004) Reconstruction du "3D" par interférométrie radar haute résolution. PhD dissertation, IRIT
Quartulli M, Dactu M (2001) Bayesian model based city reconstruction from high-resolution ISAR data. In: IEEE/ISPRS joint workshop on remote sensing and data over urban areas
Quartulli M, Dactu M (2003a) Information extraction from high-resolution SAR data for urban scene understanding. In: 2nd GRSS/ISPRS joint workshop on "data fusion and remote sensing over urban areas", May 2003, pp 115–119
Quartulli M, Datcu M (2003b) Stochastic modelling for structure reconstruction from high-resolution SAR data. In: IGARSS'03, vol 6, pp 4080–4082
Rosen P, Hensley S, Joughin I, Li F, Madsen S, Rodrìguez E, Goldstein R (2000) Synthetic aperture radar interferometry. In: Proc IEEE 88(3):333–382
Soergel U, Schulz K, Thoennessen U, Stilla U (2000a) 3D-visualization of interferometric SAR data. In: EUSAR 2000, pp 305–308
Soergel U, Thoennessen U, Gross H, Stilla U (2000b) Segmentation of interferometric SAR data for building detection. Int Arch Photogram Remote Sens 33:328–335
Soergel U, Thoennessen U, Stilla U (2003) Iterative building reconstruction from multi-aspect InSAR data. In: ISPRS working group III/3 Workshop, vol XXXIV
Tison C, Nicolas J, Tupin F, Maître H (2004a) A new statistical model of urban areas in high resolution SAR images for Markovian segmentation. IEEE Trans Geosci Remote Sens 42(10):1674 2046–2057
Tison C, Tupin F, Maître H (2004b) Retrieval of building shapes from shadows in high-resolution SAR interferometric images. In: IGARSS'04, vol III, pp 1788–1791
Tison C, Tupin F, Maître H (2007) A fusion scheme for joint retrieval of urban height map and classification from high-resolution interferometric SAR images. IEEE Trans Geosci Remote Sens 45(2):495–505
Tupin F (2003) Extraction of 3D information using overlay detection on SAR images. In: 2nd GRSS/ISPRS joint workshop on "data fusion and remote sensing over urban areas, pp 72–76
Tupin F, Bloch I, Maître H (1999) A first step toward automatic interpretation of SAR images using evidential fusion of several structure detectors. IEEE Trans Geosci Remote Sens 37(3): 1327–1343

Chapter 8
Building Reconstruction from Multi-aspect InSAR Data

Antje Thiele, Jan Dirk Wegner, and Uwe Soergel

8.1 Introduction

Modern space borne SAR sensors like TerraSAR-X and Cosmo-SkyMed provide geometric ground resolution of one meter. Airborne sensors (PAMIR [Brenner and Ender 2006], SETHI [Dreuillet et al. 2008]) achieve even higher resolution. In data of such kind, man-made structures in urban areas become visible in detail independently from daylight or cloud coverage. Typical objects of interest for both civil and military applications are buildings, bridges, and roads. However, phenomena due to the side-looking scene illumination of the SAR sensor complicate interpretability (Schreier 1993). Layover, foreshortening, shadowing, total reflection, and multi-bounce scattering of the RADAR signal hamper manual and automatic analysis especially in dense urban areas with high buildings. Such drawbacks may partly be overcome using additional information from, for example topographic maps, optical imagery (see corresponding chapter in this book), or SAR acquisitions from multiple aspects.

This chapter deals with building detection and 3d reconstruction from InSAR data acquired from multiple aspects. Occlusions that occur in single aspect data may be filled with information from another aspect. The extraction of 3d information from urban scenes is of high interest for applications like monitoring, simulation, visualisation, and mission planning. Especially in case of time critical events, 3d

A. Thiele (✉)
Fraunhofer-IOSB, Sceneanalysis, 76275 Ettlingen, Germany
and
Karlsruhe Institute of Technology (KIT), Institute of Photogrammetry and Remote Sensing (IPF), 76128 Karlsruhe, Germany
e-mail: antje.thiele@iosb.fraunhofer.de; antje.thiele@kit.edu

J.D. Wegner and U. Soergel
IPI Institute of Photogrammetry and GeoInformation, Leibniz Universität Hannover, 30167 Hannover, Germany
e-mail: wegner@ipi.uni-hannover.de; soergel@ipi.uni-hannover.de

U. Soergel (ed.), *Radar Remote Sensing of Urban Areas*, Remote Sensing and Digital Image Processing 15, DOI 10.1007/978-90-481-3751-0_8,

reconstruction from SAR data is very important. The active sensor principle and long wavelength of the signal circumvent disturbances due to signal loss in the atmosphere as experienced by passive optical or active laser systems.

The following section provides an overview of current state-of-the-art approaches for building reconstruction from multi-aspect SAR data. Subsequently, typical building features in high-resolution InSAR are explained and their potential for 3d reconstruction is high-lighted. Thereafter, we describe in detail an approach to detect buildings and reconstruct their 3d structure based on both magnitude and phase information. Finally, results are discussed and conclusions are drawn.

8.2 State-of-the-Art

A variety of building reconstruction methods have lately been presented in literature. In this section, the focus is on recent developments in the area of object recognition and reconstruction from multi-aspect SAR data. All approaches are characterized by a fusion of information from different aspects on a higher semantic level than pixel level in order to cope with layover and shadowing.

8.2.1 Building Reconstruction Through Shadow Analysis from Multi-aspect SAR Data

Building height and dimension may be derived by analysis of the corresponding shadow from a single image (Bennett and Blacknell 2003). However, such measurements may be ambiguous because different roof types have to be considered, too. Shadow analysis from multi-aspect SAR images of the same scene may help to resolve such ambiguities in order to come up with a more robust reconstruction of buildings. In Moate and Denton (2006), Hill et al. (2006) and Jahangir et al. (2007) object recognition and reconstruction based on multiple active-contours evolving simultaneously on all available SAR images of the scene is proposed. Parameterized wire-frame building models are used to simulate the building's appearance in all images of the scene. Building parameters are continuously adjusted until an optimal segmentation of the building shadow in all images is achieved.

In general, building reconstruction methods based merely on shadow analysis are limited to rural areas or suburban areas. Reconstruction from shadows alone delivers satisfying results if the shadows are cast on flat terrain and no interferences with other objects exist. Approaches making use of additional information besides the RADAR shadow have to be developed if dealing with densely populated urban areas with high buildings.

8.2.2 Building Reconstruction from Multi-aspect Polarimetric SAR Data

An approach for automatic building reconstruction from multi-aspect polarimetric SAR data based on buildings reconstructed as cuboids is presented in Xu and Jin (2007). As a first step, edges are extracted in images of four different aspects and a local Hough transform is accomplished to extract parallel line segments. Parallelograms are generated, which contain the bright scattering from layover areas caused by building façades. Subsequently, such building façades are parameterized. A classification takes place in order to discriminate parallelograms caused by direct reflection of façades from parallelograms that are due to double-bounce signal propagation and shadow. Building parameters are described probabilistically and normal distributions are assigned to all parameters. The corresponding variances are estimated based on the variance of the detected edge points in relation to the straight line fitted through them by the Hough transform. A maximum likelihood method is adopted to match all multi-aspect façade images and to reconstruct buildings three-dimensionally.

A prerequisite for this approach are detached buildings. Interfering façade images from multiple high buildings will lead to imprecise reconstruction results.

8.2.3 Building Reconstruction from Multi-aspect InSAR Data

In Bolter and Leberl (2000) multi-aspect InSAR data are used to detect and reconstruct buildings based on InSAR height and coherence images.

A maximum decision strategy is deployed to combine four different views of a village consisting of small houses. First, the maximum height value of all four acquisitions is chosen and the resulting height map is smoothed with a median filter. Thereafter, a binary mask with potential building regions is generated by subtracting bare earth from the original height map. Minimum bounding rectangles are fit to regions of interest after some morphological filter operations have been applied. Differentiation between buildings and other elevated vegetation is done based on mean and standard deviation of the regions' coherence and height map. Furthermore, simple building models with either a flat roof or a symmetric gabled roof are fit to the segmented building regions. Inside the minimum bounding rectangle planes are fit to the height map using least squares adjustment.

This approach is further extended in Bolter (2001) including information from corresponding SAR magnitude data. Optimal results are achieved if measurements from building shadow analysis are combined with hints from the InSAR height map. Based on the RADAR shadows, building positions and outlines can be estimated while height information is deduced from InSAR heights. Moreover, a simulation step is proposed to refine reconstruction results. A SAR image is simulated using the previously reconstructed 3d hypothesis as input. Subsequently, based on a comparison of real and simulated data the 3d hypothesis is adjusted and refined to minimize the differences.

Problems arise if buildings stand closely together and if they are higher than the ambiguity height of the InSAR acquisition since this approach very much relies on the InSAR height map.

8.2.4 Iterative Building Reconstruction Using Multi-aspect InSAR Data

Iterative building reconstruction from multi-aspect InSAR data is carried out in two separate steps: building detection and building generation (Soergel 2003). For building detection, the InSAR data is first pre-processed in order to reduce speckle. Subsequently, primitive objects are extracted by applying a segmentation of the slant range data. Edge and line structures are detected in intensity data while objects with a significant elevation above ground are segmented in height data. Building hypotheses are set up by generating complex objects from primitive objects.

Thereafter, such hypotheses are projected from slant range geometry to ground range geometry in order to prepare for building generation. Model knowledge is introduced in this step. Buildings are reconstructed as elevated objects with three different kinds of parametric roof models (flat, gabled, and pent roofs) as well as right-angled footprints. More complex building structures are addressed introducing right-angled polygons as footprints and allowing different heights of adjacent building parts (prismatic model). Building heights and roof types are estimated by an analysis of the shadow and by fitting planes to the height data.

In order to fill occluded areas and to compensate for layover effects, building candidates from multiple aspects of the same scene are fused. They are used as input for a simulation to detect layover and shadow regions. In the next step, the simulated SAR data are re-projected to slant range geometry and compared to the original SAR data. In case differences are detected, false detections are eliminated and new building hypothesis are created. The entire procedure is repeated iteratively and is expected to converge towards a description of the real 3d scene. Criteria for stopping the process are either a maximum number of iterations or a threshold of the root mean square error between simulated and real world DEM. These works from Soergel (2003) and Soergel et al. (2003) have been further developed and extended by Thiele et al. (2007a) which will be described in much more detail in the following sections.

8.3 Signature of Buildings in High-Resolution InSAR Data

In this section we focus on the analysis of the building signature in high-resolution InSAR data. Thereby, the characteristic of well known SAR phenomena such as layover, multi-bounce reflection, and shadow (Schreier 1993) is discussed for the examples of a flat-roofed and a gable-roofed building model. Furthermore, the influence of different viewing geometries and building dimensions is shown for both magnitude and phase data. Example images depicted in this section have been recorded in X-band by airborne and space borne systems.

8.3.1 Magnitude Signature of Buildings

In this section the magnitude signature of buildings is discussed considering two different roof types and orthogonal viewing directions. Corresponding illustrations are displayed in Fig. 8.1. The appearance of buildings highly depends on the side-looking viewing geometry of the SAR sensors and range measurements. Received signal of DSM points of same distance to the sensor (e.g., ground, building wall and roof) is integrated in the same image cell. This so called layover effect is shown schematically in the first column of Fig. 8.1. It usually appears bright due to superposition of the various contributors. Comparing the layover of flat- and gable-roofed buildings (Fig. 8.1 second and fourth row), a subdivision of the layover area is observable depending on building dimensions and illumination geometry. This effect was discussed thoroughly for flat-roofed buildings in Thiele et al. (2007a) and for gable-roofed buildings in Thiele et al. (2009). With a decreasing building width, stronger roof pitch, or decreasing off-nadir angle θ such subdivision of the layover signature becomes more pronounced. In both cases, a bright line appears in the flat- and gable-roofed signature. It is caused by a dihedral corner reflector spanned by the ground and building wall, which leads to a sum of all signals that hit the structure and undergo double-bounce propagation back to the sensor. This line, called corner line from now on, is a characteristic part of the building footprint and can be distinguished from other lines of bright scattering using the InSAR phases (see next section).

The subsequent single reflection signal of the building roof may appear as bright or dark area in the SAR magnitude image, depending on the average height variety of the roof structure in proportion to the wavelength and the illumination geometry. In case the roof structure is smooth in comparison to the wavelength of the recording SAR system, the building roof acts like a mirror. All signals are reflected away from the sensor and the roof appears dark. In contrast, a relatively rough roof surface shows Lambertian backscattering and thus appears brighter. Additionally, superstructures on the roof like chimneys can lead to regular patterns or irregular signatures. The ground behind the roof signature is partly occluded by the building shadow appearing as a dark region in the magnitude image. A real magnitude signature of a building can actually differ from this theoretically described building signature because backscatter signal from adjacent objects such as trees and buildings may often interfere.

Figure 8.1 illustrates the variation of the magnitude signature due to illumination direction and building geometry. Real magnitude images of a flat-roofed and a gable-roofed building, acquired by the airborne SAR sensor AeS-1 (Schwaebisch and Moreira 1999) under nearly orthogonal viewing conditions, are displayed

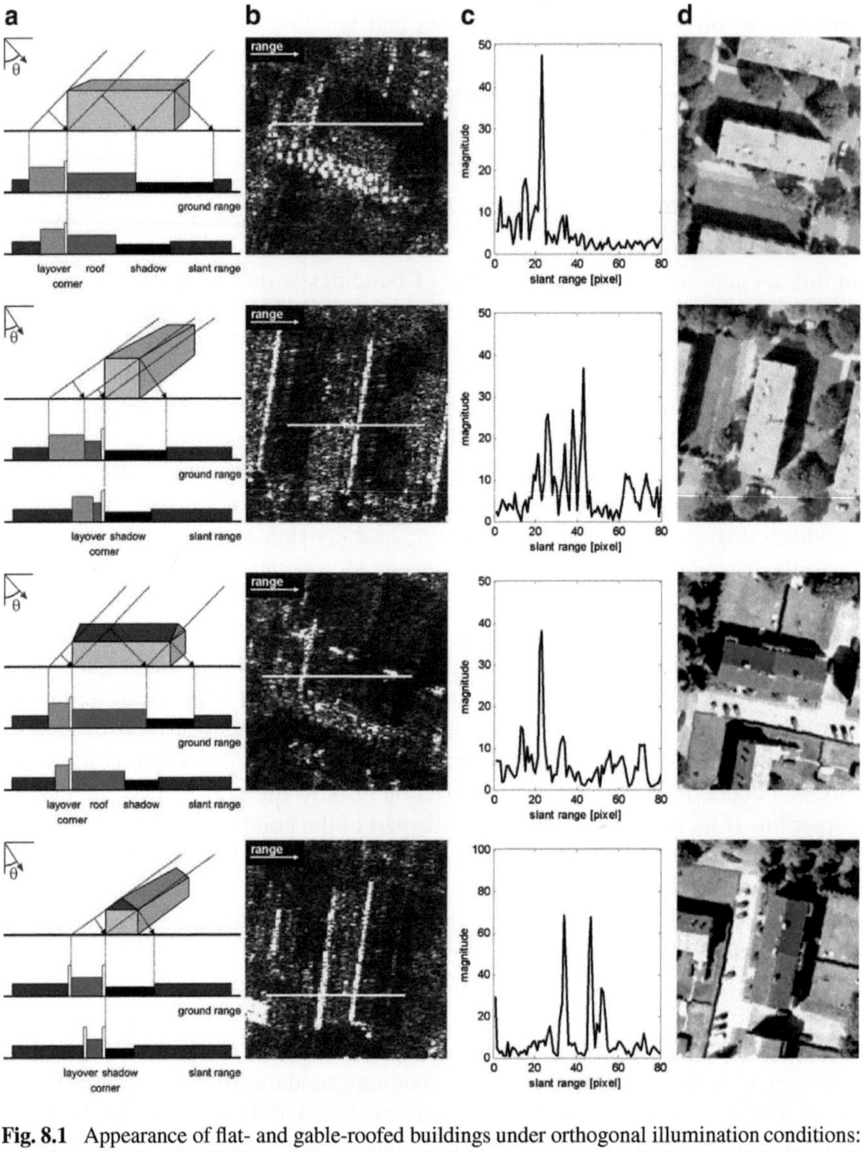

Fig. 8.1 Appearance of flat- and gable-roofed buildings under orthogonal illumination conditions: (**a**) schematic view, (**b**) SAR magnitude data, (**c**) slant range profile of SAR magnitude data, (**d**) corresponding optical image

(Fig. 8.1b). A detailed view of the magnitude slant range profiles corresponding to the white lines in the magnitude images is provided in Fig. 8.1c. Additionally, optical images of the scene are shown in Fig. 8.1d.

In the first row a flat-roofed building (width × length × height, 12 × 36 × 13 m) is facing the sensor with its short side. A small off-nadir angle θ and a large building height result in a long layover area. On the contrary, a larger off-nadir angle would lead to a smaller layover area, but at the cost of a bigger shadow area. In the real SAR data, a bright layover region, dominated by façade structures, occurs at the long building side because the building is not perfect in line with the range direction of the SAR sensor. The corner line appears as short bright line, oriented in azimuth direction. Next, a homogenous area resulting from single-bounce roof signal followed by a shadow area can be seen. Corresponding magnitude values are displayed in the range profile.

The second row shows the same building imaged orthogonally by the SAR sensor. Its appearance changes radically compared to the case described previously. The entire signal of the roof is obscured by layover which is, above all, due to the small building width. Furthermore, the layover region and the corner line show-up more clearly, which is caused by less occlusion of the building front by trees (see corresponding optical image). The shadow area is less developed because of interfering signal from nearby trees and the neighbouring building. Such effects of interfering reflection signals often occur in densely populated residential areas complicating image interpretation.

A gable-roofed building ($11 \times 33 \times 12$ m) facing the sensor with its short side is displayed in the third row of Fig. 8.1. Layover and direct reflection from the roof are less strong compared to the flat-roofed building. This is caused by the building geometry in general and the local situation. Both, the slope of the roof and its material, define the reflection properties. In the worst-case scenario the entire signal is reflected away from the sensor. In the example image the appearance of layover is hampered by a group of trees situated in front of the building. The corner line is clearly visible in the magnitude image and in the corresponding profile.

In the fourth row of Fig. 8.1 the same building as in row three is imaged orthogonally by the SAR sensor. Its magnitude signature shows two significant peaks. The first one is part of the layover area and results from direct reflection of the tilted roof. Width and intensity of this first maximum depend on the incidence angle between the roof plane normal and the off-nadir angle (θ). The brightest signal appears if the off-nadir angle equals the slope of the roof (i.e., zero incidence angle). Under such a configuration all points of the sensor facing roof plane have the same distance to the sensor and are mapped to one single line. Moreover, with increasing span angle between ridge orientation and azimuth direction the signature resembles more a flat-roofed building. However, strong signal occurs for certain angles due to constructive interference at regular structures (Bragg resonance), for example, from the roof tiles. An area of low intensity between the two peaks originates from direct reflection of ground and building wall. The second peak is caused by the double-bounce signal between ground and wall. It appears as one long line along the entire building side. Single response from the building roof plane facing away from the sensor is not imaged due to high roof slope compared to the off-nadir angle. Thus, a dark region caused by the building shadow occurs behind the double peak signature.

Besides illumination properties and building geometry, the image resolution of a SAR system defines the appearance of buildings in SAR imagery. In Fig. 8.2 magnitude images acquired by airborne and space borne sensors showing the same building group are displayed. Both images in column b of Fig. 8.2 were acquired by

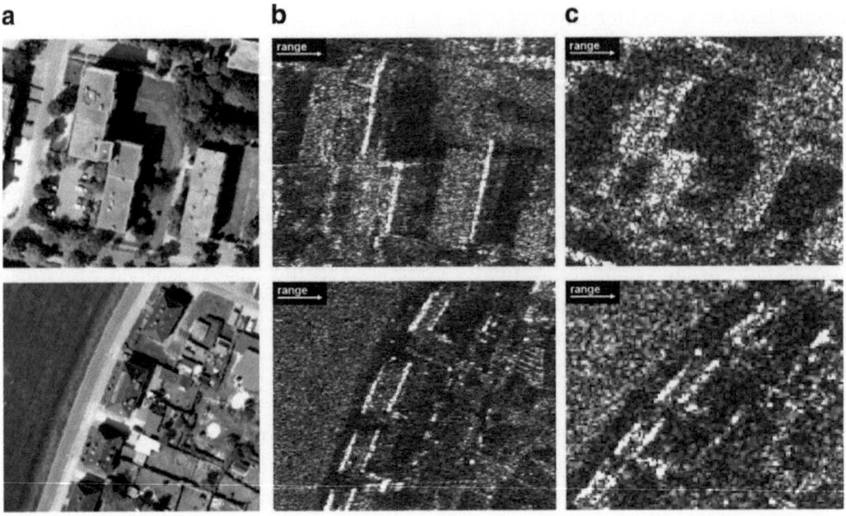

Fig. 8.2 Appearance of flat- and gable-roofed buildings in optical (**a**), AeS-1 (**b**), and TerraSAR-X (HS) data (**c**) (Courtesy of Infoterra GmbH)

the airborne sensor AeS-1 with a resolution of 38 cm in range and 16 cm in azimuth direction. Column c of Fig. 8.2 shows space borne high-resolution spotlight data of TerraSAR-X of approximately 1 m resolution in range direction.

Focusing first on the group of flat-roofed buildings, a layover area is observable in data of both the AeS-1 sensor and the TerraSAR-X satellite. Corner lines are clearly detectable in the AeS-1 data, but less developed in such of TerraSAR-X whereas a shadow region is visible in both data sets. The analysis of the building group depicted in the second row, characterised by hip roofs, shows the previously described "double line" signature. Two maxima occur in both data sets. However, line widths and line continuities are different. Possible explanations for such differences may be slightly different illumination directions and specifics of the SAR data processing like the filtering window.

In summary it can be stated that corner lines are the most stable and dominant building features. They appear in data of all four illumination and building configurations of Fig. 8.1 and especially in high-resolution airborne and space borne data (Fig. 8.2). Hence, building recognition and reconstruction is often based primarily on such corner lines. We will consider this fact in the following sections.

8.3.2 Interferometric Phase Signature of Buildings

Beside the magnitude pattern, the interferometric phase signature of buildings is characterized by the SAR effects layover, multi-bounce reflection, direct reflection from the roof and shadow, too. In Fig. 8.3 the variation of the InSAR phase signature

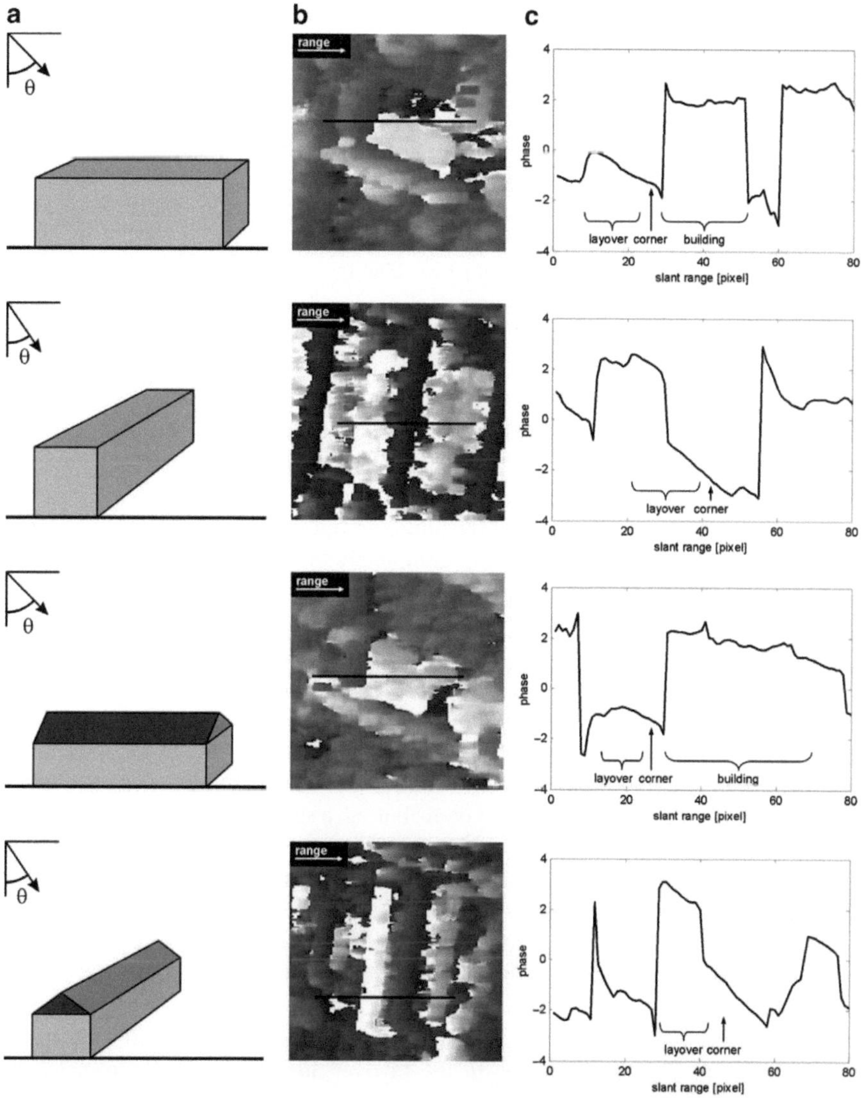

Fig. 8.3 Imaging flat- and gable-roofed building under orthogonal illumination directions: (**a**) schematic view, (**b**) real InSAR phase data, (**c**) slant range profile of InSAR phase data

due to different illumination properties and building geometries is illustrated by means of a schematic view (a), InSAR phase data (b), and slant range profiles (c).

In general, the phase values of a single range cell result, just as the magnitude values, from a mixture of the signal of different contributors. The final interferometric phase value considering across track configurations is proportional to the contributor heights. Hence, the InSAR height derived from an image pixel is a function of the heights from all objects contributing signal to the particular range cell. For example, heights from terrain, building wall, and roof contribute to the final InSAR height of a building layover area. Consequently, the shape of the phase profiles is defined among others by illumination direction and building geometry.

The first row of Fig. 8.3 shows the phase signature of a flat-roofed building oriented in range direction. It is characterised by a layover region, also called front-porch region (Bickel et al. 1997), and a homogenous roof region. These two regions are marked in the corresponding interferometric phase profile, as well as the position of the described significant corner line. The layover profile shows a downward slope, which is caused by two constant (ground and roof) and one varying (wall) height contributor. Hence, the longer the range distance to the sensor becomes, the lower the local height of the reflecting point at the wall will get. The corner line position in the magnitude profile shows in the phase profile a phase value nearly similar to local terrain phases. This is caused by the sum of the double-bounce reflections between ground and wall, which have the same signal run time as a direct reflection at building corner point.

Thereafter, the single response of the building roof leads to a constant trend in the phase profile. If the first layover point completely originates from response of the building roof, then this maximum layover value is equal to the phase value from the roof. Examples for real and simulated InSAR data are shown in Thiele et al. (2007b). In the subsequent shadow region no signal is received so that the phase is only characterized by noise.

The second row of Fig. 8.3 shows the same flat-roofed building illuminated from an orthogonal perspective. Its first layover point, corresponding to the maximum, is dominated by the response of the roof and thus by the building height. Due to the mixture of ground, wall, and roof contributors, a subsequent slope of the phases occurs. Differences to the first row of Fig. 8.3 are caused by the smaller off-nadir angle at this image position leading to a smaller 2π unambiguous elevation interval. Hence, a higher phase difference is equal to the same height difference. Furthermore, a single reflection of the roof cannot be seen due to the small width of the building. Hence, after the layover area the shadow area begins and the corner line separates both.

In the third row of Fig. 8.3 the InSAR signature of a gable-roofed building is depicted. The phase values in the layover area are mainly dominated by the backscattering of ground and building wall. Reasons for less developed response of the building roof were mentioned in the previous section. Phase values at the corner line position are again corresponding to terrain level. Single response of the roof starts at high level and shows a weak trend downwards. This effect appears because the building is not completely oriented in range direction. In addition, the choice of the profile position in the middle of the building plays a role. With a ridge orientated precisely in range direction of the sensor, the phase profile would show a constant trend, such as for the flat-roofed building.

The orthogonal imaging configuration of the gable-roofed building is depicted in the fourth row of Fig. 8.3. In comparison to the previously described illumination configuration, the resulting phase is dominated by backscattering of the building roof which was also observable in the magnitude profile. As a consequence, the layover maximum is much higher. The shape of the layover phase profile is determined by the off-nadir angle, the eave, and the ridge height. For example, a strong steep slope leads to a high gradient in the phase profile. Higher phase differences between ground and roof are again caused by the smaller 2π unambiguous elevation interval. Single backscatter signal of the roof is not observable due to the small width of the building and the roof plane inclination.

Geometric information of a building is mainly contained in its layover region. Therefore, the analysis of the phase profile of gable-roofed buildings is very helpful especially for 3d reconstruction purposes. Results of this analysis are used later on for the post-processing of building hypotheses.

8.4 Building Reconstruction Approach

An introduction to building detection and building reconstruction based on multi-aspect SAR data was given in Section 8.2. All briefly outlined algorithms began with the extraction of building hypotheses based on a single aspect. The subsequent fusion of multi-aspect information was realized by a comparison of single building hypotheses. These procedures are restricted to buildings, which are detectable and can be reconstructed based on a single view. However, the signature of small residential buildings is extremely sensitive to changes of illumination geometry (refer to building examples in Section 8.2). Therefore, the extraction of such buildings very often is not successful based on merely one single aspect (Thiele et al. 2007a).

In the following, an approach is described considering multi-aspect building signatures to generate initial building hypotheses. Additionally, prior knowledge of the buildings that are reconstructed is introduced. First, buildings are assumed to have rectangular footprints. Second, a minimum building extension of $8 \times 8 \times 4$ m (width $\times$ length $\times$ height) is expected. Third, buildings are presumed to have vertical walls and a flat or gable roof. The recorded InSAR data have to consist of acquisitions from at least two aspects spanning an angle of 90° in the optimal case in order to benefit from complementary object information.

8.4.1 Approach Overview

The approach can be subdivided in two main parts, which consist of the analysis of magnitude and interferometric data, respectively. Based on the findings presented in Section 8.3, the approach focuses on corner lines. Building detection as well as the generation of the first building hypotheses mainly rely on the analysis of magnitude data. Calculation of building heights and post-processing of the building hypotheses primarily exploits the interferometric phases.

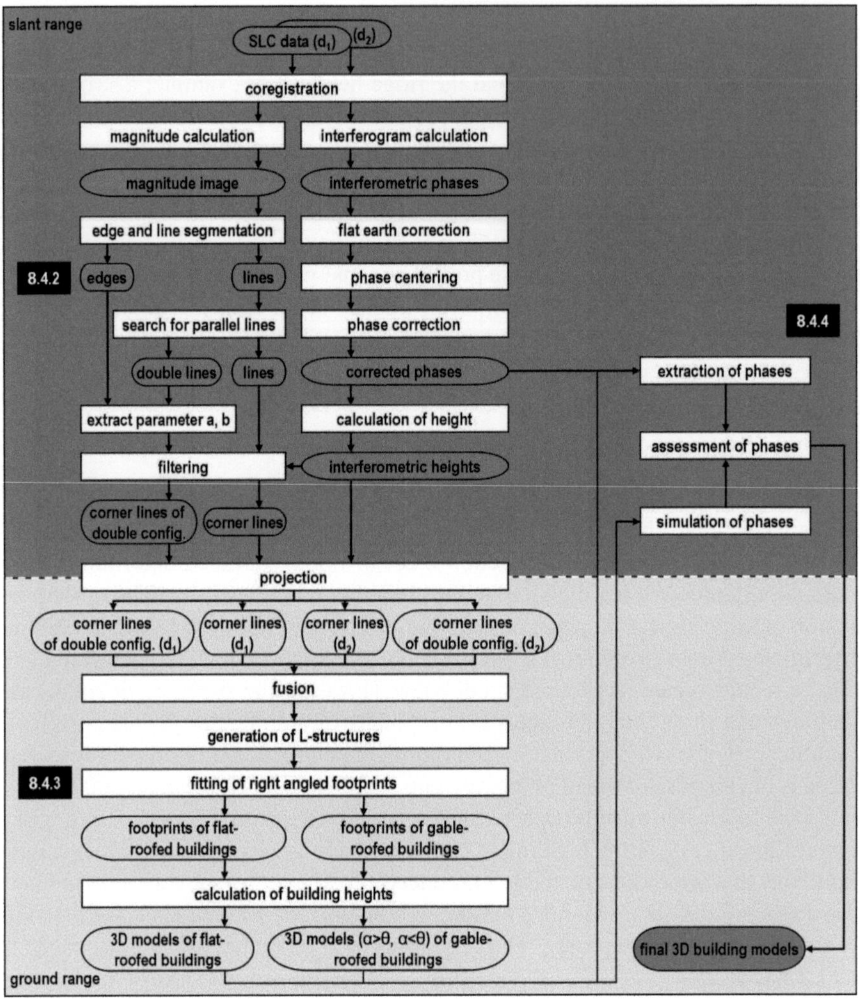

Fig. 8.4 Workflow of algorithm

In the following, a brief description introduces the algorithm shown schematically in Fig. 8.4. More detailed information is presented in subsequent sections.

Processing starts in slant range geometry with sub-pixel registration of the interferometric image pairs as a prerequisite for interferogram generation. This interferogram generation includes multi-look filtering, followed by flat earth compensation, phase centring, phase correction, and height calculation. Since these processing steps are well-established in the field of InSAR analysis, no detailed description will be provided.

Based on the calculated magnitude images, the detection and extraction of building features is conducted. Low-level segmentation of primitives (edges and lines), high-level generation of "double line" signatures, and extraction of geometric building parameters is done. Thereafter, the filtered primitives of each aspect are projected from their individual slant range geometry to the common ground range geometry. This transformation allows the fusion of primitives of all aspects for building hypotheses generation. Subsequently, height estimation is conducted. Results of "double line" segmentation are used to distinguish between flat- and gable-roofed building hypotheses. The resulting 3d building hypotheses are post-processed in order to improve the building footprints and to solve ambiguities in the gable-roofed height estimation. Post-processing consists of interferometric phase simulation and extraction of the corresponding real interferometric phases. Eventually, the real interferometric phases are compared to the simulated phases during an assessment step and the final 3d building results are created. All previously outlined processing steps will be explained in much detail in the following sections.

8.4.2 Extraction of Building Features

The extraction of building features is restricted to slant range InSAR data of a single aspect. Hence, this part of the workflow is accomplished separately for each view. The subsequent step of building hypotheses generation requires the projection of features to a common coordinate system based on interferometric heights.

8.4.2.1 Segmentation of Primitives

As described in Section 8.3.1 the segmentation of primitives exploits only bright lines, which are mainly caused by direct reflections and double-bounce propagation.

Different kinds of edge and line detectors may be used for corner line extraction. Two main categories exist: Detectors that are specifically designed for the statistics of SAR imagery and detectors designed for optical data. Examples for non SAR specific operators are the Canny-operator (Canny 1986) and the Steger-operator (Steger 1998), needing radiometrically pre-processed SAR magnitude images (e.g., speckle reduction and logarithmic rescaling). SAR specific operators have been developed, for instance, by Touzi et al. (1988) and Tupin et al. (1998) considering the statistics of original magnitude images. These template detectors determine the probability of a pixel to belong to an edge or line.

In the presented case, the two referred SAR specific operators are used considering eight different template orientations. Line detection is based on a template consisting of a central region and two neighbouring regions of equal size. The edge detection template has only two equally sized windows. In Fig. 8.5 the steps of line detection from a SAR magnitude image showing a gable-roofed building (a) are displayed. One out of eight probability images resulting from a vertical template orientation is provided in Fig. 8.5b. The fusion of the eight probability images conducted in the original approach (Tupin et al. 1998) is not done in this case. Since

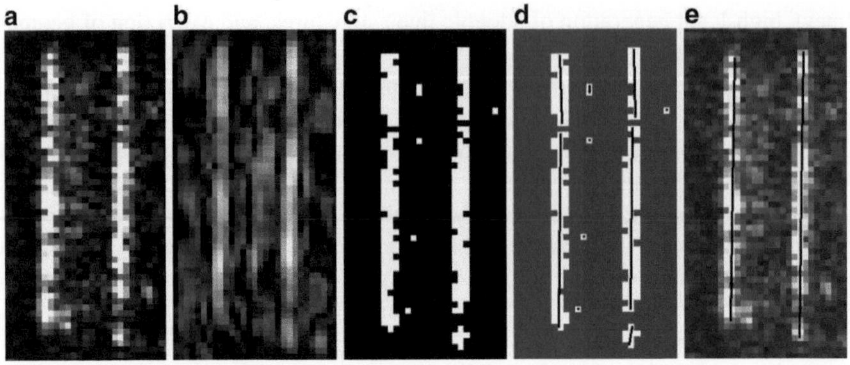

Fig. 8.5 Example of gable-roofed signature in magnitude data (**a**), one corresponding probability image of line detection (**b**), the binary image of line hints (**c**), the binary image overlaid with line segments (**d**) and final result of line segmentation after the prolongation step (**e**)

buildings are assumed to be rectangular objects, edges, and lines are supposed to be straight. Additionally, they are believed to show their maximum in that probability image whose respective window orientation is the closest to the real edge or line orientation. Fusion of the probability images is necessary only for applications considering curved paths such as road extraction.

Subsequently, both, a magnitude and a probability threshold are applied. The magnitude threshold facilitates to differentiate between bright and dark lines. Figure 8.5c shows exemplarily one resulting binary image, which includes line hints. Additionally, straight lines and edges are fitted to this binary image, respectively (see Fig. 8.5d). Moreover, small segments are connected to longer ones as shown in Fig. 8.5e. Criteria for this prolongation step are a maximum distance between two adjacent segments and their orientation. In a final filtering step, the orientation of the resulting lines and edges has to match the window orientation of the underlying probability image.

8.4.2.2 Extraction of Building Parameters

The extraction of building features at this stage of the approach mainly supports the reconstruction of gable-roofed buildings. In the first step, pairs of parallel lines are detected from the previously extracted corner lines. In order to be grouped to a pair of parallel lines, candidate lines have to meet certain requirements with respect to distance, orientation, and overlap. During the second step, edges enclosing the extracted bright lines are extracted. Based on this constellation of lines and edges, two parameters are determined. The first parameter, a, is defined as distance between the two lines whereas the second parameter b is the width of the layover maximum as shown in Fig. 8.6a.

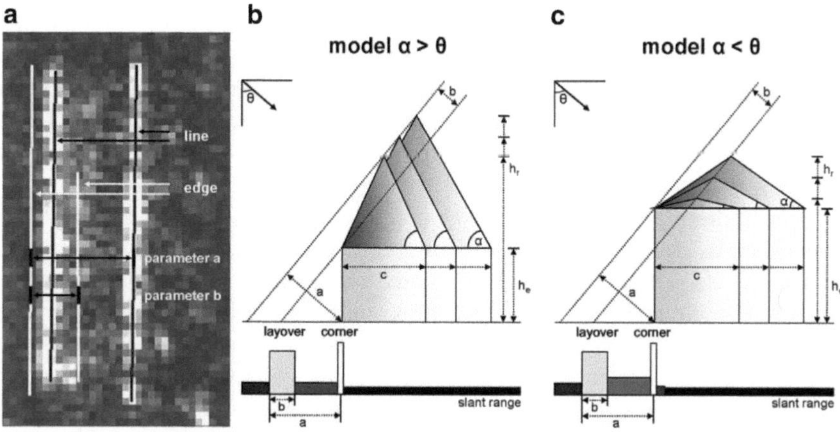

Fig. 8.6 Extraction schema of parameter a and b in the magnitude data (**a**) and groups of gable-roofed building hypotheses showing comparable magnitude signature (**b,c**)

These two parameters allow the generation of two groups of gable-roofed building hypotheses, which show a comparable magnitude signature. The layover maximum of the first building group (Fig. 8.6b), defined by a roof pitch angle α greater than the off-nadir angle θ, results from direct signal reflection from roof and ground. A second group of buildings (Fig. 8.6c) leading to the same magnitude signature as the first one, is characterized by α smaller than θ. The result is a combination of signal from roof, wall, and ground. Both groups of hypotheses can be reduced to only one hypothesis for each of them by considering another aspect direction enabling the extraction of the building width. In Fig. 8.6b, c this building width is marked with the parameter c and the appropriate extraction is described in the following section.

8.4.2.3 Filtering of Primitive Objects

The aim of the filtering step is to find reliable primitive objects to assemble flat- and gable-roofed buildings. Inputs are all previously segmented line objects, useful features are calculated from the interferometric heights (see the workflow in Fig. 8.4).

A flat-roofed building as well as a not ridge-azimuth parallel oriented gable-roofed building is characterized by a corner line. These lines have to be distinguished from other lines, for example, resulting from direct reflection. A gable-roofed building showing ridge-azimuth parallel orientation is characterized by a pair of parallel lines if the incidence angle is small enough. The sensor close line results from direct reflection and the sensor far line from double-bounce propagation. Hence, the single corner lines as well as the described double line constellations have to be separated from all other lines. Filtering is possible based on the interferometric heights at line positions. The previous analysis of the InSAR phases at building locations pointed out, that due to the double-bounce propagation between ground and wall, the interferometric phase value at corner position is similar to local terrain phase. In comparison, the layover maximum of gable-roofed buildings is dominated by direct signal reflection from the roof leading to heights that are higher than the terrain height.

Hence, filtering works like a production rule using the interferometric heights of the lines as decision criterion to derive corner line objects from the initial set of line objects. The mean height in an area enclosing the line is calculated and compared to the local terrain height. First, only lines whose height differences pass a low height threshold are accepted as building corner lines and as reliable hint for a flat or gable-roofed building. Second, line pairs which show both a sensor close line with a height clearly above the local terrain height and a sensor far line fitting the corner line constraints are accepted as hint for a gable-roofed building. The sensor far corner line is marked as candidate for a gable-roofed building.

8.4.2.4 Projection and Fusion of Primitives

The step of projection, also known as geo-coding or orthorectification, enables the fusion of multi-aspect and multi-sensor information in a common coordinate system. All extracted corner line objects of each aspect are transformed from slant range geometry to the common world coordinate system. For this transformation, which has to be carried out individually for each corner line, the previously calculated InSAR heights in the enclosing area of the line are used.

In Fig. 8.7 a LIDAR DSM is overlaid with the projected corner lines. The data set contains lines from two aspects enclosing an angle of approximately 90°. The corner lines of the first flight direction, corresponding to top-down illumination, are marked in black, the corner lines of the second direction in white. The set union of the corner lines from both directions reveals the benefit of orthogonal views for object recognition with SAR sensors. Both views complement one another resulting in much more accurately detected parts of the building outlines.

8.4.3 Generation of Building Hypotheses

This section is split up in two parts. First, the step of generating building footprint objects exploiting the previously detected corner line objects is described. Second, height information is extracted making use of the parameters a and b and the calculated InSAR heights, to finally achieve 3d building hypotheses.

Fig. 8.7 LIDAR-DSM overlaid with projected corner lines (*black* – direction 1, *white* – direction 2)

8.4.3.1 Building Footprint

The generation of building footprints exploits the frequently appearing constellations of corner lines spanning an L-structure in a single aspect or in the ground projection of multi-aspect data. A schematic illustration of the combined feature analysis and the resulting building hint in ground geometry is given in Fig. 8.8a.

First, a simplified magnitude signature of a flat- and a gable-roofed building under orthogonal viewing directions is shown in slant range geometry. Second, as described previously, only corner line objects (in Fig. 8.8a labelled with "corner d_1" and "corner d_2") are projected to a common coordinate system in ground range geometry. At the bottom centre of Fig. 8.8a the L-structure object is generated by the corner line objects from two orthogonal viewing directions can be seen. Based on this constellation, building footprints are generated. The exploitation of such simple geometric structures was also published in Simonetto et al. (2005), and Xu and Jin (2007).

The reconstruction of the building footprint starts with the generation of L-structure objects comprising the search of pairs of corner line objects, which must meet angle, bridging, and gap tolerances. Furthermore, only extracted lines that appear on the sensor facing side of a building are actually real corner lines. In dense urban areas, where many building corners are located closely, it may happen that corner lines of different buildings are combined to L-structure objects. In that case, it is possible to eliminate this kind of L-structures by exploiting the different sensor flight directions. In detail, using orthogonal flight directions for example,

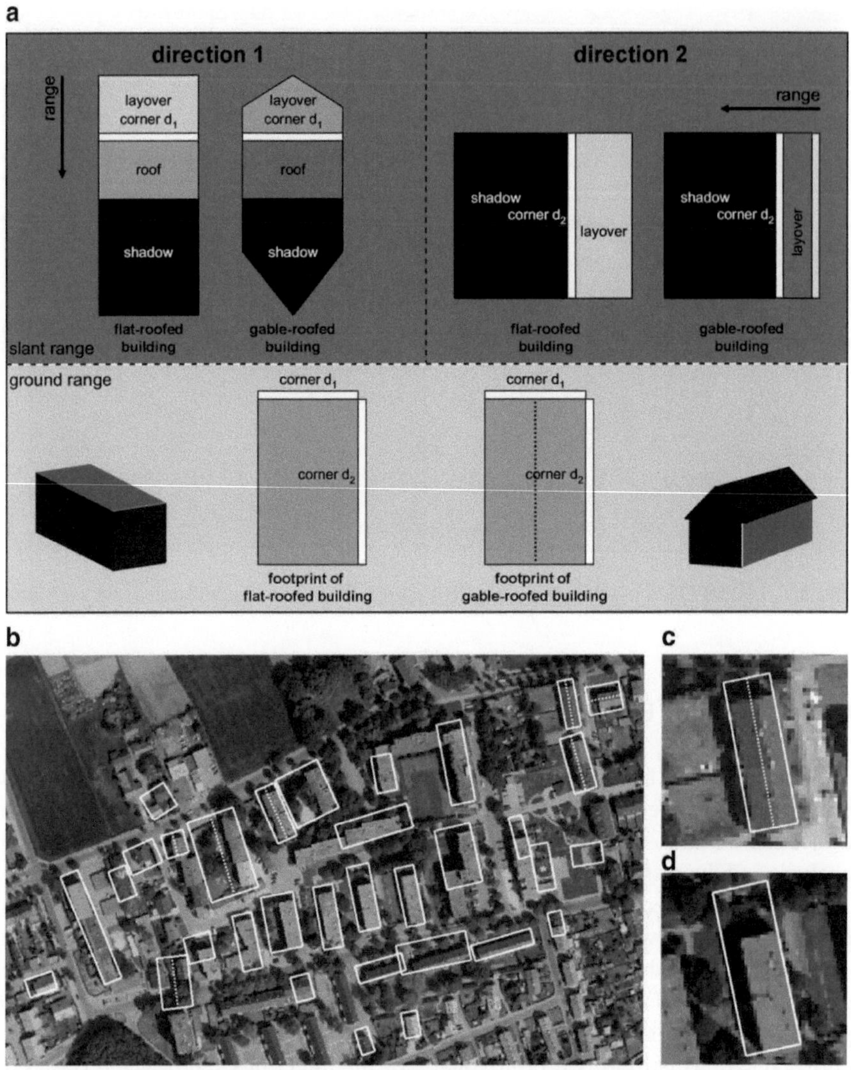

Fig. 8.8 Schematic illustration of building recognition based on multi-aspect primitives (**a**), orthophoto overlaid with resulting building hypotheses (**b**), gable-roofed building hypothesis (**c**), and flat-roofed building hypothesis (**d**)

only those L-structures are suitable, which form an L facing with the exterior to the two flight paths. This is shown in more detail in Thiele et al. (2007a).

In the next step, parallelogram objects are derived from the filtered L-structures. Since most of the generated L-structure objects are not forming an ideal L-structure as illustrated in Fig. 8.8a, filtering of the generated parallelograms is conducted afterwards. In this step the mean InSAR height and the standard deviation of the InSAR heights inside the parallelogram are used as decision criteria.

Furthermore, the span area of the L-structure has to pass a threshold to avoid misdetections resulting from crossing corners. The definition of these decision parameters depends on the assumed building roof type and the fitting accuracy of model assumptions and local architecture. For example, the expected standard deviation of InSAR heights inside a parallelogram of a gable-roofed building is much higher than that of a flat-roofed building. These steps all together were presented in more detail and with example threshold values in Thiele et al. (2007a).

In general, the remaining parallelograms still overlap. Hence, the ratio of average height and standard deviation inside the competing parallelograms is computed and the one with the highest ratio is kept. In the last step, a minimum bounding rectangle is determined for each final parallelogram. It is considered as the final building footprint. In Fig. 8.8b the footprint results of a residential area, based on the segmented corner lines shown in Fig. 8.7, are shown. All building footprint objects generated from corner lines which are part of a parallel line pair are hypotheses for gable-roofed building objects. They are marked with a dotted ridge line in Fig. 8.8b. A detailed view of results of gable- and flat-roofed buildings is provided in Fig. 8.8c, d. The gable-roofed hypothesis (Fig. 8.8c) fits quite well the orthophoto signature of the building. On the contrary, the hypothesis of the flat-roofed building shows higher differences to the optical building signature and post-processing becomes necessary. This issue will be described and discussed in the following section.

8.4.3.2 Building Height

In addition to 2d information, a complete building reconstruction also includes height estimation. In order to properly reconstruct buildings three-dimensionally, their roof type has to be considered. For a flat-roofed building the height h_f is determined by calculating the difference between the mean InSAR height inside the generated right-angle footprint h_b and the mean local terrain height h_t around the building as shown in Eq. (8.1).

$$h_f = \overline{h_b} - \overline{h_t} \tag{8.1}$$

In order to determine the height of gable-roofed buildings an ambiguity problem has to be solved. Two different building hypotheses can be generated based on the same magnitude signature (Fig. 8.6b, c). Using the extracted parameters a and b (see Fig. 8.6a), the width of the building (parameter c), and the local off-nadir angle θ at the position of the parallel line pair, three important parameters for 3d building reconstruction can be calculated: The eave height h_e, the ridge height h_r, and the pitch angle α of the hypotheses. Applying Eq. (8.2), the first hypothesis shows a greater α than θ and this results in a lower eave height h_e but in a higher overall height h_r.

$$\alpha > \theta, \quad h_e = \frac{(a-b)}{\cos\theta} \quad h_r = h_e + \frac{c}{2}\cdot\tan\alpha \quad \tan\alpha = \tan\theta + 2\frac{b}{c\cdot\cos\theta} \tag{8.2}$$

$$\alpha < \theta, \quad h_e = \frac{a}{\cos\theta} \quad h_r = h_e + \frac{c}{2}\cdot\tan\alpha \quad \tan\alpha = \tan\theta - 2\frac{b}{c\cdot\cos\theta} \tag{8.3}$$

The second hypothesis (Eq. 8.3) assumes a smaller α than θ leading to a higher h_e but lower total height h_r. The existing ambiguity cannot be solved at this stage of the processing. It will be part of the post-processing of the building hypotheses described in the following section.

8.4.4 Post-processing of Building Hypotheses

Post-processing of building hypotheses focuses on solving the ambiguity of gable-roofed building reconstruction and correcting oversized building footprints. Its main idea is a detailed analysis of the InSAR phases at the position of the building hypotheses supported by simulated interferometric phases based on these hypotheses. Simulation takes the current 3d building hypotheses, as well as the sensor, and scene parameters of the InSAR data as input parameters. Our process of interferometric phase simulation was presented in Thiele et al. (2007b). It takes into account that especially at building locations a mixture of several contributions can define the interferometric phase of a single range cell. A ground range height profile of each building hypothesis is generated taking into account azimuth and range direction. The ground range profile in range direction is split up into connected linear components of constant gradient. Afterwards, for each range cell, certain features are calculated from these segments, such as normal vector, local incidence angle, range distance differences and phase differences. The simulation is carried out according to the slant range grid of the real measured interferogram. Finally, the interferometric phase of each single range cell is calculated by summing up all contributions (e.g., from ground, building wall and roof).

The subsequent assessment of the similarity of simulated and real InSAR phases is based on the correlation coefficient and delivers a final hypothesis. In the following, the post-processing, based on two reconstruction results, is described in more detail.

8.4.4.1 Ambiguity of the Gable-Roofed Building Reconstruction

The ambiguity of the gable-roofed building reconstruction process can theoretically be solved by a high-precision analysis of the magnitude or phase signature. The analysis of the magnitude signature would start with the ridge-perpendicular orientation of the building. Due to the different building heights h_e and h_r of model $\alpha > \theta$ and $\alpha < \theta$, the shape of layover and shadow area would show differences. Such an analysis would suppose a clear shape of the areas without any interference from other objects, but this condition is usually not met in dense urban areas. Furthermore, the magnitude signature of the ridge-parallel configuration would also show variations caused by the different signal contributors (ground, wall and roof). However, a prerequisite of this potential magnitude analysis is that all relevant parameters (e.g., wall and roof materials) are given, which is not practicable in reality.

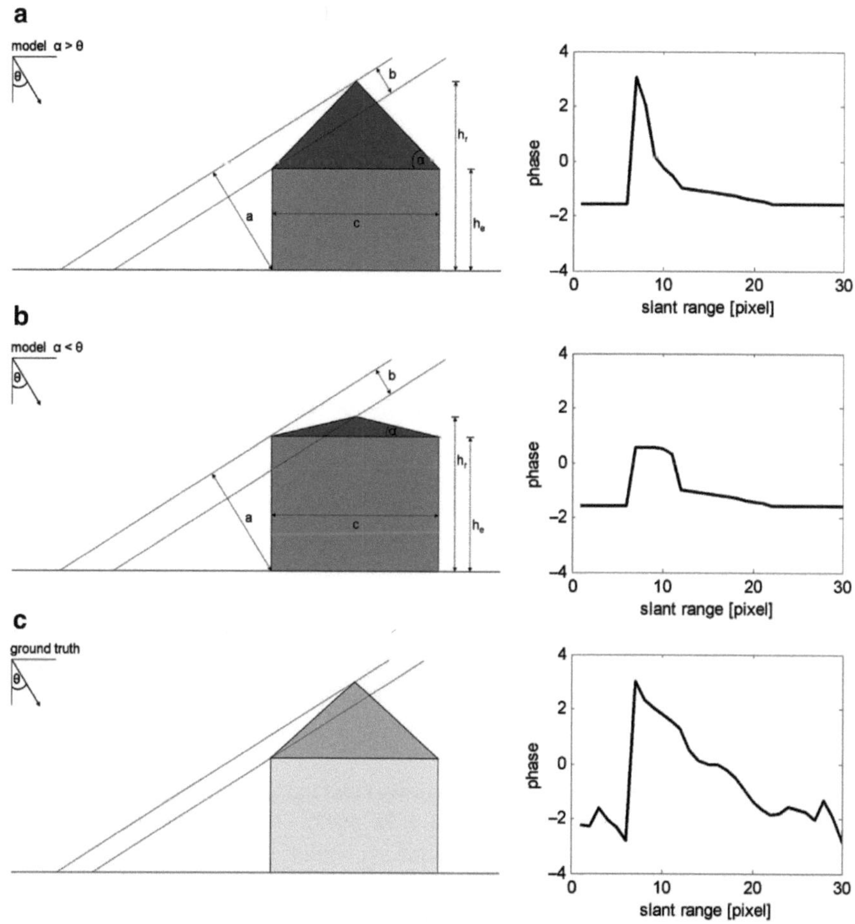

Fig. 8.9 Ambiguity of the reconstruction of gable-roofed buildings: schematic view of a building and its corresponding simulated phase profile of model $\alpha > \theta$ (**a**) and $\alpha < \theta$ (**b**); schematic view of real building and real measured InSAR phase profile (**c**)

An analysis of the phase signature is more promising. Due to different geometries of the two remaining building hypotheses, the interferometric phase in the layover area is dominated by different groups of contributors resulting in different phase shapes. This effect is observable in the simulation of the interferometric phases shown in Fig. 8.9a, b, which is carried out for a range line by using the calculated building parameters (e.g., width, h_e, h_r) as well as the scene parameters (e.g., off-nadir angle, flight altitude), and sensor parameters (e.g., wave length, baseline configuration) as input.

In Fig. 8.9, the first phase values of the layover areas of the two hypotheses are divergent, due to different h_r and the different distance from sensor to building eave. Focusing first on model $\alpha > \theta$, h_r is higher and the ridge point is the closest building point to the sensor. Hence, the first backscatter information of the building contains the maximal height and leads to the highest point of the layover shape. Additionally, the first layover point allows the direct extraction of h_r if we assume dominant reflection of the roof in comparison to the ground. The second model $\alpha < \theta$ shows a lower phase value at the beginning of the layover. Thus, the eave point has the smallest distance to the sensor. As a consequence, h_e affects the first point of the profile. Depending on the ratio between α and θ, a weak downtrend, a constant trend (Fig. 8.9b), or an uptrend of the phase profile, caused by stronger signal of the ridge point, occurs. This trend depends on the mixture of the signal of the three contributors, roof, wall, and ground. In comparison to model $\alpha > \theta$, the direct extraction of h_r based on the first layover value is not possible in this case.

In addition to the previously described differences at the start point of the phase profiles, the subsequent phase shape shows different descents (Fig. 8.9a, b). This effect is caused by the mixture of heights of the different contributors. The layover part, marked by the parameter b, of hypothesis $\alpha > \theta$ is governed by signal contributions of roof and ground. Therefore, the height contribution of the roof is strongly decreasing whereas the ground stays constant. In comparison, the same layover part of hypothesis $\alpha < \theta$ is caused by the response of roof, wall, and ground. The height information of the roof is slightly increasing; the one of the wall is decreasing and the one of the ground again stays constant. The mixture of these heights can show a nearly constant trend up to the ridge point position. Alternatively, a decreasing or increasing trend may occur because the decreasing trend of the wall can or cannot compensate the increasing trend of the roof. Generally, the phase profile descent of model $\alpha < \theta$ is weaker than the descent of model $\alpha > \theta$ due to the interacting effects of multiple contributors.

The remaining part of the layover area between the two maxima is characterized by the two contributors, wall and ground. It begins at slant range position 12 pixel in the phase profiles in Fig. 8.9a, b and shows a similar trend for both models. The phase value at the corner position (slant range position 22 pixel) is a little higher than the terrain phases in the simulated profiles. Due to the radar shadow behind the building, the phase shape behind the layover area contains no further information for the example provided here.

The real InSAR signature is calculated by the steps multi-look filtering, flat earth compensation, phase centring, and phase correction, which are described in more detail in Thiele et al. (2007a). Finally, we obtain a smooth InSAR profile shifted to $-\pi/2$ at terrain level to avoid phase jumps at building location. The same shifting is done with the simulated phase profiles, which allows direct comparison between both of them. A real single range phase profile of the building simulated in Fig. 8.9a, b is given in Fig. 8.9c. Comparing the schematic views (left column of Fig. 8.9), the real building parameters (h_e, h_r, and α) show a higher similarity with hypothesis $\alpha > \theta$ than with hypothesis $\alpha < \theta$. This similarity is also observable in the corresponding phase profiles (right column of Fig. 8.9). The very high phase value of both profiles is nearly identical in position and absolute value because both times α is larger than θ and thus the signal reflection at the beginning of the layover area is dominated by the ridge point of the roof. The strong uptrend in the simulation of model $\alpha > \theta$ is less pronounced in the real phase profile, due to multi-look filtering of the real InSAR phases. Furthermore, our simple simulation does not consider direct and double-bounce reflection resulting from superstructures of the building façade, which are of course affecting the real InSAR phases. The position and the absolute phase value at the corner position are again similar in simulated and real phase profile.

During post-processing of the gable-roofed building hypotheses, the previously described differences of the layover shapes are investigated and exploited in order to choose the final reconstruction result. Based on the detected corner line, real InSAR phases are extracted to assess the similarity between simulated and real interferometric phases. According to the model assumptions of our simulation process, which are mentioned above and given in Thiele et al. (2007b), only simulated interferometric phases unequal zero are considered for the calculation of the correlation coefficient. This assumption is fulfilled by layover areas and areas of direct reflection from the roof. Finally, the hypothesis which shows the highest correlation coefficient is chosen as final reconstruction result of the gable-roofed building object. The result and the comparison to ground truth data are presented in the following section.

8.4.4.2 Correction of Oversized Footprints

As pointed out before, some of the reconstructed building footprints are oversized, which is mainly caused by signal contributions of adjacent walls, fences, or trees. In addition, the estimated building height is affected by this phenomenon, because surrounding terrain height values contribute to building height estimation. This leads to underestimated building heights. Similar to the post-processing step of gable-roofed buildings, reconstruction results of flat-roofed buildings can be improved comparing simulated and real InSAR phases. In Fig. 8.10 the post-processing is visualized in detail for a building hypothesis of the flat-roofed building already shown in Fig. 8.8d.

The process begins with the simulation of the hypothesis based on the extracted building width, length, and height (Fig. 8.10a). A schematic view in the left column illustrates the illumination situation for an oversized hypothesis and the idealized position of the two extracted building corners d_1 and d_2. The centre column displays the simulated interferometric phases of this oversized hypothesis. In front of the building the L-shaped layover area is observable, followed by the constant phase area resulting from the single-bounce reflection of the building roof (light grey). Based on this simulation result and the current building footprint, corresponding real InSAR phases are extracted (Fig. 8.10c, last column). The differences between simulated and real (Fig. 8.10c, right column) phases are given in the right column of Fig. 8.10a. Only a small part of the simulated phases corresponds to the real phase signature of a building (shown in mean to dark grey). The oversized part of the hypothesis shows grey values from mean to light grey. Furthermore, the overlap between simulated and real phases is brightness coded darker than zero level, due to the underestimated building height mentioned before.

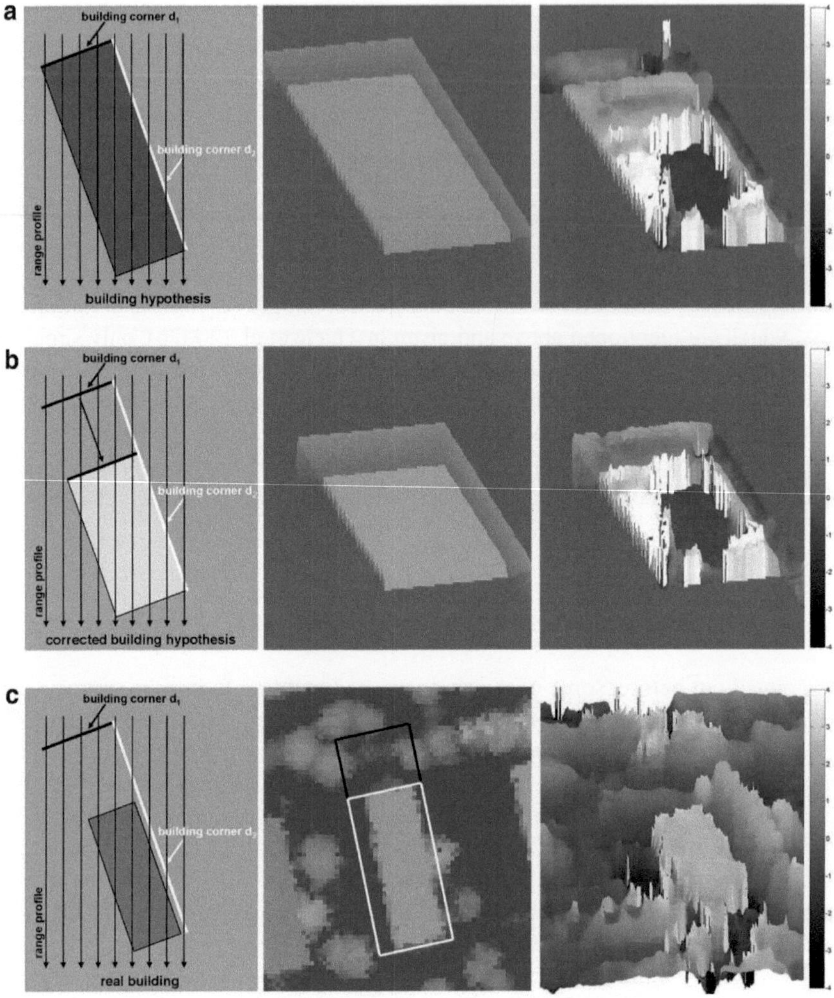

Fig. 8.10 Oversized hypothesis: schematic view, simulated phases and differences between simulated and real phases (**a**), corrected hypothesis: schematic view, simulated phases and differences between simulated and real phases (**b**), real building: schematic view, LIDAR-DSM overlaid with oversized (*black*) and corrected (*white*) hypothesis footprint and extracted real phases (**c**)

In order to improve the result, a correction of the building corner position is necessary. The updating of the position is realized by a parallel shift of corner d_1 along corner d_2 (Fig. 8.10b) in discrete steps. At each new corner position the geometric parameters width, length, and height of the building are recalculated and used for a new phase simulation. Based on this current simulation results and the extracted real InSAR phases, differences and correlation coefficient between both of them are calculated.

The final position of the building corner d_1 (Fig. 8.10b, left column) is defined by the maximum of the correlation coefficients. The centre column shows the corresponding simulated phase image and the right column the differences between simulated and real InSAR phases. Due to the smaller building footprint and the recalculated building height, smaller difference areas and lower height differences occur. Compared to the start situation (Fig. 8.10a), the grey values at the right layover area and the inner part of roof area show lighter grey values closer to zero level. The layover area at the upper part of the building still shows light grey values indicating high differences. This effect is caused by a weakly developed building layover in the real InSAR data. A group of adjacent trees and local substructures avoid the occurrence of well pronounced building layover as well as building corners, and led to the oversized building footprint. The LIDAR-DSM, provided in Fig. 8.10c (centre column), shows this configuration. Furthermore, the oversized hypothesis (black) and the corrected hypothesis (white) are overlaid. The validation of post-processing is given in the following section.

8.5 Results

The presented approach of building reconstruction based on InSAR data exploits different aspects to extract complementary object information. A dense urban area in the city of Dorsten (Germany), characterized mainly by residential flat- and gable-roofed buildings, was chosen as test site. All InSAR data were acquired by the Intermap Technologies X-Band sensor AeS-1 (Schwaebisch and Moreira 1999) in 2003 with an effective baseline of $B \approx 2.4$ m. The data have spatial resolution of about 38 cm in range and 16 cm in azimuth direction; they were captured with an off-nadir angle θ ranging from 28° to 52° over swath. Furthermore, the InSAR data were taken twice from orthogonal viewing directions.

All detected footprints of building hypotheses based on this data set are shown in Fig. 8.8b. The majority of the buildings in the scene are well detected and shaped. Additionally, most of the building roof types are detected correctly. Building recognition may fail if trees or buildings are located closely to the building of interest resulting in a gap of corner lines at this position. Furthermore, too close proximity of neighbouring buildings also results in missing L-structures. Some of the reconstructed footprints are larger than ground truth, due to too long segmented corner lines caused by signal contributions of adjacent trees. Hence, much attention has to be paid to the post-processing results.

The detected footprints of a gable-roofed and a flat-roofed building were shown in Fig. 8.8c, d superimposed onto an orthophoto. Their magnitude and phase signatures were described in Sections 8.3.1 and 8.3.2 because they show similar geometric dimensions. Numerical reconstruction results and the corresponding ground truth data of both buildings are summarized in Table 8.1. Cadastral maps provided ground truth building footprints and a LIDAR-DSM their heights as well as the roof-pitch angle of gable-roofed buildings.

Table 8.1 Reconstruction results of gable- and flat-roofed building compared to ground truth data

	Gable-roofed building			Flat-roofed building		
Building parameter	Ground truth	Model $\alpha > \theta$	Model $\alpha < \theta$	Ground truth	Intermediate result	Corrected (final) result
Off-nadir angle θ (°)	33.5	33.5	33.5	45.3	45.3	45.3
Length (m)	33	35.9	35.9	36	50.7	36.9
Width (m)	11	10.3	10.3	12	17.6	17.6
Height h_f (m) (std.)	–	–	–	13	9.8 (4.0)	11.4 (3.3)
Eave height h_e (m)	9	7.6	8.9	–	–	–
Ridge height h_r (m)	12	12.4	11.1	–	–	–
Pitch angle α (°)	29	43	22	–	–	–

The footprint of the gable-roofed building is well detected showing differences to ground truth of 2.9 and 0.7 m. Post-processing the different hypotheses, concerning the comparison of simulated and real InSAR phases, delivered a higher correlation with model $\alpha < \theta$ (marked in grey at Table 8.1). The estimated eave height is closer to ground truth than the ridge height because pitch angle α was estimated to small.

The result of the flat-roofed building reconstruction shows the high potential of the post-processing of the preliminary footprints. The building length is well corrected from 50.7 m down to 36.9 m. Furthermore, the building height and the height standard deviation inside the building footprint are strongly improved.

Summarizing the results, there is still room for improvement. One possibility would be to consider all building sides in the post-processing step instead of just one. Furthermore, the completeness of building recognition can be enhanced by combining more than only two aspects, to compensate or relieve occlusion effects caused by intersection effects between neighbouring trees and buildings.

8.6 Conclusion

In this chapter an approach for the reconstruction of flat-roofed and gable-roofed buildings from multi-aspect high-resolution InSAR data was presented. We focused especially on small buildings, units typical for residential areas with a minimum extension of 8 × 8 × 4 m (width × length × height). First, the signatures of flat- and gable-roofed buildings in magnitude and phase data were discussed with focus on particular effects due to different illumination geometries. Second, the reconstruction approach benefiting from the exploitation of multi-aspect data was described and intermediate results were shown for several processing steps. The main steps are:

- Segmentation of primitives based on original magnitude images
- Extraction of gable-roofed building parameters
- Filtering and fusion of primitives considering local InSAR heights

- Generation of 3D building hypotheses
- Post-processing of building hypotheses comparing real and simulated InSAR phases

Finally, the reconstruction results were discussed and evaluated by comparing them to cadastral and LIDAR data.

The reconstruction results of this approach show the great benefit of using multi-aspect data and, in particular, of orthogonal views. The completeness of the building recognition could be enhanced combining more than only two aspects, in order to compensate occlusion effects. Furthermore, the benefit of exploiting InSAR phases was described, especially to solve the ambiguity problem in the reconstruction of gable-roofed buildings and to overcome the oversized building footprints.

References

Bennett AJ, Blacknell D (2003) The extraction of building dimensions from high-resolution SAR imagery. In: IEEE Proceedings of the international radar conference, pp 182–187

Bickel DL, Hensley WH, Yocky DA (1997) The effect of scattering from buildings on interferometric SAR measurements. In: Proceedings of IGARSS, vol 4, pp 1545–1547

Bolter R (2001) Buildings from SAR: detection and reconstruction of buildings from multiple view high-resolution interferometric SAR data. Dissertation, University of Graz, Austria

Bolter R, Leberl F (2000) Detection and reconstruction of human scale features from high resolution interferometric SAR data. In: IEEE Proceedings of the international conference on pattern recognition, pp 291–294

Brenner AR, Ender JHG (2006) Demonstration of advanced reconnaissance techniques with the airborne SAR/GMTI sensor PAMIR. In: IEE Proceedings radar, sonar and navigation, vol 153, No. 2, pp 152–162

Canny J (1986) A computational approach to edge detection. IEEE Trans Pattern Anal Mach Intell 8(6):679–698

Dreuillet P, Bonin G, du Plessis OR, Angelliaume S, Cantalloube H, Dubois-Fernandez P, Dupuis X, Coulombeix C (2008) The new ONERA multispectral airborne SAR system. In: Proceedings of IEEE international geoscience and remote sensing symposium, vol 4, pp IV-165–IV-168

Hill RD, Moate CP, Blacknell D (2006) Urban scene analysis from SAR image sequences. In: Proceedings of SPIE algorithms for synthetic aperture radar imagery XIII, vol 6237

Jahangir M, Blacknell D, Moate CP, Hill RD (2007) Extracting information from shadows in SAR imagery. In: IEEE Proceedings of the international conference on machine vision, pp 107–112

Moate CP, Denton L (2006) SAR image delineation of multiple targets in close proximity. In: Proceedings of 6th European conference on synthetic aperture radar, VDE Verlag, Dresden

Schreier G (1993) Geometrical properties of SAR images. In: Schreier G. (ed) SAR geocoding: data and systems. Wichmann, Karlsruhe, pp 103–134

Schwaebisch M, Moreira J (1999) The high-resolution airborne interferometric SAR AeS-1. In: Proceedings of the fourth international airborne remote sensing conference and exhibition, pp 540–547

Simonetto E, Oriot H, Garello R (2005) Rectangular building extraction from stereoscopic airborne radar images. IEEE Trans Geosci Remote Sens 43(10):2386–2395

Soergel U (2003) Iterative Verfahren zur Detektion und Rekonstruktion von Gebäuden in SAR- und InSAR-Daten. Dissertation, Leibniz Universität Hannover, Germany

Soergel U, Thoennessen U, Stilla U (2003) Iterative building reconstruction in multi-aspect InSAR data. In: Maas HG, Vosselman G, Streilein A (eds) 3-D reconstruction from airborne laserscanner and InSAR data, IntArchPhRS, vol 34, part 3/W13, pp 186–192

Steger C (1998) An unbiased detector of curvilinear structures. IEEE Trans Pattern Anal Mach Intell 20(2):113–125

Thiele A, Cadario E, Schulz K, Thoennessen U, Soergel U (2007a) Building recognition from multi-aspect high-resolution InSAR data in urban area. IEEE Trans Geosci Remote Sens 45(11):3583–3593

Thiele A, Cadario E, Schulz K, Thoennessen U, Soergel U (2007b) InSAR phase profiles at building locations. In: Proceeding of ISPRS photogrammetric image analysis, vol 36, part 3/W49A, pp 203–208

Thiele A, Cadario E, Schulz K, Soergel U (2009) Analysis of gable-roofed building signatures in multiaspect InSAR data. IEEE Geoscience and Remote Sensing Letters, Digital Object Identifier: 10.1109/LGRS.2009.2023476, online available

Touzi R, Lopes A, Bousquet P (1988) A statistical and geometrical edge detector for SAR images. IEEE Trans Geosci Remote Sens 26(6):764–773

Tupin F, Maitre H, Mangin JF, Nicolas JM, Pechersky E (1998) Detection of linear features in SAR images: application to road network extraction. IEEE Trans Geosci Remote Sens 36(2):434–453

Xu F, Jin YQ (2007) Automatic reconstruction of building objects from multiaspect meter-resolution SAR images. IEEE Trans Geosci Remote Sens 45(7):2336–2353

Chapter 9
SAR Simulation of Urban Areas: Techniques and Applications

Timo Balz

9.1 Introduction

The simulation of synthetic aperture radar (SAR) data is a widely used technique in radar remote sensing. Using simulations, data from sensors which are still under development can be synthesized. This provides data for developing image interpretation algorithms before the real sensor is launched. Simulations can further create simulated images from precisely defined scenes. They can deliver simulated images of any object of interest from various orbits, at a wide range of angles, using different wavelengths.

In the long history of SAR simulation, many variants of SAR simulation tools have been developed for different applications. In urban areas, SAR simulators are primarily used for mission planning, for the scientific analysis of the complex backscattering, and for geo-referencing. More broadly, simulators are used for sensor design, algorithm development, and for training & education. The different applications and their different requirements have lead to the development of several SAR simulation techniques. Common methods and models will be presented in this chapter, concentrating on the simulation of urban scenes.

Many simulators are based on methods developed by computer graphics which are adapted for SAR simulation. When this is the case, the special geometry and radiometry of SAR images have to be considered. For SAR simulation, the radar Eq. (9.1) is most important. The power received by the radar antenna P_R depends on the power of the sender P_S, the antenna gain G, the wavelength λ, the distance between the antenna and the target r_O, and the radar cross section (RCS) σ.

$$P_R = \frac{P_S \cdot G^2 \cdot \lambda^2}{(4\pi)^3 \cdot r_O^4} \cdot \sigma \tag{9.1}$$

T. Balz (✉)
State Key Laboratory of Information Engineering in Surveying, Mapping and Remote Sensing, Wuhan University
e-mail: balz@lmars.whu.edu.cn

U. Soergel (ed.), *Radar Remote Sensing of Urban Areas*, Remote Sensing and Digital Image Processing 15, DOI 10.1007/978-90-481-3751-0_9,

The transmitted power, the antenna gain, and the wavelength, directly depend on the sensor properties. The calculation of the distance between sensor and the radar target is trivial. Therefore, the RCS is the most important parameter for SAR simulation. The RCS can be expressed in a form where σ is described by the energy scattered back from the target $\vec{E}_s$ and the energy intercepted by the target $\vec{E}_i$ (Knott et al. 2004).

$$\sigma = 4\pi \cdot \frac{\left|\vec{E}_s\right|^2}{\left|\vec{E}_i\right|^2} \tag{9.2}$$

σ is an area and is expressed in m^2. It is used for point targets, whereas the backscattering from these areas is described by the backscattering coefficient σ°. σ° has no dimension and is defined as the radar cross section of an area A normalized by A (Ulaby and Dobson 1989).

$$\sigma^\circ = \frac{\sigma}{A} \tag{9.3}$$

The RCS for a certain polarization σ° can be expressed as the product of two functions: the function describing the surface roughness $f_s(\theta_i)$ and the function of the dielectric properties of the material $f_p(\varepsilon_r \theta_i)$ with ε_r as relative permittivity of the material.

$$\sigma^o(\theta_i) = f_p(\varepsilon_r \theta_i) \cdot f_s(\theta_i) \tag{9.4}$$

(Weimann et al. 1998)

In Section 9.2, the development of SAR simulation systems will be discussed and SAR simulation systems will be differentiated and classified. Section 9.3 will present different models for SAR simulations which will lead to the requirements for the 3D models used as input data for simulations as discussed in Section 9.4. In Section 9.5, examples for applications will be shown, and finally conclusions are drawn.

9.2 Synthetic Aperture Radar Simulation Development and Classification

9.2.1 Development of the SAR Simulation

By the early 1960s, the first SAR simulators had already been developed to optimize the acquisition of SAR images for radar stereo analysis (La Prade 1963) as well as for analyzing the radar backscattering of Venus and Moon (Muhleman 1964). The point scattering model developed by Holtzman et al. (1978) is still the basis for many simulators today. This model divides images into cells, each containing reflectivity and height information used to calculate the SAR image intensity. The speckling is added at the end as Rayleigh-distributed noise.

The SAR raw data simulator SARSIM (Pike 1985) was used for sensor design as well as for the development of SAR processors. It has been widely applied for simulating ERS data. Ray tracing by shooting and bouncing rays was used by Ling et al. (1989) for the RCS calculation of complex objects. SARAS (Franceschetti et al. 1992) was the first extended scene simulator. SARAS calculated the reflection based on the electromagnetic properties of the materials and the local incident angle, derived from a digital elevation model (DEM). This simulation demonstrated a good overall correlation with real ERS SAR images (Franceschetti et al. 1994). For the simulation of artificial structures such an approach cannot be applied as the rationales behind scattering and radar models for natural and man-made scenes are completely different (Franceschetti et al. 2003).

SAR image simulation systems directly simulate SAR images. As a result they are not only common in research and development but they also have commercial applications. Examples include the SE-RAY-EM from OKTAL-SE, a SAR image simulation system based on shooting and bouncing rays (Mametsa et al. 2002), and the commercial PIRDIS image simulation system which supports the simulation of moving target indication (Meyer-Hilberg et al. 2008).

SARViz (Balz and Stilla 2009) simulates extended scenes in fractions of a second. This is possible by simplifying the simulation and using the rasterization approach, which utilizes the flexible programmability of modern Graphics Processing Units (GPU).

9.2.2 Classification of SAR Simulators

Different types of SAR simulators are used for different applications. To differentiate between the types of simulations, SAR simulators can be classified according to different schemes. At present there is no standard for the classification of SAR simulators available. Classification schemes can be differentiated in classifications based on the input or output of the SAR simulator.

An output-based classification has been presented by the Marconi Research Center. The Marconi Research Center differentiates between three types of SAR simulators (Marconi 1984):

1. *System simulators*, which simulate a raw radar signal using a DEM and land use data. The simulated raw data then undergoes a SAR processing to get a SAR image.
2. *Image simulators*, which simulate only the signal parts and basic SAR effects necessary to produce a realistic looking SAR image. Normally the speckling is calculated statistically, and various SAR image effects are not simulated.
3. Simulations based on real SAR images of comparable scenes.

Another output-based classification has been developed by Franceschetti et al. (1995). They differentiate between SAR raw data simulators and SAR image simulators. Image simulators directly produce a final SAR magnitude image, whereas

SAR raw data simulators simulate the raw data of a sensor which has to be SAR processed in order to get a SAR image. Furthermore, these raw data simulators are divided into point simulators and extended scene simulators. Point simulators mainly model the sensor, whereas scene simulators are intended to simulate landscapes and scenes realistically, focusing on the more natural appearance of the radar backscattering.

Simulators can be further distinguished by the way they calculate the backscatter coefficient, $\sigma°$. The input-based classification from Leberl (1990) distinguishes between three types:

1. Simulators that determine $\sigma°$ using look-up tables.
2. Approaches where $\sigma°$ is derived from real SAR images of areas with comparable land coverage and a DEM.
3. Simulators where the backscattering coefficient is directly calculated based on physical models.

Another input-based means of classifying simulation tools is their model handling. Four types can be identified, although only the final three are SAR simulators:

1. *Radar target simulators* – calculating the RCS of single targets.
2. *SAR background simulators* – simulating natural landscapes, often based on 2.5D DEM data and look-up tables.
3. *SAR target-background simulators* – separating the background from the target. The model applied to the background is often simplified and the target simulation may or may not be based on radar target simulators using shooting and bouncing rays.
4. *Integrated SAR simulators* – not differentiating between the background and the targets. All objects are simulated in 3D and both inter- and intra-object interactions are supported, covering multiple objects in an extended scene.

Table 9.1 provides an overview of the SAR simulation classification. For simplification and coherency only the simulation of the target is taken into account for SAR

Table 9.1 SAR simulation system classification matrix

		Simulation output				
		SAR raw data simulator		SAR image simulator		
		Look-up tables	Physical model	Simplified	Look-up tables	Physical model
Simulation Input	SAR background simulator	Holtzman et al. 1978	SARAS (Franceschetti et al. 1992)	Sheng and Alsdorf 2005	SARViz (Balz 2006)	
	SAR target-background simulator		GRECOSAR (Margarit et al. 2006)			
	Integrated SAR simulator				SARViz (Balz and Stilla 2009)	SE-RAY-EM (Mametsa et al. 2002)

target-background simulators. Often SAR target-background simulators simulate the background using look-up tables, but implement a complete physical model for the backscattering of the target.

9.3 Techniques of SAR Simulation

9.3.1 Ray Tracing

Many SAR simulators use ray tracing or a derived technique. Ray tracing has been developed for the visualization of scenes in the optical dimension of light, following geometrical optics. In ray tracing, the path of each "ray" is followed from the SAR sensor, throughout the scene, and eventually back to the sensor. The number of rays followed depends on the implementation and a trade-off between the desired degree of realism and the need to minimize the computational load. At least one ray per slant-range resolution cell has to be followed, but many simulators divide the scene into smaller cells, thus following more rays throughout the scene. This method, then, allows the superposition of signal contributions from many individual scattering objects located in the cell to be modeled.

The effects of multiple reflections, shadows, and occlusions are automatically considered using ray tracing and the simulation of the complex radar signal, including phase information, is possible.

Because the paths of millions of rays have to be calculated for an entire image, ray tracing is a computationally intensive technique. Every ray has to be checked for possible interactions with each polygon of every 3D model included in the scene. If a ray is hitting more than one surface, the closest hit has to be found. By optimizing the data organization for a fast search, the calculation speed of ray tracing can be drastically increased (Glasner 1984). Typically, Octrees (Whang et al. 1995), binary space partitioning (Paterson and Yao 1990), or Kd-trees (Wald and Havran 2006; Tao et al. 2008) are used.

For target RCS simulation, the shooting and bouncing rays technique (Ling et al. 1989) is often applied. In shooting and bouncing rays, each ray is traced through a target, according to the law of geometrical optics, until it leaves the target. Afterwards, physical optics is applied in order to address diffraction, refraction, etc. Similar concepts are also used by SAR simulators (Mametsa et al. 2002).

9.3.2 Rasterization

Besides ray tracing, a second technique, known by many as rasterization, can be used to visualize scenes with lower computational load. This technique is widely used in real-time visualization applications, even though producing less realistic results.

Rasterization does not rely on tracing rays. Instead, the scene elements (i.e., object primitives in vector format, such as triangles) are transformed from world coordinates into the image coordinate system and undergo a rasterization step to a given grid before they are finally drawn. When compared to ray tracing, this process is usually faster, because instead of tracking millions of rays, only thousands of objects have to be transformed by a series of multiplication operations. Recently this process has become hardware accelerated, further increasing the visualization speed. To visualize occlusions correctly, the primitives are either sorted in range direction or a second method, known as Z-buffer technique (Catmull 1978), is applied to prevent primitives closer to the virtual camera from being overdrawn by primitives which are farther away from the virtual camera. This is done by computing a depth value corresponding to the distance between the viewer and each pixel. If a new pixel is about to be written, the depth values are compared against the current depth value of already processed pixels. Only pixels closer to the viewer and possessing a smaller depth-value are drawn, replacing the current value in the Z-buffer.

The radar target simulation GRECO (Graphical Electromagnetic Computing) relies on rasterization to determine the visible parts of a target in order to speed up the RCS calculation (Rius et al. 1993). GRECO visualizes a 3D model using Graphics Processing Units (GPU). Instead of saving RGB color information the face normal directions are saved and interpreted later during the post-processing carried out from the Central Processing Unit (CPU). The concept was later extended for the SAR simulator GRECOSAR (Margarit et al. 2006). Today's graphics hardware is even more powerful, allowing for the GPU to be exclusively used for both the calculation and visualization of SAR images (Balz and Stilla 2009).

9.3.3 Physical Models Used in Simulations

Ray tracing and rasterization are two ways to determine how the simulated sensors and the virtual scene interact. The effect of every interaction of the "travelling rays" is calculated by the physical model. A wide variety of different physical models are applied, ranging from very simple to the most realistic possible. The desired complexity depends on the application.

Simple simulations used for layover and occlusion calculation do not need a physical model and may rely on geometric constraints alone. After implementing the SAR geometry, simple layover and occlusion detection can be realized by setting the basic reflection value of each "ray-" or each "fragment-interaction" to a basic constant reflection value r, such as $r = 1$ as an example. In the final image shadow regions will have a reflection value of $r = 0$, whereas layover areas coincide with values $r > 1$.

Other simplified simulations are based on the Lambertian reflection model. The backscattering is always supposed to be of Lambertian nature, thus the backscatter has the same strength in all directions. Under this assumption the observed backscattering intensity depends only on the local incidence angle θ_i and the reflectivity, r, determined, for example, by look-up tables.

$$\sigma^o = r \cdot \cos(\theta_i) \qquad (9.5)$$

Similar to Phong (1975) shading used in Computer Graphics, the backscattering can be divided into the sum of two contributions: a Lambertian part and a specular part (ambient and emissive part can be neglected in SAR simulation). Calculating diffuse and specular reflections separately and combining them afterwards for the overall reflection strength has, for example, been done before by Arnold-Bos et al. (2007) based on previous work in the field of underwater sound scattering (APL-UW 1994).

Other simulators implement a complete physical model. There is a wide variety of electromagnetic models used for this purpose (Fung et al. 1992; Fung 1994; Long 2001). To calculate the backscatter, various variables are needed to describe the interaction of signal and material. Even for simplified models at least the di-electrical properties and parameters describing the surface roughness must be known.

The backscatters of different objects do influence each other, yet in rasterization, as well as in ray tracing, the contribution of each ray or pixel is calculated separately. Therefore, the effects of mutually influencing SAR scatterers have to be calculated in a post-processing step. This step, however, can be rather time consuming depending on the number of scatter centers included in the calculation for each SAR image pixel, and is often neglected by simulators requiring fast simulation results. Simulators trying to achieve a very precise simulation of the wave propagation avoid ray-tracing and rasterization for this reason and instead simulate the wave propagation in a 3D voxel space.

The speckle simulation is often simplified by assuming Rayleigh-distributed speckle and by multiplying the calculated backscatter intensity with a number generated randomly according to the Rayleigh distribution (Hammer et al. 2008). Regarding high-resolution SAR images, the basic assumption for a Rayleigh distribution, a high-number of independent scatterers of comparable signal strength in each cell, can be wrong, especially for artificial structures. Depending on the surface, various other probability distribution functions, including the Rician inverse Gaussian distribution (Eltoft 2005), K-distribution (Jakeman and Pussey 1976), and various others (Nadarajah and Kotz 2008), can describe the speckle more precisely. Besides approximating the appropriate probability density function, the multiplication of the random value with the calculated backscatter intensity is an additional simplification of the speckle calculation. A more realistic speckle model can be achieved by distributing point scatterers randomly in each voxel (volume elements) and coherently adding all signal contributions of these point scatterers.

9.4 3D Models as Input Data for SAR Simulations

9.4.1 3D Models for SAR Simulation

The quality of the 3D models used for simulation is crucial for the overall simulation quality. Even the most sophisticated physical model is useless if the simulated 3D models are erroneous. The required quality and complexity of the model geometry depends on the desired simulation quality and resolution. The parameters needed to describe the material behavior depend on the physical model used for the simulation.

In principle, modeling of the scene can be distinguished between raster and vector descriptions of the models. A raster description grids the scene into a 3D voxel space in which each voxel contains information about the material of the voxel. This approach consumes a great deal of memory and is therefore used most commonly by simulators requiring a more precise simulation of the microwave propagation. For acceptable wave propagation modeling, the spatial step should be less than one-tenth of the wavelength (Dellière et al. 2007). However, for simulating extended urban scenes the memory requirements of the raster description are simply too high.

Alternatively, a complex scene can be described based on the full set of primitive objects located therein, which are stored in an appropriate vector format. This kind of approach is also referred to as symbolic representation. Scene primitives can include a wide variety of objects ranging from triangles to curves or spheres. However for the sake of easier processing, most simulators only support flat primitives such as polygons or triangles. By combining numerous triangles, complex shapes can be built.

9.4.2 Numerical and Geometrical Problems Concerning the 3D Models

Small angular variations while acquiring the SAR data can lead to huge differences in the resulting SAR images. To simulate these effects correctly, models used for the simulation have to be very rich in detail. Such models can, for instance, be acquired by terrestrial laser scanning. But simulating extended scenes containing hundreds or even thousands of such comprehensive models would largely exceed the memory and the computational capacity of standard computers.

Furthermore, when using such high-resolution models, the simulation of curved structures is still problematic. After triangulation, a curved structure will consist of many flat triangles, yet such a structure will still reflect differently than a structure represented by a boundary consisting of planar facets.

Using interpolated normal directions for calculating the reflection can reduce the problem. But in rasterization, as well as in ray tracing, the amount of reflections analyzed per resolution cell is finite. Therefore, the problem still exists but can be reduced to a minimum by using models with a spatial resolution near that of the radar wavelength and an internal simulation resolution of approximately half of the wavelength. However, this again increases the memory usage and calculation time.

The importance of the models for the simulation cannot be overestimated. For many applications the availability of good 3D models is more important than the simulation technique. The analysis of the TerraSAR-X image of the pyramids of Giza in the following section provides an excellent example of the necessity of such models.

9.5 Applications of SAR Simulations in Urban Areas

9.5.1 Analysis of the Complex Radar Backscattering of Buildings

By comparing simulated SAR images with real SAR images, the interpretation of SAR images can be improved. Due to various disturbing effects in SAR images, such as speckle, layover, shadows, and so forth, their interpretation is rather difficult. Simulations can help to explain certain structures in SAR images and they can be used to verify assumptions about the image content (Guida et al. 2008).

The appearance of the pyramids of Giza in one of the first TerraSAR-X images is only explainable by the double-bounce reflections between the ground and the pyramids (Bamler and Eineder 2008), which can be verified by simulations (Auer et al. 2008a). A subset of the TerraSAR-X spotlight image from the pyramids is depicted in Fig. 9.1a.

The TerraSAR-X image in Fig. 9.1a is an HH polarized high-resolution spotlight image acquired from a descending orbit with a ground resolution of 1.4×1.4 m and an incidence angle of 53°. The simulation in Fig. 9.1b is showing a SARViz (Balz and Stilla 2009) simulation, only including single-bounce reflection. The front side of the Cheops pyramid is fore-shortened to a very small area containing the first-order backscattering contributions of the pyramid front. The reflections from the triangular front portion have to be caused by double-bounce reflections. The combined single- and double-bounce simulation in Fig. 9.1c verifies this assumption.

The double-bouncing occurs between the ground and the pyramid. Figure 9.1d shows only the double-bounce. The simulation without a surrounding ground plane is shown in Fig. 9.1e. Without the surrounding ground, the front side cannot be seen. Hence, the double-bouncing occurs between the ground and the pyramid, as shown by Bamler and Eineder (2008).

The double-bouncing only occurs because of the step-like structure of the pyramid. Simulating a simplified model, a mathematically perfect pyramid, results in no double-bouncing and the simulation result matches the single-bounce simulation in

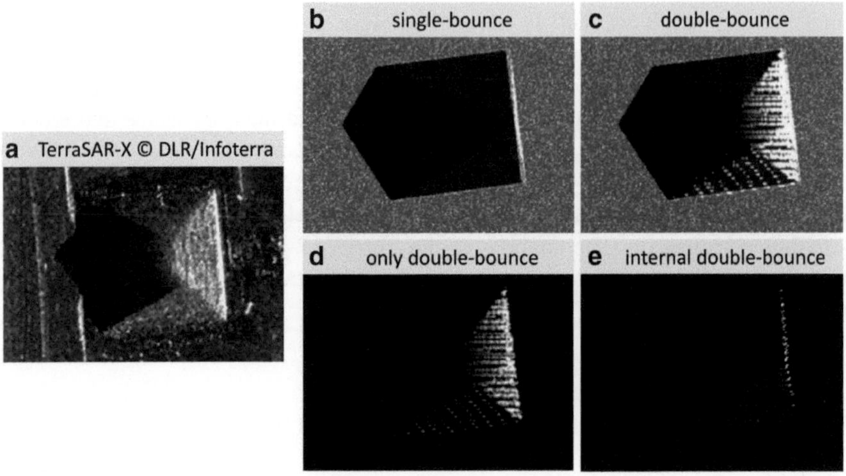

Fig. 9.1 (**a**) TerraSAR-X image of the pyramid © DLR/Infoterra, (**b**) single-bounce simulation of the pyramid model, (**c**) combined single- and double-bounce simulation, (**d**) double-bounce simulation, and (**e**) double-bounce simulation of the pyramid model without a surrounding ground

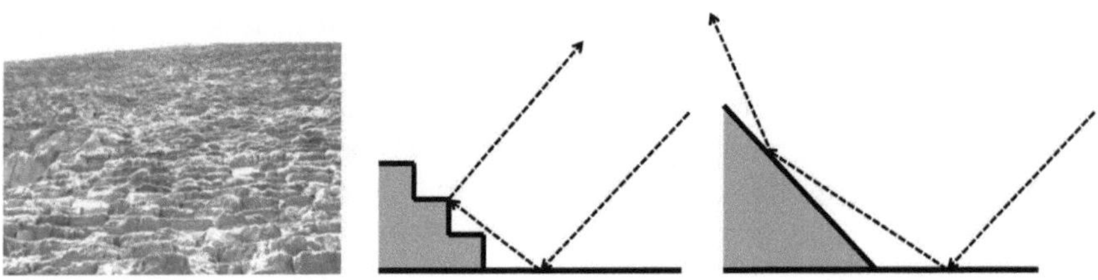

Fig. 9.2 Photo of the pyramid front (*left*), sketches of the double-bouncing of different pyramid shapes (*middle* and *right*)

Fig. 9.1b. As depicted in Fig. 9.2, the forward scattered energy from the ground is not scattered back to the sensor by a perfect pyramid. Only because of the step-like structure of the pyramids of Giza is energy backscattered.

The experiences gained from the analysis of pyramids and other artificial structures are valuable for other SAR image interpretation tasks. The analysis of the appearance of collapsed buildings in high-resolution SAR images, such as those acquired after the devastating Wenchuan Earthquake on May 12, 2008, requires a comprehensive understanding of SAR. SAR simulations can directly support image interpretation in these cases and can provide a deeper understanding of the backscattering of collapsed and partly collapsed buildings (Shinozuka et al. 2000; Wen et al. 2009), which is fundamental for automated or semi-automated damage assessment tools.

SAR simulations can help to understand SAR effects and support the analysis of SAR images. Using simulations certain effects can be simulated separately and the mutual influence can be analyzed in detail. However, the 3D models used for the simulations must be chosen properly.

9.5.2 SAR Data Acquisition Planning

The first SAR simulation tool was developed 1963 to improve mission planning for acquiring stereoscopic radar images (La Prade 1963). Sophisticated mission planning is crucial for successful data acquisition and data of a certain area must be acquired at a certain time and at low cost. Depending on the application, the importance of the time delay and cost differ. For example, military applications or applications in disaster management are time-critical and therefore higher costs are acceptable.

Further limitations including the sensor properties, orbit parameters, weather conditions, etc., must also be considered by mission planning. But most important is the visibility of the area of interest. SAR is a side-looking system and occlusions can hinder the analysis of SAR images, especially in occlusion-rich environments such as those found in urban areas. When this is the case a simple shadow and layover analysis is suitable (Soergel et al. 2003). The acquisition of usable data for multi-aspect object analysis or bi-static SAR acquisition is even more complex and requires sophisticated preparations and simulation assisted mission planning (Gebhardt et al. 2008).

To simulate the occlusions and layover areas, a digital surface model (DSM) is necessary. For high-resolution SAR data acquisition, a 3D city model can be used. Though trees are not included in most 3D city models, they can occlude a considerable area in urban SAR images and their absence in the simulated models can hinder the occlusion analysis (Stilla et al. 2003). DSMs derived from LIDAR or InSAR measurements are therefore preferred.

Reducing occlusions is a trade-off between shadow and layover. Steeper looking angles will reduce the amount of shadows in the acquired SAR image but increase the layover area and vice versa. SAR simulations can assist the mission planning and can determine the best acquisition parameters. This is extremely important for time-critical operations in urban areas. In military applications or disaster management, an optimized data acquisition can be a matter of life and death.

9.5.3 SAR Image Geo-referencing

If the information gathered from SAR images needs to be localized precisely and the SAR data is about to be used together with other spatial information, the SAR image needs to be spatially referenced. For geo-referencing and mosaicking SAR images, the local topography has to be taken into account. SAR simulators used for assisting the geo-referencing of SAR images are typically simplified in that they either simulate SAR geometry alone, or implement simplified SAR radiometry calculations; an example being constantly assuming Lambertian reflection (Gelautz et al. 1998). This was the case when Sheng and Alsdorf (2005) simulated SRTM DEM to improve the geo-referencing and mosaicking of JERS-1 images over the Amazon basin.

To improve the geo-referencing of high-resolution SAR images in urban areas, the spatial resolution of SRTM data is too low. Road network information is better suited. Roads do not backscatter a lot of energy. They appear dark in SAR images. Getting a simplified SAR simulated image of the road structure is possible by using GIS road network data and SAR simulating the roads, while assuming their backscatter is low relative to the surrounding backscatter. The geometry of the road structure has to be SAR simulated using available DEMs. By automatically comparing the simulation around road junctions with the real SAR image, tie-points between the GIS data and the real SAR image can be found automatically (Balz 2004).

9.5.4 Training and Education

SAR simulators are important tools for training and education and can increase knowledge of the physical mechanisms of SAR (Nunziata et al. 2008). Only through the experience of examining a variety of SAR images can students begin to develop "SAR eyes", allowing them to be able to immediately recognize objects of interest contained within a scene. Due to their ability to deliver defined SAR images of objects of interest, SAR simulators are widely used to educate trainees on the on the possibilities of this kind of imaging.

SAR simulators offer a cost-effective way to generate thousands of simulated SAR images of certain objects as it allows for a multitude of looking and azimuth angle combinations to be demonstrated at low or no cost. SAR simulations further offer the possibility to turn certain SAR-effects, such as speckling, on or off as demonstrated in Fig. 9.3. These possibilities make SAR simulations a huge advantage both for the training and education of students as well for the analysis and interpretation of SAR data as previously discussed.

Often a set of training databases, containing different SAR simulated images, are created and used for training purposes. Learning by using these training databases of simulated images is normally not very effective. A fast SAR image simulator should be used in place of such databases as they allow students to "play" with the sensor properties and explore the results with immediate feedback.

As demonstrated in Fig. 9.4, it should be possible to compare the results of the SAR simulation to 3D visualizations of the models used for the simulation, as well as to real SAR and aerial images.

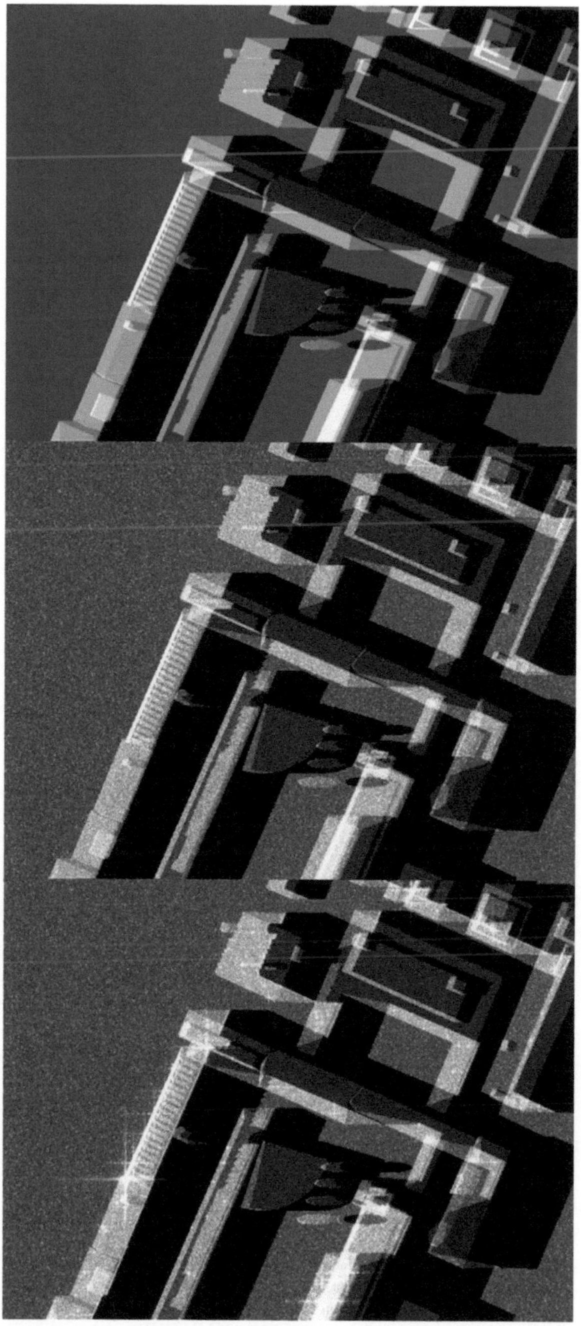

Fig. 9.3 SAR simulated images without speckling (*top*), without double-bouncing (*middle*), with double-bouncing and side lobes (*bottom*)

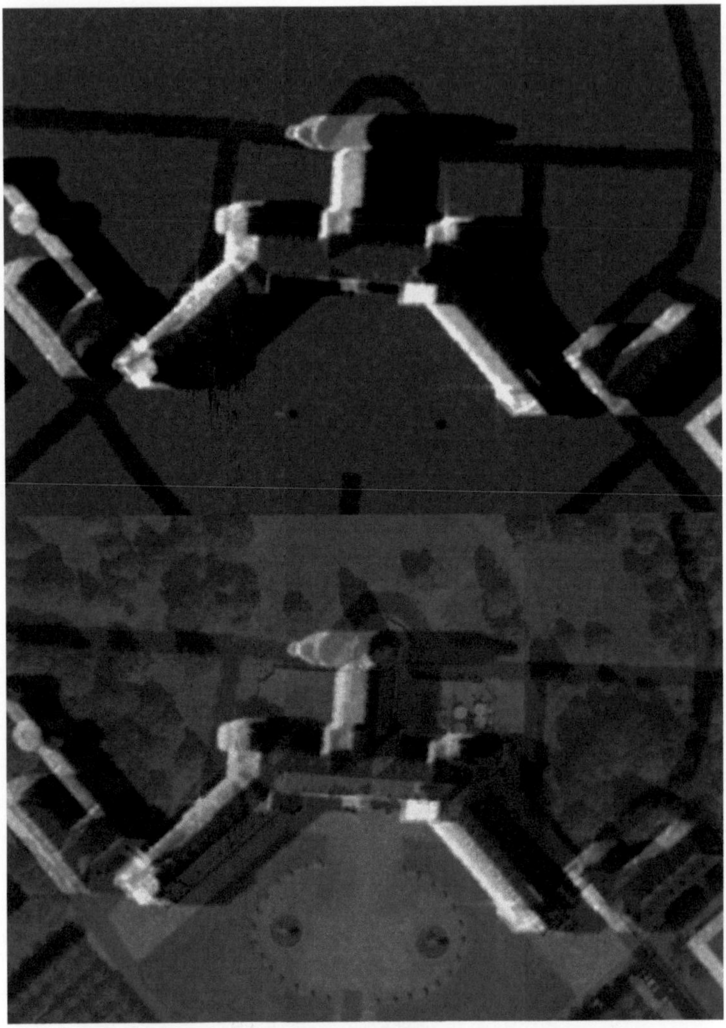

Fig. 9.4 SAR simulated image and SAR simulated image overlaid with optical image

9.6 Conclusions

SAR simulators are used for a wide variety of applications, each having different requirements. No simulator will be usable for every application due to the fact that some application requirements are mutually exclusive to others. The wide variety of SAR simulation types and applications for SAR simulations makes comparisons between SAR simulators difficult. SAR simulators are used for a special purpose and must fulfill the requirements of that purpose. Basic simulations for geo-referencing have to fulfill different requirements than simulations for algorithm or sensor design.

The techniques used for implementing a SAR simulation, as well as the physical models, depend on the needs of the application the simulation is used for.

To this end, an enduring yet seldom discussed problem is the availability of usable 3D models for simulation. The simulation of high-resolution SAR images requires 3D models with a very high level of detail. Furthermore, the materials must be modeled which becomes time consuming for extended scenes. There is still no practical solution available to generate 3D city models for SAR simulation in a practicable and productive way.

SAR simulators are constantly reinvented and re-implemented in companies and research institutes all over the world. This hinders the development of SAR simulators and their broader use. A widely supported modular open-source SAR simulator could be of great use for the scientific community. Using open-source ray tracers and adapting them for SAR simulations (Auer et al. 2008b) can become a remarkable way of reducing development overhead.

SAR simulations are important tools for various applications, but they are not ends in themselves. Simulations never represent the reality in every detail, but they are instead a simplification of reality. Although this is, of course, a drawback of all simulations, it can be rather advantageous for many applications. SAR simulations provide simplified, controllable images from defined scenarios. For algorithm design and testing, as well as for education and scientific analysis, this simplification can be simulations most prevalent benefit.

Acknowledgement This work was supported by the Research Fellowship for International Young Scientists of the National Natural Science Foundation of China under Grant 60950110351.

References

APL-UW (1994) High-frequency ocean environmental acoustic models handbook. Advanced Physics Laboratory, University of Washington, Seattle, WA. Technical Report TR 9407

Arnold-Bos A, Khenchaf A, Martin A (2007) Bistatic radar imaging of the marine environment – Part I: theoretical background. IEEE Trans Geosci Remote Sens 45:3372–3383

Auer S, Gernhardt S, Hinz S, Adam N, Bamler R (2008a) Simulation of radar reflection at man-made objects and its benefits for persistent scatterer interferometry. In: Proceedings of the 7th European conference on SAR (EUSAR 2008), Friedrichshafen, Germany

Auer S, Hinz S, Bamler R (2008b) Ray tracing for simulating reflection phenomena in SAR images. In: Proceedings of IGARSS 2008, Boston, MA

Balz T (2004) SAR simulation based change detection with high-resolution SAR images in urban environments. In: IAPRS 35, Part B, Istanbul

Balz T (2006) Real time SAR simulation on graphics processing units. In: Proceedings of the 6th European conference on SAR (EUSAR 2006), Dresden, Germany

Balz T, Stilla U (2009) Hybrig GPU based single- and double-bounce SAR simulation. IEEE Trans Geosci Remote Sens 47:3519–3529

Bamler R, Eineder M (2008) The pyramids of Gizeh seen by TerraSAR-X – a prime example for unexpected scattering mechanisms in SAR. IEEE Geosci Remote Sens Lett 5:468–470

Catmull E (1978) A hidden-surface algorithm with anti-aliasing. In: Proceedings of the 5th annual conference on computer graphics and interactive techniques SIGGRAPH '78, Atlanta

Dellière J, Maître H, Maruani A (2007) SAR measurement simulation on urban structures using a FDTD technique. In: Proceedings of urban remote sensing joint event, Paris, France
Eltoft T (2005) The Rician inverse Gaussian distribution: a new model for non-Rayleigh signal amplitude statistics. IEEE Trans Image Process 14:1722–1735
Franceschetti G, Migliaccio M, Riccio D, Schirinzi G (1992) SARAS: a synthetic aperture radar (SAR) raw signal simulator. IEEE Trans Geosci Remote Sens 30:110–123
Franceschetti G, Migliaccio M, Riccio D (1994) SAR raw signal simulation of actual ground sites in terms of sparse input data. IEEE Trans Geosci Remote Sens 32:1160–1169
Franceschetti G, Migliaccio M, Riccio D (1995) The SAR simulation: an overview. In: Proceedings of IGARSS '95, quantitative remote sensing for science and application, Florence, Italy
Franceschetti G, Iodice A, Riccio D, Ruello G (2003) SAR raw signal simulation for urban structures. IEEE Trans Geosci Remote Sens 41:1986–1995
Fung AK (1994) Microwave scattering and emission models and their applications. Artech House: Norwood, MA
Fung AK, Li Z, Chen KS (1992) Backscattering from a randomly rough dielectric surface. IEEE Trans Geosci Remote Sens 40:356–369
Gebhardt U, Loffeld O, Nies H (2008) Hybrid bistatic SAR experiment TerraPAMIR – Geometric description and point target simulation. In: Proceedings of the 7th European Conference on Synthetic Aperture Radar (EUSAR 2008), Friedrichshafen, Germany
Gelautz M, Frick H, Raggam J, Burgstaller J, Leberl F (1998) SAR image simulation and analysis of alpine terrain. ISPRS J Photogramm Remote Sens 53:17–38
Glasner AS (1984) Space subdivision for fast ray tracing. IEEE Comput Graph Appl 4(10):15–22
Guida R, Iodice A, Riccio D, Stilla U (2008) Model-based interpretation of high-resolution SAR images of buildings. IEEE J Select Topics Appl Earth Obs Remote Sens 1:107–119
Hammer H, Balz T, Cadario E, Soergel U, Thoennessen U, Stilla U (2008) Comparison of SAR simulation concepts for the analysis of high-resolution SAR data. In: Proceedings of the 7th European conference on SAR (EUSAR 2008), Friedrichshafen, Germany
Holtzman JC, Frost VS, Abbott JL, Kaupp VH (1978) Radar image simulation. IEEE Trans Geosci Electron 16:297–303
Jakeman E, Pussey PN (1976) A model for non-Rayleigh sea echo. IEEE Trans Antennas Propag 24:806–814
Knott EF, Shaeffer JF, Tuley MT (2004) Radar cross section, 2nd edn. SciTech Publishing: Raleigh, NC
La Prade GL (1963) An analytical and experimental study of stereo for radar. Photogramm Eng 29:294–300
Leberl FW (1990) Radargrammetric image processing. Artech House, Norwood, MA
Ling H, Chou RC, Lee SW (1989) Shooting and bouncing rays: calculating the RCS of an arbitrarily shaped cavity. IEEE Trans Antennas Propag 37:194–205
Long MW (2001) Radar reflectivity of land and aea. Artech House, Norwood, MA
Mametsa HJ, Rouas F, Berges A, Latger J (2002) Imaging radar simulation in realistic environment using shooting and bouncing rays technique. In: Proceedings of SPIE: 4543. SAR Image Analysis, Modeling and Techniques IV, Toulouse, France
Marconi (1984) SAR simulation concept and tools, Final Report. In: Report MTR 84/34, Marconi Research Centre, UK
Margarit G, Mallorqui JJ, Rius JM, Sanz-Marcos J (2006) On the usage of GRECOSAR, an orbital polarimetric SAR simulator of complex targets, to vessel classification studies. IEEE Trans Geosci Remote Sens 44:3517–3526
Meyer-Hilberg J, Neumann C, Senkowski H (2008) GMTI systems simulation using the SAR simulation tool PIRDIS. In: Proceedings of the 7th European Conference on Synthetic Aperture Radar (EUSAR 2008), Friedrichshafen, Germany
Muhleman DO (1964) Radar scattering from Venus and the Moon. Astronom J 69:34–41
Nadarajah S, Kotz S (2008) Intensity models for non-Rayleigh speckle distributions. Int J Remote Sens 29:529–541

Nunziata F, Gambardella A, Migliaccio M (2008) An educational SAR sea surface waves simulator. Int J Remote Sens 29:3051–3066

Paterson MS, Yao FF (1990) Efficient binary space partitions for hidden-surface removal and solid modeling. Discrete Comput Geom 5:485–503

Phong BT (1975) Illumination for computer generated pictures. Commun ACM 18:311–317

Pike TK (1985) SARSIM, a synthetic aperture radar system simulation model. In: DFVLR-Mitt, 85–11, Oberpfaffenhofen

Rius JM, Ferrando M, Jofre L (1993) High-frequency RCS of complex radar targets in real-time. IEEE Trans Geosci Remote Sens 31:1306–1319

Sheng Y, Alsdorf DE (2005) Automated georeferencing and orthorectification of Amazon basin-wide SAR mosaics using SRTM DEM data. IEEE Trans Geosci Remote Sens 43:1929–1940

Shinozuka M, Ghanem R, Houshmand B, Mansouri B (2000) Damage detection in urban areas by SAR imagery. J Eng Mech 126:779–777

Soergel U, Schulz K, Thoennessen U, Stilla U (2003) Event-driven SAR data acquisition in urban areas using GIS. GIS J Spat Info Decis 16(12):32–37

Stilla U, Soergel U, Thoennessen U (2003) Potential and limits of InSAR data for building reconstruction in built-up areas. ISPRS J Photogramm Remote Sens 58:113–123

Tao YB, Lin H, Bao HJ (2008) Kd-tree based fast ray-tracing for RCS prediction. Prog Electromagn Res 81:329–341

Ulaby FT, Dobson MC (1989) Handbook of radar scattering statistics for terrain. Artech House, Norwood, MA

Wald I, Havran V (2006) On building fast kd-Trees for ray tracing, and on doing that in O(N log N). In: Proceedings of IEEE symposium on interactive ray tracing 2006, Salt Lake City, UT, pp 61–69

Weimann A, von Schoenermark M, Schumann A, Joern P, Gunther R (1998) Soil moisture estimation with ERS-1 SAR data in East-German loess soil area. Int J Remote Sens 19:237–243

Wen XY, Zhang H, Wang C (2009) The high-resolution SAR image simulation and analysis of the damaged building in earthquake (in Chinese). J Remote Sens 13:19–176

Whang KY, Song JW, Chang JW, Kim JY, Cho WS, Park CM, Song IY (1995) Octree-R: an adaptive octree for efficient ray tracing. IEEE Trans Vis Comput Graph 1:343–349

Chapter 10
Urban Applications of Persistent Scatterer Interferometry

Michele Crosetto, Oriol Monserrat, and Gerardo Herrera

10.1 Introduction

This chapter reviews the urban applications of Persistent Scatterer Interferometry (PSI), the most advanced type of differential interferometric Synthetic Aperture Radar techniques (DInSAR) based on data acquired by spaceborne SAR sensors. The standard DInSAR techniques exploit the information contained in the radar phase of at least two complex SAR images acquired at different times over the same area generating interferograms or interferometric pairs. For a general review of SAR interferometry, see Rosen et al. (2000) and Crosetto et al. (2005). A large part of the DInSAR results obtained in the 1990s has been achieved by using the standard DInSAR configuration, which in some cases is the only one that can be implemented due to the limited SAR data availability.

A remarkable improvement in the quality of the DInSAR results is given by the advanced DInSAR methods that make use of large sets of SAR images acquired over the same deformation phenomenon. These techniques represent an outstanding advance with respect to the standard ones, both in terms of deformation modelling capabilities and quality of the deformation estimation. Different DInSAR approaches based on large SAR datasets have been proposed, starting from the late 1990s. However, a fundamental step was the publication of the so-called Permanent Scatterers technique by Ferretti et al. (2000). As it is discussed later in this section, different new techniques have been proposed in the last years following this approach. They were initially named Permanent Scatterers techniques, while now all these techniques, including the original Permanent Scatterers technique, are called Persistent Scatterer Interferometry (PSI) techniques. Note that the term Permanent

M. Crosetto (✉) and O. Monserrat
Institute of Geomatics, Av. Canal Olímpic s/n, 08860 Castelldefels (Barcelona), Spain
e-mail: michele.crosetto@ideg.es; oriol.monserrat@ideg.es

G. Herrera
Instituto Geológico y Minero de España (IGME), Rios Rosas 23, 28003 Madrid, Spain
e-mail: g.herrera@igme.es

U. Soergel (ed.), *Radar Remote Sensing of Urban Areas*, Remote Sensing and Digital Image Processing 15, DOI 10.1007/978-90-481-3751-0_10,

Scatterers is directly associated with the original technique patented by the Politecnico di Milano (Italy), and which is licensed to TRE (www.treuropa.com), a spin-off company of this university.

What is the key difference between DInSAR and PSI techniques? As already mentioned, the first difference is the redundant number of SAR images needed. A second substantial difference is that PSI techniques implement suitable data modelling procedures that make the estimation of different parameters possible. It is worth noting that the estimation is based on appropriate statistical treatments of the available redundant DInSAR observations. The estimated parameters are briefly discussed below. The first one is the time series of the deformation, which can provide information on the temporal evolution of the displacement. The deformation time series and the map of the average displacement rates are the two key products of a PSI analysis, as shown in Fig. 10.1. Another parameter is the so-called residual

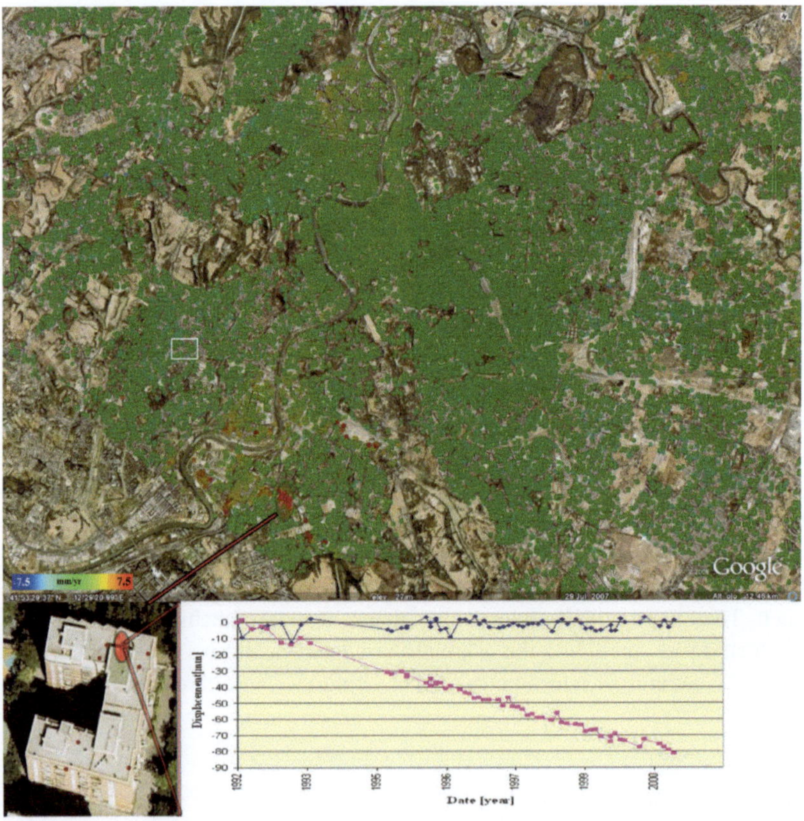

Fig. 10.1 Example of PSI deformation velocity map over the city of Rome, which has been geocoded and imported in Google Earth (*above*). *Below*, on the *left*, it is shown a zoom of the velocity map over a deformation area. *Below*, on the *right*, are shown the deformation time series of a PS located in the zoom area and a PS belonging to a stable area. The white frame covers the area shown in Fig. 10.4

Fig. 10.2 3D visualization with Google Earth of PS over a portion of Barcelona. The coloured circles represent the measured PS, which have been geocoded using the so-called residual topographic error. The colour of each PS represents its estimated residual topographic error

topographic error, which is the difference between the true height of the scattering phase centre of a given point, and the height given by the employed digital elevation (DEM) model in this point, see Fig. 10.2. This parameter plays an important role only for two specific goals: modelling purposes, that is proper estimation of the residual topographic component, separating it from the deformation component, and for geocoding purposes.

The standard geocoding methods simply employ the same DEM used in the DInSAR processing to geocode the DInSAR products, that is they use an approximate value of the true height of the scattering phase centre of a given pixel, which results in a location error in the geocoding. By using the residual topographic error this kind of error can be largely reduced, thus achieving a more precise geocoding: this may considerably help the interpretation and the exploitation of the results. An example of advanced geocoding is shown in Fig. 10.3. An additional parameter is the atmospheric phase component of each image of the used SAR stack. The estimation of this component is fundamental to properly estimate the deformation contribution.

As mentioned above, different PSI techniques have been proposed in the last years. Some of the most relevant works are briefly discussed below. The original Permanent Scatterers approach (Ferretti et al. 2000, 2001; Colesanti et al. 2003a) was followed by several other authors. The Small Baseline Subset (SBAS) technique

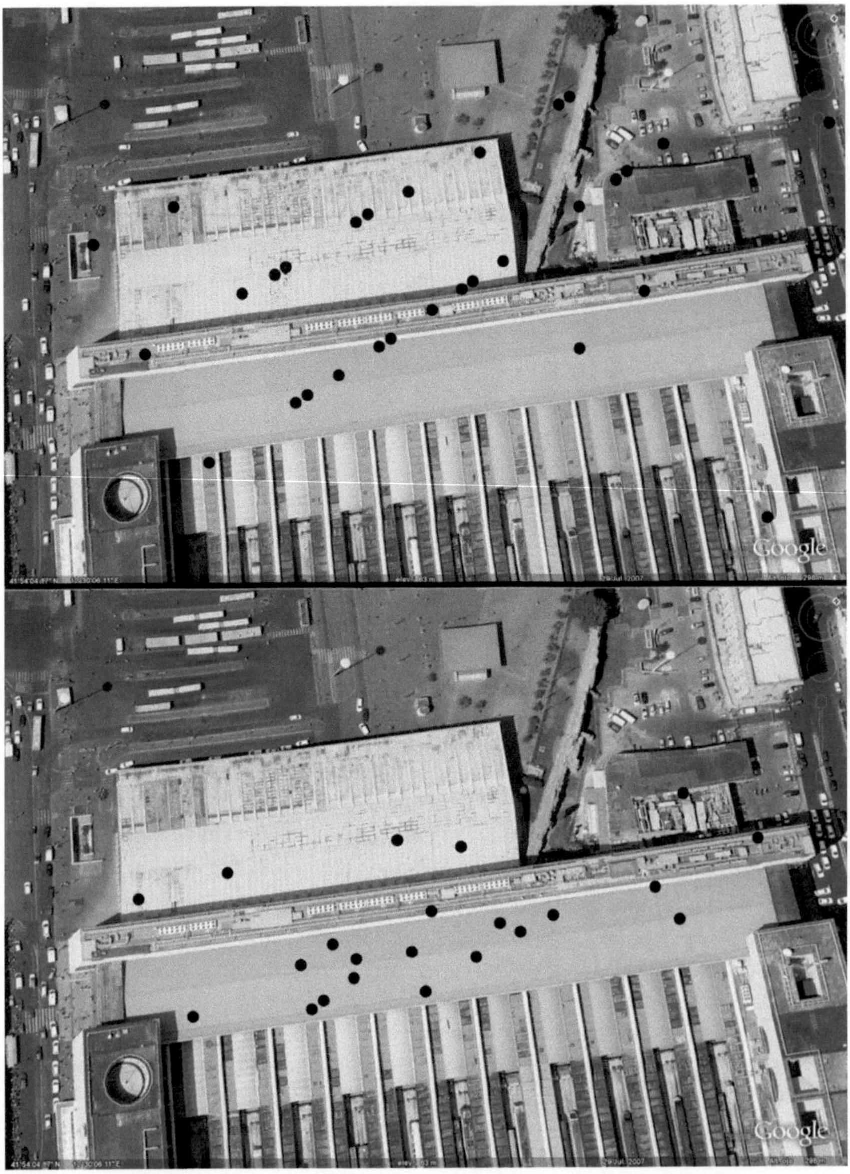

Fig. 10.3 Example of advanced geocoding of PSI results. PS geocoded without (*above*) and with (*below*) the correction based on the so-called residual topographic error. The geocoded points are visualized in Google Earth

is one of the most important and well documented PSI approaches (Berardino et al. 2002; Lanari et al. 2004; Pepe et al. 2005, 2007). A similar approach was proposed by Mora et al. (2003). Two companies that provide PSI services, Gamma

Remote Sensing (www.gamma-rs.ch) and Altamira Information (www.altamira-information.com), described their approaches in Werner et al. (2003) and Arnaud et al. (2003), respectively. Hooper et al. (2004) described a procedure useful in geophysical applications. Crosetto et al. (2005) proposed a simplified approach based on stepwise linear functions for deformation and least squares adjustment. Crosetto et al. (2008) described a PSI chain, which includes an advanced phase unwrapping approach. Finally, further relevant contributions include Kampes and Hanssen (2004), which adapted the LAMBDA method used in GPS to the problem of PSI, and Van Leijen and Hanssen (2007), which described the use of adaptive deformation models in PSI.

This chapter is organized as follows. In Section 10.2, the major advantages and the most important open technical issues related to PSI urban applications are discussed. Then, the most important PSI urban applications are reviewed. Finally the paper describes the results of the main validation activities carried out to prove the quality of the PSI-derived deformation estimates. Conclusions follow.

10.2 PSI Advantages and Open Technical Issues

This section discusses the major advantages and the most important open technical issues of PSI urban applications. The advantages of PSI are manifold. PSI offers wide-area coverage associated with a relatively high spatial resolution. This allows us to study a whole metropolitan area, thus getting a global outlook of their deformation phenomena, keeping at the same time the capability to measure individual structures and buildings. Another important advantage of PSI is its sensitivity to small deformations, which are of the order of 1 mm/year. Since it is based on spaceborne sensors, PSI exploits periodic and relatively low-cost data acquisitions. An unmatched capability is given by the ability to measure past deformation phenomena. This is possible by using the huge SAR image archives, which in case of ERS starts in 1991. This unique aspect of the technique means that it is possible to study ground motions that occurred in the past and for which no other survey data are available. Additionally, by using PSI analyses it is possible to get a potential reduction in the amount of ground-based observations, achieving simplified logistics operations and reducing personnel time and costs. The PSI technique provides two deformation measurement products. The first product, the average displacement rates, allows a user to quickly identify areas of motion that may be of interest. Once the user has identified areas of interest, a more in depth analysis can be carried out using the second type of product, that is the deformation time series. The time series allow a user to examine the motion history for a time period of interest. This is key information to identify potential causes of deformation, for example by analysing the time series with respect to the schedule of underground construction works, the lowering of a water table, etc.

Some of the most relevant PSI technical open problems are discussed below. Note that most of them concern all PSI applications, and not only the urban ones.

Spatial sampling. Even though the average density of Persistent Scatterers (PS), that is the points where PSI phase is good enough to get deformation measurements, is relatively high (e.g. 560 PS/km^2 with ERS and 730 PS/km^2 with Radarsat, see www.treuropa.com/Portals/0/pdf/PSmeasures.pdf), it has to be considered that PSI is an opportunistic deformation measurement method, which is able to measure deformation only over the available PS, see Fig. 10.4. PS density is usually low in vegetated and forested areas, over low-reflectivity areas, that is very smooth surfaces, and steep terrain. By contrast, the PS are usually abundant over buildings, monuments, antennas, poles, conducts, etc. In general the location of the PSs cannot be known a priori: this affects in particular the study of areas and objects of small spatial extent, for example specific buildings, which can be under-sampled or even not sampled at all. Note that this is particularly important for sensors like those of ERS, Envisat and Radarsat, while high-resolution SAR sensors, like TerraSAR-X,

Fig. 10.4 Example of PS density over a 200 by 170 m subset of the PSI velocity map from Fig. 10.1. The ERS SAR sensor sampled this area with an approximate density of 1 sample/80 m^2, that is getting 425 samples. Seventeen out of 425 resulted to be PS useful for deformation monitoring purposes. This illustrates the opportunistic character of PSI

should considerably improve PS density (Adam et al. 2008). It is worth underlining a remarkable difference between PSI and ground-based geodetic and surveying techniques. The latter ones are based on strategically located points, that is they measure points chosen ad hoc on the objects of interest. By contrast, PSI performs a massive and opportunistic sampling, identifying PS that provide a strong and consistent radar reflectance over time. Usually PS can be located on the ground, on the side, or on the top of buildings or structures. But since some PS may show the deformation of a building and others of the ground, the direct comparison of PSI estimates with other data has to be carried out carefully.

Temporal sampling. The capability of sampling deformation phenomena over time depends on the SAR data availability, which in turn depends on the revisiting time capabilities of the SAR satellites and on the data acquisition policies. For instance, Envisat has a revisiting time of 35 days but it carries several sensors, which cannot acquire data simultaneously. The SAR satellite revisiting time has a major impact on the temporal resolution of PSI deformation monitoring: PSI can typically monitor slow deformation phenomena, which evolve over several months or years. In addition to the temporal frequency of SAR images, PSI requires a large number of SAR scenes acquired over the same area. Typically more than 15–20 images are needed. Currently this amount of images is unavailable in several locations of the world.

Line-of-sight measurements. The PSI deformation measurements are made in the line-of-sight (LOS) of the SAR sensor, that is the line that connects the sensor and the target at hand. Given a generic 3D deformation in the area of interest, PSI provides the estimate of one component of this deformation, which is obtained by projecting the 3D deformation in the LOS direction. By using ascending and descending SAR data one can retrieve the vertical and east-to-west horizontal components of deformation. For this purpose it is required the independent processing of the ascending and descending datasets. With the orbits of the current SAR systems, PSI has a very low sensitivity to the north-to-south horizontal deformations.

Fast motion and linear deformation models. Due to the ambiguous nature of the PSI observations, that is the wrapped interferometric phases, PSI suffers severe limitations in the capability to measure "fast" deformation phenomena. Since PSI measures relative deformations, this limitation depends on the spatial pattern of the deformation phenomenon at hand. As a rule of thumb, with the current revisiting times of the available C-band satellites, PSI has usually difficulties to measure deformation rates above 4–5 cm/year. An additional disadvantage is due to the fact that most of PSI approaches make use of a linear deformation model in their deformation estimation procedures. For instance, all PSI deformation products generated in the Terrafirma project (http://www.terrafirma.eu.com) are based on this model. This assumption, which is needed to unwrap the interferometric phases (one of the most important stages of any PSI technique), can have a negative impact on the PSI deformation estimates for all phenomena characterized by non-linear deformation behaviour, that is where the assumption is not valid. In areas where the deformation shows "significantly non-linear motion" and/or high motion rates the PSI products lack PSs. This lack of PSs represents an important limitation because it affects the areas where the interest to measure deformation is the highest.

Time series. The time series represent the most advanced PSI deformation product and also the most difficult one to be estimated. They are an ambitious product because they provide a deformation estimate to each of the acquisition dates of the used SAR images. The time series are particularly sensitive to phase noise. Their interpretation should take into account the above mentioned limitation related to the linear deformation model assumption. To the authors' experience the real information content of the PSI deformation time series has not been fully understood so far. Even if excellent time series examples have been published in the literature, their limitations have been not clarified. It is worth noting that very few PSI time series validation results have been published in the literature.

Geocoding. PS geocoding has a direct impact on urban applications. According to the results of the Terrafirma Validation project (www.terrafirma.eu.com/Terrafirma_validation.htm), the east-to-west PS positioning precision (1σ) is 2–4 m, and the PS height precision (1σ) ranges between 1 and 2 m. In addition to these values it is also important to consider the uncertainty in the location of the PS within a resolution cell, for example, 20 by 4 m in case of ERS SAR imagery. Even though the above values are certainly good if one considers that they are derived from satellite-based data, they limit the interpretation and exploitation possibilities of PSI results. This is particularly important for all applications related to the deformation of single buildings or structures.

Deformation tilts or trends. Tilts or trends in the PSI deformation velocity maps have to be considered with particular care. In fact they can occur due to uncompensated orbital errors and low frequency atmospheric effects. Therefore, a tilt in a given deformation velocity map could be due to the above error sources, or to a real geophysical signal. With a standard PSI processing it is not possible to estimate (subtle) low-frequency geophysical deformation signals. Two opposite situations may happen. First, one may get tilts in the PSI products that are interpreted as geophysical signals, while in fact they are simple residual processing errors. Second, we may get a product without any tilt, which is interpreted by a geophysicist as no signal, for example quiescence of a given phenomenon, while in fact the site may have undergone significant geophysical low-frequency deformations that have been removed during the PSI processing. If this is so, one should clearly communicate to the end user that the given PSI deformation products do not include the deformations characterized by low spatial frequencies.

10.3 Urban Application Review

This section reviews the most important PSI urban applications. The references provided below are intended to be relevant examples and by no means form an exhaustive reference list. The majority of the examples are based on SAR data acquired by ERS-1/2 and Envisat, which represent the most important PSI data sources. In the forthcoming years it is expected to have an important increase of the urban applications based on high-resolution TerraSAR data, for example see Strozzi et al. (2008). The increment will be mainly driven by the increased spatial resolution, which could open several applications related to the monitoring of single structures or buildings. Another important factor will be the shorter revisiting time capability of the new systems. On the other hand, one has to consider that data availability could be a limiting factor. In fact, from one side the current high-resolution SAR system can only cover a fraction of the entire globe, and from the other the cost of the data could represent a limit to the development of some types of applications.

The deformation analysis over entire urban or metropolitan areas is one of the most powerful PSI urban applications. This type of analysis, which fully exploits the key advantages of PSI, that is wide-area coverage, measure of past deformation phenomena, and low cost, allows to get a global outlook of the deformation phenomena occurring in the area of interest. This type of analysis can be used to detect and measure deformation generated by different mechanisms, including unknown deformation phenomena. The best available collection of this type of analysis is given by the Terrafirma project, funded by the European Space Agency (ESA). Table 10.1 lists the cities analysed during the first stage of this project. Another rich set of European cities was analysed during its second stage. A wide collection of PSI results is available in the project webpage www.terrafirma.eu.com, following the link "Products/Stage 1/2 results". In addition, this page offers comprehensive information on project partners, products and documentation.

Table 10.1 PSI analyses over metropolitan areas performed in the Stage 1 of the Terrafirma project, see the deformation maps at www.terrafirma.eu.com/stage_1_results.htm

Product	Country	Covered area (km^2)	Period studied	Number of SAR scenes	Satellite data	Number of PS	PS density (PS/km^2)
Amsterdam	The Netherlands	1.600	1992–2002	91	ERS1/2	326,630	204
Athens	Greece	900	1992–1999	38	ERS1/2	98,111	109
Berlin	Germany	533	1995–2000	56	ERS1/2	446,893	837
Brussels	Belgium	900	1992–2003	74	ERS1/2	221,273	246
Haifa	Israel	900	1992–2000	47	ERS1/2	35,064	39
Istanbul	Turkey	1.000	1992–2002	49	ERS1/2	116,404	116
Lisbon	Portugal	800	1992–2003	55	ERS1/2	200,196	250
Lyon	France	2.310	1992–2000	50	ERS1/2	462,282	1605
Moscow	Russia	550	1992–2000	27	ERS1/2	166,439	302
Palermo	Italy	150	1992–2003	57	ERS1/2	108,398	722
Sofia	Bulgaria	800	1992–2003	45	ERS1/2	37,399	48
Sosnowiec	Poland	1.200	1992–2003	79	ERS1/2	122,926	102
St. Petersburg	Russia	550	1992–2004	45	ERS1/2	47,028	86
Stoke-on-Trent	UK	920	1992–2003	70	ERS1/2	178,109	194

PSI is currently used to monitor subsidence and uplift phenomena in several cities around the world. A relevant example is described in Dixon et al. (2006), which show a PSI-derived subsidence map of New Orleans. This map reveals that parts of the city underwent rapid subsidence in the 3 years before the Hurricane Katrina disaster occurred in 2005. An interesting study of the natural and anthropogenic subsidence that affects the south-eastern Po Plain (Italy), which includes the city of Bologna, is described in Zerbini et al. (2007). The authors describe an analysis which combines different techniques to extract information on the spatial and temporal variability of the subsidence: GPS, gravity and PSI. Other interesting applications are described in Lanari et al. (2004), Ferretti et al. (2004), Herrera et al. (2007), Crosetto et al. (2008) and Vallone et al. (2008).

The monitoring of deformation caused by water, gas and oil extraction represents one of the most important PSI applications. Different private companies such as Telerilevamento Europa (www.treuropa.com), Altamira Information (www.altamira-information.com), Gamma Remote Sensing (www.gamma-rs.ch) and Fugro NPA Ltd (www.npagroup.com) offer monitoring services related to them. Examples of PSI studies related to groundwater pumping are discussed in Tomás et al. (2005) and Bell et al. (2008). An interesting study of a gas extraction area, where four independent PSI analyses were carried out, is described in the Terrafirma Validation Project (www.terrafirma.eu.com/Terrafirma_validation.htm, follow the link "Product Validation Report"). Different interesting results can be found in webgis.irea.cnr.it, which publishes on-line the complete PSI results (velocity maps and time series) over different cities.

In some cases PSI has provided key information to study seismic faults in urban areas, for example see Bürgmann et al. (2006) and Funning et al. (2007). Both works concern the San Francisco Bay. The latter one is based on the joint analysis of PSI and GPS data. In addition, PSI has revealed important characteristics of the land deformation induced by volcanic activity in the area of Naples (Italy), for example, see Lanari et al. (2003).

The study of landslide phenomena in urban areas is another important type of PSI application. Due to the deformation rate limitation discussed in the previous section, PSI is useful to study slow-moving deformation landslides. An example is provided by Hilley et al. (2004).

As already mentioned in the previous section, an advantage of PSI is the capability to measure whole metropolitan areas with a spatial resolution that, in some cases, allows us to measure individual structures and buildings. In this context it is important to recall the limitation of spatial sampling density mentioned in the previous section. An example of infrastructure monitoring is described in Crosetto et al. (2008), which concerns the main dike of the port of Barcelona (Spain). Another very interesting example of dike monitoring, which concerns the assessment of the safety of water defence systems, a crucial activity in low-lying countries such as the Netherlands, is described in Hanssen and van Leijen (2008). They show that, over the Netherlands, PSI can be used to obtain weekly updates on dike stability for a significant part of all dikes in the country. An example of study of buildings in the city of Rome (Italy) can be found in Manunta et al. (2008).

Finally it is worth mentioning an additional PSI application, which exploits the so-called residual topographic error. Using the topographic error, Perissin and Rocca (2006) assess the possibility to derive urban DEMs. An important limitation of this application is the relatively low PS density which can be achieved in urban area.

10.4 PSI Urban Applications: Validation Review

This section reviews some of the most important PSI validation results, which concern in particular the monitoring of urban areas. Any new deformation measurement technique needs to demonstrate the quality of its measurements. This is fundamental to increase its acceptability and establish a long-term market. For this purpose, in the last years some important efforts have been made to study the quality of PSI results. The next section describes the outcomes of a major validation project founded by the ESA. Afterwards the most important validation results published in the literature are being discussed.

10.4.1 Results from a Major Validation Experiment

The newest and most important PSI validation results come from the Validation Project in which is part of the Terrafirma project. It addressed key issues, like PSI quality assessment, assessment of performances, estimation of precision and accuracy and evaluation of the consistency of PSI results coming from different providers. This project was focused on the four Terrafirma PSI providers, that is Telerilevamento Europa, Altamira Information, Gamma Remote Sensing and Fugro NPA. It included two main parts: process validation and product validation. The process validation involved the inter-comparison of the different providers' processed outputs and the analysis of their intermediate results. The product validation was based on PSI products generated over two test sites: Alkmaar and Amsterdam. The Alkmaar area, which includes a spatially correlated deformation field due to gas extraction, was studied using ERS-1/2 and Envisat data. Ground truth data on this site are available from levelling campaigns. The city of Amsterdam, which includes autonomous and mainly spatially uncorrelated movements, was studied using Envisat data and ground truth covering the area of the North-South metro line.

The inter-comparison activity generated useful global statistics, which concern large sets of PSs and provide information on the global inter-comparison behaviour of velocities, time series, topographic errors and PS geocoding. These values, which are summarized in Table 10.2, can be used to derive error bars to indicate the quality of the estimates derived by PSI. In addition to them, we briefly mention the validation results over the Amsterdam test site. Due to geocoding errors, it was not possible to make a perfect one-to-one comparison between PSs and ground truth. This affected negatively the validation results. A more in depth analysis can be

Table 10.2 PSI validation: summary of the main results coming from the Terrafirma Validation project

Parameter	Validation result	Estimated range	Comments
Deformation velocity	Standard deviation of the deformation velocity	$\sigma_{VELO} =$ 0.4–0.5 mm/year	Statistics derived over sites largely dominated by zero or very moderate deformation rates
Deformation time series	Standard deviation of the deformation time series	$\sigma_{TSeries} = 1.1$–4 mm	
Topographic error	Standard deviation of the topographic error	$\sigma_{TOPO} = 0.9$–2 m	The topographic error has a direct impact on the PS geocoding
Geocoding	Standard deviation of the geocoding	$\sigma_{GEOCOD} = 2.1$–4.7 m	These values roughly affect the east to west direction
Velocity validation	Standard deviation of the difference PSI velocity vs. the reference velocity	$\sigma_{VELO} =$ 0.8–0.9 mm/year	Validation based on tachymetry data. In general the PSI data show a reasonably good correlation with them
Time series validation	Average RMS errors of single deformation measurements	RMS = 4.2–5.5 mm	

found at www.terrafirma.eu.com/Terrafirma_validation.htm. It is worth noting that the statistics of Table 10.2 were derived over sites largely dominated by zero or very moderate deformation rates. Therefore they are representative of all PSI studies that concern areas with similar characteristics.

10.4.2 PSI Validation Results

Assessing the PSI validation results published in the literature, one has to consider that PSI performances vary as a function of different factors, like SAR image availability, PS availability (spatial sampling), PS quality, temporal deformation behaviour, deformation rates and spatial extent of the analysed area. The evaluation of any validation result should always consider the characteristics of the validation experiment. Any extrapolation to different PSI conditions should be avoided. It is worth mentioning that most of the available validation results concern the PSI deformation velocity, while the other key product, that is deformation time series, is much less studied. Further research is needed to study the quality of the PSI time series.

This section concisely discusses some examples of PSI validation. Crosetto et al. (2008) describe the validation of PSI measurements over a dike of the port of Barcelona, which was based on levelling data. A good agreement between the PSI estimations and the reference values was achieved: the maximum difference of the deformation velocities was 0.7 mm/year. The same paper describes an example of thermal dilation of an industrial building. Even though it is not a validation example, it is useful to appreciate the sensitivity of PSI, which is able to sense millimetre-level deformations. Herrera et al. (2008) analyse the subsidence of Murcia exploiting the PSI time series. They perform a comparison of PSI with extensometers and a comparison between two different PSI techniques. Teatini et al. (2007) analyse the area of Venice using PSI results. They describe the comparison of PSI and levelling, and provide an interesting example of PSI interpretation in urban area. Colesanti et al. (2003b) describe the validation results over a landslide phenomenon close to the city of Ancona (Italy), which was based on levelling data. Finally, two additional validation exercises, where relatively negative PSI results were achieved, are worth mentioning. The first one is PSIC4, a major ESA project devoted to PSI validation (see earth.esa.int/psic4). In this project the results of eight different PSI chains were analysed and validated. The poor PSI performances were mainly due to the big deformation rates of the analysed area due to mining extraction activity. These results illustrate the PSI limitations with fast motion and linear deformation models, which are discussed in Section 10.2. The second example is given by the Jubilee Line (London) validation analysis performed in the Terrafirma project (see www.terrafirma.eu.com/JLE_intercomparison.htm). The analysis was focused on the deformation induced by tunnel construction works. The relatively poor validation results in this case where caused by the highly non-linear deformation and the relatively poor temporal and spatial PS sampling with respect to the deformation phenomena of interest.

10.5 Conclusions

In this chapter the deformation monitoring in urban areas based on the PSI technique has been discussed. The key characteristics of this SAR-based technique have been described, highlighting the differences between the classical DInSAR and the PSI. The main products of a PSI analysis have been briefly described, and the most important PSI approaches have been concisely reviewed, providing a comprehensive list of references.

The major advantages of PSI deformation monitoring have been considered and an extended list of the most important open technical issues has been provided. Examples of open PSI issues are: the spatial and temporal sampling, the problems with fast motion and non-linear deformation, geocoding errors, and the tilts in the deformation velocity maps. The latter one limits the PSI capability to analyse geophysical deformation phenomena characterized by low spatial frequency behaviour. Despite being a relatively new technique, PSI has undergone a fast development and has been applied in a wide number of different applications The most important PSI urban applications have been reviewed, which include analyses at entire urban or metropolitan areas, subsidence and uplift phenomena, deformation caused by water, gas and oil extraction, seismic faults in urban areas, landslides, and the monitoring of infrastructures and single buildings. Even though the majority of the examples provided are based on SAR data acquired by ERS-1/2 and Envisat, in the near future it is expected a remarkable increase of the applications based on high-resolution TerraSAR-X data. Finally, the main PSI validation activities have been described. Proving the quality of any new technique is necessary for its acceptability and for establishing a long term market. In recent years major PSI validation projects have been funded by ESA. Their major outcomes have been discussed in this paper.

References

Adam N, Eineder M, Yague-Martinez N, Bamler R (2008) High-resolution interferometric stacking with TerraSAR-X. In: Proceedings of the International Geoscience Remote Sensing Symposium, IGARSS 2008, Boston, MA

Arnaud A, Adam N, Hanssen R, Inglada J, Duro J, Closa J, Eineder M (2003) ASAR ERS interferometric phase continuity. In: IGARSS 2003, Toulouse, France, 21–25 July 2003

Bell JW, Amelung F, Ferretti A, Bianchi M, Novali F (2008) Permanent scatterer InSAR reveals seasonal and long-term aquifer-system response to groundwater pumping and artificial recharge. Water Resour Res 44:1–18

Berardino P, Fornaro G, Lanari R, Sansosti E (2002) A new algorithm for surface deformation monitoring based on small baseline differential SAR interferograms. IEEE Transactions on Geoscience and Remote Sensing, 40(11):2375–2383 November 2002

Bürgmann R, Hilley G, Ferretti A, Novali F (2006) Resolving vertical tectonics in the San Francisco Bay Area from Permanent Scatterer InSAR and GPS analysis. Geology 34(3):221–224

Colesanti C, Ferretti A, Novali F, Prati C, Rocca F (2003a) SAR monitoring of progressive and seasonal ground deformation using the permanent scatterers technique. IEEE Trans Geosci Remote Sens 41(7):1685–1701

Colesanti C, Ferretti A, Prati C, Rocca F (2003b) Monitoring landslides and tectonic motions with the permanent scatterers technique. Eng Geol 68:3–14

Crosetto M, Crippa B, Biescas E, Monserrat O, Agudo M, Fernández P (2005) Land deformation monitoring using SAR interferometry: state-of-the-art. Photogramm Fernerkundung Geoinfo 6:497–510

Crosetto M, Biescas E, Duro J, Closa J, Arnaud A (2008) Quality assessment of advanced interferometric products based on time series of ERS and Envisat SAR data. Photogramm Eng Remote Sens 74(4):443–450

Dixon TH, Amelung F, Ferretti A, Novali F, Rocca F, Dokkas R, Sella G, Kim SW, Wdowinski S, Whitman D (2006) Subsidence and flooding in New Orleans. Nature 441:587–588

Ferretti A, Prati C, Rocca F (2000) Nonlinear subsidence rate estimation using permanent scatterers in differential SAR interferometry. IEEE Trans Geosci Remote Sens 38(5):2202–2212

Ferretti A, Prati C, Rocca F (2001) Permanent scatterers in SAR interferometry. IEEE Trans Geosci Remote Sens 39(1):8–20

Ferretti A, Novali F, Bürgmann R, Hilley G, Prati C (2004) InSAR permanent scatterer analysis reveals ups and downs in San Francisco bay area. EOS 85(34):317–324

Funning GJ, Bürgmann R, Ferretti A, Novali F, Fumagalli A (2007) Creep on the Rodgers Creek fault, northern San Francisco Bay area from a 10 year PS-InSAR dataset. Geophys Res Lett 34:L19306, doi:10.1029/2007GL030836

Hanssen RF, van Leijen FJ (2008) Monitoring water defense structures using radar interferometry, Radar Conference, 2008. In: Radar '08. IEEE, Rome, Italy, 26–30 May 2008

Herrera G, Tomás R, López JM, Delgado J, Mallorquí JJ, Duque S, Mulas J (2007) Advanced DInSAR analysis on mining areas: La Union case study (Murcia, SE Spain). Eng Geol 90: 148–159

Herrera G, Toms R, Lopez-Sanchez JM, Delgado J, Vicente F, Mulas J, Cooksley G, Sanchez M, Duro J, Arnaud A, Blanco P, Duque S, Mallorqui JJ, De la Vega-Panizo R, Monserrat O (2008) Validation and comparison of advanced differential interferometry techniques: Murcia metropolitan area case study. ISPRS J of Photogrammetry & Remote Sensing 64(5):501–512, September 2009, doi:10.1016/j.isprsjprs.2008.09.008

Hilley GE, Bürgmann R, Ferretti A, Novali F, Rocca F (2004) Dynamics of slow-moving landslides from permanent scatterer analysis. Science 304(5679):1952–1955, doi:10.1126/science.1098821

Hooper A, Zebker H, Segall P, Kampes B (2004) A new method for measuring deformation on volcanoes and other natural terrains using InSAR persistent scatterers. Geophys Res Lett 31:L23611, doi:10.1029/2004GL021737

Kampes BM, Hanssen RF (2004) Ambiguity resolution for permanent scatterer interferometry. IEEE Trans Geosci Remote Sens 42(11):2446–2453

Lanari R, Berardino P, Borgström S, Gaudio CD, Martino PD, Fornaro G, Guarino S, Ricciardi GP, Sansosti E, Lundgren P (2003) The use of IFSAR and classical geodetic techniques for caldera unrest episodes: application to the Campi Flegrei uplift event of 2000. Volcanol J Geothermal Res 133:247–260

Lanari R, Zeni G, Manunta M, Guarino S, Berardino P, Sansosti E (2004) An integrated SAR/GIS approach for investigating urban deformation phenomena: The city of Napoli (Italy) case study. Int J Remote Sens 25:2855–2862

Manunta M, Marsella M, Zeni G, Sciotti M, Atzori S, Lanari R (2008) Two-scale surface deformation analysis using the SBAS-DInSAR technique: a case study of the city of Rome, Italy. Int J Remote Sens 29(6):1665–1684, doi:10.1080/01431160701395278

Mora O, Mallorquí JJ, Broquetas A (2003) Linear and nonlinear terrain deformation maps from a reduced set of interferometric SAR images. IEEE Trans Geosci Remote Sens 41(10): 2243–2253

Pepe A, Sansosti E, Berardino P, Lanari R (2005) On the generation of ERS/ENVISAT DInSAR time-series via the SBAS technique. IEEE Trans Geosci Remote Sens Lett 2:265–269

Pepe A, Manunta M, Mazzarella G, Lanari R (2007) A space-time minimum cost flow phase unwrapping algorithm for the generation of persistent scatterers deformation time-series. In: Proceedings of IGARSS 2007, Barcelona, Spain, 23–27 July 2007

Perissin D, Rocca F (2006) High-accuracy urban DEM using permanent scatterers. IEEE Trans Geosci Remote Sens 44(11):3338–3347

Rosen PA, Hensley S, Joughin I (2000) Synthetic aperture radar interferometry. Proc of the IEEE 88(3):333–382

Strozzi T, Tosi L, Teatini P, Wegmüller U (2008) Monitoring land subsidence in the Venice lagoon with TerraSAR-X. In: 3rd TerraSAR-X science team meeting, Oberpfaffenhofen, Germany, 25–26 November 2008

Teatini P, Strozzi T, Tosi L, Wegmüller U, Werner C, Carbognin L (2007) Assessing short- and long-time displacements in the Venice coastland by synthetic aperture radar interferometric point target analysis. J Geophys Res 112:F01012, doi:10.1029/2006JF000656

Tomás R, Márquez Y, Lopez-Sanchez JM, Delgado J, Blanco P, Mallorquí JJ, Martinez M, Herrera G, Mulas J (2005) Mapping ground subsidence induced by aquifer overexploitation using advanced differential SAR interferometry: Vega media of the Segura River (SE Spain) case study. Remote Sens Environ 98(2–3):269–283

Vallone P, Crosetto M, Giammarinaro MS, Agudo M, Biescas E (2008) Integrated analysis of differential SAR interferometry and geological data to highlight ground deformations occurring in Caltanissetta city (Central Sicily, Italy). Eng Geol 98:144–155

Van Leijen F, Hanssen RF (2007) Persistent scatterer interferometry using adaptive deformation models. In: Proceedings of Envisat Symposium 2007, Montreux, Switzerland, 23–27 April 2007

Werner C, Wegmüller U, Strozzi T, Wiesmann A (2003) Interferometric point target analysis for deformation mapping. In: Proceedings of IGARSS 2003, Toulouse, France, 21–25 July 2003

Zerbini S, Richter B, Rocca F, van Dam T, Matonti F (2007) A combination of space and terrestrial geodetic techniques to monitor land subsidence: case study, the Southeastern Po Plain, Italy. J Geophys Res 112:B05401, doi:10.1029/2006JB004338

Chapter 11
Airborne Remote Sensing at Millimeter Wave Frequencies

Helmut Essen

11.1 Introduction

Advanced radar sensors are able to deliver highly resolved images of the earth surface with considerable information content, as polarimetric information, 3-D-features and robustness against changing environmental and operational conditions. This is possible also under adverse weather conditions, where electro-optical sensors are limited in their performance.

Typical applications cover the control of agricultural activities, the survey of traffic during special events or even the regular monitoring of motorways. A special utilization for easily deployable imaging sensors are all kinds of natural or man-made environmental disasters, as the monitoring of volcanic activities, the survey of pipelines or of accidents like that at Chernobyl, where radiation hazard or other dangers are given for monitoring by humans.

All these utilizations require sensors, which have to cope with a high variability of atmospheric conditions while supplying complete information about the status of the earth surface. Millimeter wave SAR is able to serve these demands with best possible results and ease of operation as long as only short or medium ranges are required. Especially the latter condition can be fulfilled due to the unique properties of millimeter wave SAR, which are roughly described by short aperture length for given resolution, inherently low speckle, low blasting of strong scattering centers and simple processing.

H. Essen (✉)
FGAN- Research Institute for High Frequency Physics and Radar Techniques,
Department Millimeterwave Radar and High Frequency Sensors (MHS),
Neuenahrer Str. 20, D-53343 Wachtberg-Werthhoven, Germany
e-mail: essen@fgan.de

U. Soergel (ed.), *Radar Remote Sensing of Urban Areas*, Remote Sensing and Digital Image Processing 15, DOI 10.1007/978-90-481-3751-0_11,

11.2 Boundary Conditions for Millimeter Wave SAR

11.2.1 Environmental Preconditions

The electromagnetic wave, which is transmitted by the radar, scattered at the target of interest and the surrounding and than reflected back to the radar is influenced by the atmosphere. The propagation medium may be described by its refraction index and absorption by molecules in the atmosphere (clear air propagation) at one hand and influences of weather or other environmental conditions, as that is the presence of hydrometeors or dust.

11.2.1.1 Transmission Through the Clear Atmosphere

The millimeter wave region exhibits considerably different propagation properties if compared with classical radar bands (Skolnik 1980). This is due to resonance absorption at these frequencies, which is related to energy levels of vibration and rotation states of molecules in the atmosphere, like water vapor and oxygen.

For radar applications mainly the transmission windows around 35 GHz (Ka-Band) and 94 GHz (W-Band) are employed. It has however to be noted, that relatively high propagation losses are inhibitive for long range applications of millimeter wave radar (>10 km).

11.2.1.2 Attenuation Due to Rain

A severe influence on millimeter wave propagation is given by hydrometeors with a high density or, even worse, with a dropsize in the order of magnitude of the electromagnetic wavelength. The latter phenomenon is again due to resonance, where the drop is acting as an antenna, absorbing the energy of the resonant electromagnetic wave and is the determining factor for attenuation in the millimeter wave region (Marshall and Palmer 1948).

11.2.1.3 Propagation Through Snow, Fog, Haze and Clouds

For remote sensing applications the propagation through snow, fog, haze and clouds is determined by the same physical interactions as for the IR- and visible frequency region of the electromagnetic spectrum. While in the EO region the drop size within fog and clouds is in an order of magnitude, that interactions are most likely (Weiβ-Wrana et al. 1995), this does not apply in a comparable amount for millimeter waves. The effects are of much minor importance as long as the density of droplets is not too high. Snow has only marginal influence on millimeter wave propagation as long as the liquid water content is not excessively high (Kendra et al. 1995).

11.2.1.4 Propagation Through Sand, Dust and Smoke

The use of airborne sensors is essential for any mission of humanitarian or assisting nature in disaster areas. Besides darkness and adverse weather, dust and sand storms impose most critical conditions for remote sensing. Dust clouds in contrary to ordinary dust storms possess a wide spectrum of sand and dust particle sizes (Nüßler et al. 2007). The bigger particles may sometimes even have diameters in the order of magnitude of the wavelength related to the upper millimeter wave region. A further reason for propagation loss in the atmosphere is the smoke of the burning savannah or of volcanic eruptions. Only radar sensors in the microwave or millimeter wave region offer the capability of sufficient transmission to cope with the described environmental conditions (Skolnik 1980).

Concerning sand and dust, simulations have been conducted (Wikner 2008; Brooker et al. 2007) which start from the precondition that dust particles are almost spherical in shape and that their forward scattering can be described by Mie scattering. The results fit well with experiments and can be used for an estimation of propagation loss (Rangwala et al. 2007; Hägelen et al. 2008).

Smoke consists of even smaller particles if compared with dust. Experimental data are available (Essen and Baars 1986) which show the low attenuation of any type of smoke for millimeter waves.

11.2.2 Advantages of Millimeter Wave Signal Processing

As at millimeter waves the wavelength is extremely short in comparison with classical radar bands, the related phase is changing very rapidly. One might suspect that this would be a disadvantage for any algorithm, which, like SAR, is based upon the evaluation of the phase of the backscattered signal. However, the contrary is true. This is partly due to the geometry for millimeter wave SAR, which is typical short range, and partly due to the specific scattering mechanism, which is dominated by a relatively rougher surface, scaled by a factor of 10 compared with X-band.

In addition imaging errors inherent to SAR processing are of minor importance. One of the major advantages is the short aperture length, which for equal cross-range resolution is also scaled by a factor of 10 in comparison to X-band and thus makes millimeter wave SAR more robust against uncontrolled movements of the carrier aircraft. In the following a short survey on general properties of the millimetre wave SAR is given. More details are presented during the description of a typical SAR system, the MEMPHIS radar (Boehmsdorff and Essen 1998).

11.2.2.1 Roughness Related Advantages

Roughness of surfaces gives reason for diffuse scattering, while smooth surfaces are resulting in specular reflection processes. Roughness, however, is not an absolute criterion, but related to the wavelength of the illuminating signal. At millimeter wave frequencies most surfaces appear rough, and the diffuse scattering dominates imaging with SAR. Diffuse scattering leads to an averaging, which has a similar effect as multilook processing. The consequence for the imaging process is, that the inherent speckle within scenes with equal surface structure is lower at millimeter wave frequencies than at X-band for an equal amount of multi look processing. Another effect is due to a higher requirement upon rectangularity of angles between perpendicular surfaces for a perfect corner reflector effect. The phase state of the electromagnetic wave incident on a flat plate has to be constant over the total surface area for a coherent superposition. If this is not the case, destructive interference between waves reflected at different loci of the surface will occur and thus a rapid decrease in the overall RCS. The consequence for SAR images is, that a strong overemphasis of corners and edges, which may give reason for processing lobes for point scattering at classical SAR bands are considerably reduced for millimeter wave SAR.

11.2.2.2 Imaging Errors for Millimeter Wave SAR

During SAR processing two main sources give reason to imaging errors: Range migration and depth of focus. A concise description of these problems is given in (Curlander and McDonough 1991). The azimuth resolution of a SAR process depends mainly on the bandwidth of the Doppler signal. The phase of the Doppler signal is given by $\Phi = 4\pi R(s)/\lambda$. If a Doppler shift is present, the range to the target must change during the observation time and consequently the compressed target response is related to different ranges for consecutive samples. This is called the "range migration". The locus of these effective range cells can be approximated by

$$R(s) = R_c + (dR/dt)(s - s_c) + (d^2R/dt^2)(s - s_c)^2/2. \tag{11.1}$$

The linear part of this equation is the range walk, while the quadratic term is the range curvature. From this equation a precondition can be deduced, under which circumstances a compensation of the imaging errors has to be performed. Under the assumption that the maximum range migration ΔR should be less than about $^1/_4$ of the range resolution cell δR the criterion can be deduced to be

$$(\delta x/\lambda)^2 > R_c/8\delta R. \tag{11.2}$$

Due to the proportionality by $1/\lambda^2$ the range for which a compensation is needed is bigger by a factor of 100 between W-band and X-band in favour for the W-band.

The second important imaging error is the depth of focus criterion. This is related to the fact, that the azimuth correlation parameters, which are denoted by f_{DC} and f_R, are dependent on range. Basically this is related to a mismatch between the azimuth chirp constant f_R if the range R_c used for the correlation differs from the range of the target. This mismatch causes a phase drift between correlator function and the signal. This gives the boundary condition (Curlander and McDonough 1991):

$$dR_c < 2(\delta x)^2/\lambda. \tag{11.3}$$

Again the depth of focus at W-band is maintained for a ten times bigger range as at X-band.

11.3 The MEMPHIS Radar

The use of millimeter waves for SAR applications is a more recent trend (Boehmsdorff et al. 2001; Edrich 2004; Almorox-Gonzlez et al. 2007) and especially suited for small UAVs. The available technology and its potential for miniaturization with additional high scientific potential, as polarimetry and interferometry, are in favor for this frequency region. Additionally specific probing possibilities related to sensitivity on small-scale structures are typical for millimeter waves.

Most of the available data in the frequency bands of 35 GHz and 94 GHz were gathered with experimental radars onboard medium size aircrafts like C-160 "Transall" or similar. In Europe the RAMSES (Radar Aeroporte Multi-spectral d'Etude des Signatures) operated by ONERA (Dreuillet et al. 2006) with capabilities up to 94 GHz has been in operation for more than a decade, as well as the MEMPHIS (Millimeter Wave Experimental Multifrequency Polarimetric High Resolution Imaging System) of FGAN-FHR (Schimpf et al. 2002). In the US numerous data sets are available gathered by the Lincoln Lab millimeter wave SAR (Henry 1991).

11.3.1 The Radar System

The radar system (Schimpf et al. 2002) employs two front-ends, one at 35 GHz the other at 94 GHz, which can be operated simultaneously. Both are controlled by a common VME-bus computer and tied to the system reference, from which all frequencies and trigger impulses are derived. The IF-signals from both front-ends are fed to the data acquisition and recording electronics. The measured data are recorded by means of a high-speed digital recording system MONSSTR with a maximum recording speed of 128 MByte/s.

The architecture of both front ends is identical. The primary frequencies of 25 GHz and 85 GHz are generated by successive multiplication and filtering of the reference frequency of 100 MHz. For both subsystems the waveform and the IF-offset are modulated onto an auxiliary signal at about 10 GHz, which, together with the primary signal, is up-converted into the respective frequency band. Figure 11.1 shows the detailed diagram of the front-end.

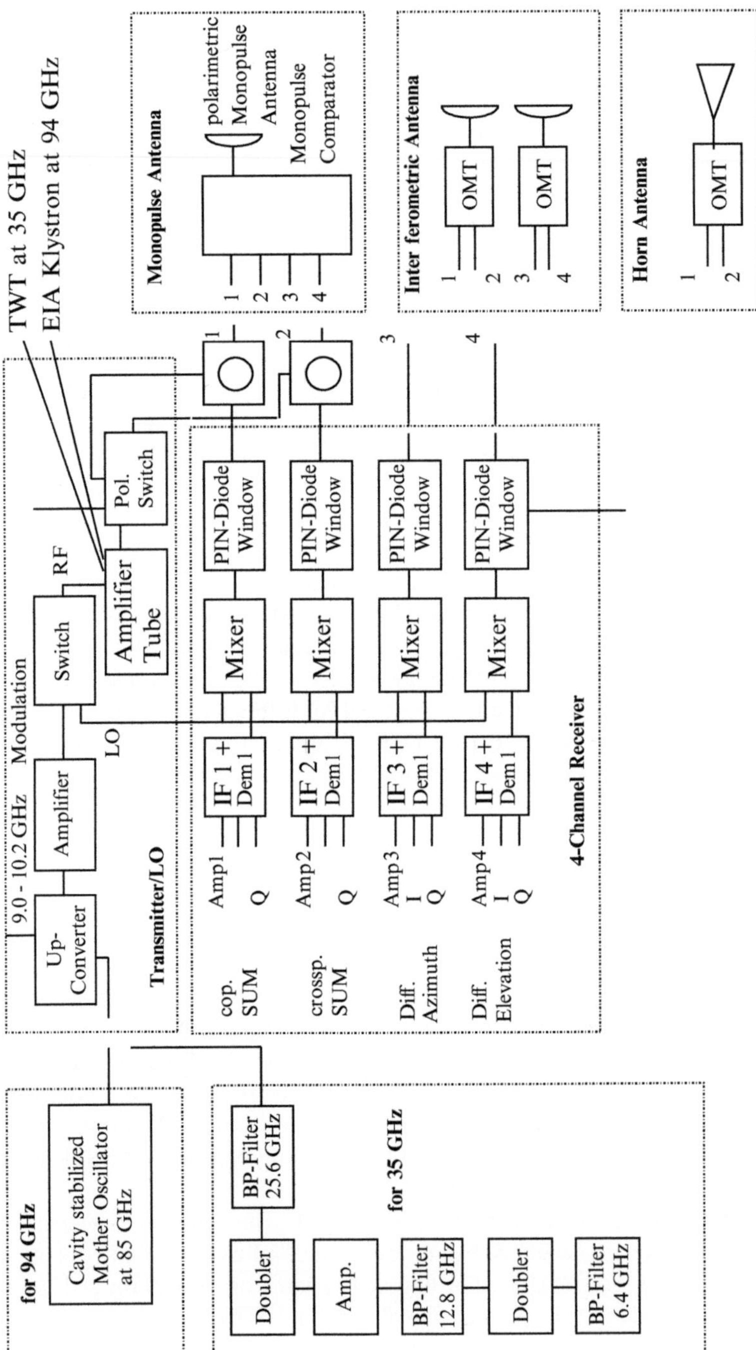

Fig. 11.1 Block diagram of the MEMPHIS millimeterwave front-end

The radar waveform is a combination of stepped frequency waveform and a FM chirp (Schimpf et al. 2004). The pulse length can be adjusted in the range of 80 ns – 2 μs. For the high-resolution mode the frequency is stepped from pulse to pulse over a bandwidth of 800 MHz in steps of 100 MHz, while at each frequency step a chirp modulation over a bandwidth of 200 MHz is done with an overlap of +50 MHz at both the lower and upper frequency limit of each successive chirp. This results in a range resolution of about 19 cm. The output power is generated by a TWT (Thales) at 35 GHz and an EAI (CPI) at 94 GHz. The transmit power is fed into a pin-switch assembly which allows to switch the transmit polarization from pulse to pulse between orthogonal components, linear horizontal or vertical or, manually switched, circular, left hand or right hand. The receiver has four channels with balanced mixers and a common local oscillator, which is coupled to the up-converter, which also supplies the transmitter stage via a SPDT-PIN switch. The down converted signals are quadrature demodulated to result in I- and Q-phase components and the logarithmically weighted amplitudes.

Depending on the application, the systems can be used with polarimetric monopulse feeds, sensing elevation and transverse deviations or an interferometric pair of antennas with orthomode transducers to sense both polarimetric components. The elevation/azimuth-asymmetry of the beam, which is generally necessary for SAR applications, is achieved by aspheric lenses in front of the feed horns. The performance data of the front-ends are summarized in Table 11.1. In addition to the radar data, inertial data from the aircraft as well as time code and GPS data are recorded.

Table 11.1 Performance data of the MEMPHIS millimeterwave front-end

	35 GHz	94 GHz
Transmitter		
Output power	500 W	750 W
PRF	2 kHz	
Pulse width	400/800 ns	
Spectral purity	> −70 dB/Hz	
Phase stability	10° RMS	
Polarization	Linear or circular H/V or R/L	
Waveform	Chirp (100/200 MHz) + stepped frequency, bandwidth 800 MHz	
Receiver		
Dynamic range	60 dB	
System noise figure	15 dB (SSB)	
Polarization	Simultaneously co and crosspolarization	
Bandwidth	100/200 MHz	
Antenna		
Typ	Dielectric lense	
Diameter	300 mm	300 mm
3 dB Beamwidth		
Azimuth	2.5°	1°
Elevation	16°	12°
Gain	29 dB	36 dB

Fig. 11.2 The MEMPHIS millimeterwave front-end

Table 11.2 Interferometric baselines of the MEMPHIS millimeterwave front-end

Channel combination	Baseline/mm
R1/R2	55
R2/R3	110
R1/R3	165
R2/R4	220
R1/R4	275

For interferometric SAR measurements the MEMPHIS radar is equipped with a multi-baseline antenna consisting of an array of six horns followed by a cylindrical lens. The complete antenna has a 3 dB beam width of 3° in azimuth and 12° in elevation. Figure 11.2 shows a photo of the 35 GHz front-end, equipped with the horn antenna array.

Due to the geometry of the horn ensemble, five independent interferograms can be generated. The possible combinations with the respective baselines are given in Table 11.2. These different interferograms are used to resolve the height ambiguity. The advantages of this multiple baseline approach for the phase unwrapping procedure is discussed in detail later on.

11.3.2 SAR-System Configuration and Geometry

For SAR applications the radar is mounted into a "Transall" aircraft looking out of a side door, as shown in Fig. 11.3. If the complete information of a specific area is required, courses with different headings are flown, covering at least the four cardinal directions.

Fig. 11.3 MEMPHIS radar in C-160 "Transall" aircraft

The radiometric calibration is based on pre- and post flight measurements against trihedral and dihedral precision corner reflectors on a pole. Sufficient height of the pole is necessary to avoid a strong influence of multipath propagation.

11.4 Millimeter Wave SAR Processing for MEMPHIS Data

11.4.1 Radial Focussing

Data are recorded with the MONSSTR system to be calibrated and evaluated by an off-line process. Images are generated by the regular SAR-process employed with the MEMPHIS data if only a linear chirp waveform with a total bandwidth of 200 MHz is used. As mentioned above, high range resolution is obtained using an LFM chirp with either 100 or 200 MHz bandwidth. Chirp length ranging between 400 and 1,200 ns can be handled by the chirp generator, which is in accordance with the required PRFs and the available duty cycles. As the required range ($>$1,000 m) is much more than the chirp length, the usual "deramp-on-receive" is not a viable technique to be implemented. Instead, the receive signal is only down-converted to the basic frequency and then the complex values sampled at a rate of 1/B (B = bandwidth of the individual chirp).

In order to increase the range resolution beyond the value of c/2B given by the chirp bandwidth, a stepped-frequency mode is implemented, using eight steps with a spacing of 100 MHz thus limiting the instantaneous bandwidth and required sample rate.

Using a synthesized chirp combining N pulses with an instantaneous bandwidth of B, post-processing is necessary to combine the individual chirps. Several methods for this processing are known, as "stepped-frequency chirp" (Levanon 2002), "frequency-jumped burst" (Maron 1990) or "synthetic bandwidth" (Berens 1999; Zhou et al. 2006). Concatenation of the individual chirps to one long chirp can be done either in the time domain (Keel et al. 1998; Koch and Tranter 1990) or in the frequency domain (Brenner and Ender 2002; William 1970; Kulpa and Misiurewicz 2006), or in a deramp-mode. The latter is used for high-resolution MEMPHIS SAR processing. Detailed results have been published in (Essen et al. 2003).

11.4.2 *Lateral Focussing*

For the test of the lateral focusing algorithm data were taken for urban areas with strong point scatterers and additionally an only weakly structured terrain with low dynamic range.

For the lateral focusing the Doppler resolution of the system is the determining parameter, which is given by:

$$F_d = 2^* f^* \; v/c^* \alpha \tag{11.4}$$

f = Frequency, v = Speed of Aircraft, c = Speed of Light, α = Squint Angle.

If the Doppler frequency within the relevant angular interval is not exceeding the PRF an unambiguous determination is possible. The unambiguous interval is given by:

$$ED = PRF^* R^* c/(2^* f^* v) \tag{11.5}$$

In the case under consideration the following parameters were relevant:

The length of the appropriate FFT is given by N = ED/R, resulting in N = 1024 for a range of R = 2 km and N = 512 for a range of R = 700 m.

The algorithm is demonstrated for an arrangement of corner reflectors of different RCS and different distances. Figure 11.4 demonstrates the arrangement and gives pseudo color representations of the respective SAR image of the reflector array.

The test arrangement was flown with different radar parameters. It turned out, that for lowest processing sidelobes the longer pulse width of 1,200 ns was the best choice.

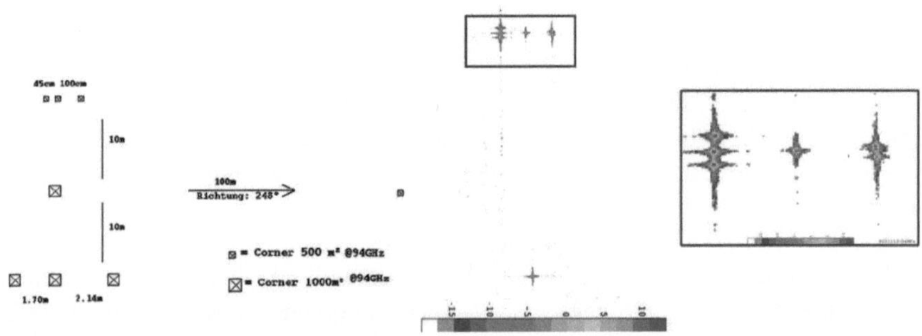

Fig. 11.4 Three Scatterers separated by 0.45 and 100 m with 0.2 and 0.8 m resolution

11.4.3 Imaging Errors

During numerous SAR flights it was observed, that generally imaging errors have a much lower importance at millimeter wave frequencies than at microwave bands, however there is also an indication, that at Ka-band the "Range-Walk" has already a slight effect. Three effects give reason for the movement of a point scatterer from one range gate to the next during one period of the Doppler FFT. These are the:

1. *Drift*: The aircraft axis is not exactly aligned to the flight direction.
2. *Beam-Width Effect*: The radar look direction covers a certain angle, given by the 3-dB-beamwidth of the antenna.
3. The *aspect angle* to the target changes during the aperture length.

Under which angle the Range-Walk is of importance demonstrate the following considerations:

The time related to an FFT of length N (Aperture Time) to give a resolution Δl is given by:

$$t_a = N/PRF \tag{11.6}$$

which is: $t_a = R^* c/(2^* f^* v^* \Delta l) \quad |t_a = 0.58\,s$

The aperture length is given by:

$$S_A = v^* t_a \tag{11.7}$$

Which is: $S_A = R^* c/(2^* f^* \Delta l) \quad |S_A = 45.7\,m$

During this period the lateral displacement (Range-walk) has to be lower than the range resolution, which results in an angle of:

$$\begin{aligned} &\beta = \Delta l/S_A \\ &\text{Which is}: \beta = \Delta l^{2*}\, 2^*\, f/(R^*\, c)|\beta = 0.23^\circ \end{aligned} \tag{11.8}$$

It is obvious, that the maximal angle increases linear with frequency and quadratic with the resolution. Table 11.3 gives some characteristic numbers.

The drift results in a range gradient linearly dependent on time. This can be compensated by shifting the start frequency of the chirp modulation, which can be done continuously, as appropriate.

The beam-width effect is not relevant at 94 GHz for a 3-dB-beamwidth of about 1. At 35 GHz, where the beam width is about 3°, this has to be taken into account. A simple solution is offered by using only part of the Doppler-FFT result,

Table 11.3 Range-walk effect at different radar frequencies, ranges and resolutions

f (GHz)	R (m)	Δl (cm)	β (°)
35	700	75	10.8
35	700	18.75	0.67
35	2,000	18.75	0.23
94	1,000	18.75	1.26
94	2,000	18.75	0.63

which is related to an evaluation of only a fraction of the full beamwidth. It has however to be considered, that for an adequate overlap of images (multilook) the data must not be shifted by a bigger portion, which leads to an increase of processing time.

The third effect caused by the aspect angle different from 90° produces non-linear (quadratic) Range-Walk due to the non-linear range gradient during the aperture time. If a circular course around a target would be flown, the range would be constant. As however, the flight course is linear, a range gradient is generated which is equal to the arch rise. For small angles the arch rise is given by:

$$h = s^2/8r \text{ (s = bow string, r = radius, h = arch rise)} \tag{11.9}$$

here: $h = \Delta l^2/(8^* R) \quad |h = 0.13\,\text{m}$
or: $h = R^* c^2/(32^* f^{2*} \Delta l^2)$

This range gradient is smaller than the resolution and thus may be neglected. Fig. 11.5a demonstrates the effect of Range Walk. The images show series of the single-look range profiles. Structures are moving through the representations from below to above. It is remarkable, that all structures appear as diagonal stripes from above left to below right. This is caused by the range-walk. The drift angle, which results in this effect, is related to the tilt angle of the single look stripes. The series of range profiles shown below has undergone a correction process. The single scatterers show now a horizontally aligned pattern, as obvious from Fig. 11.5b.

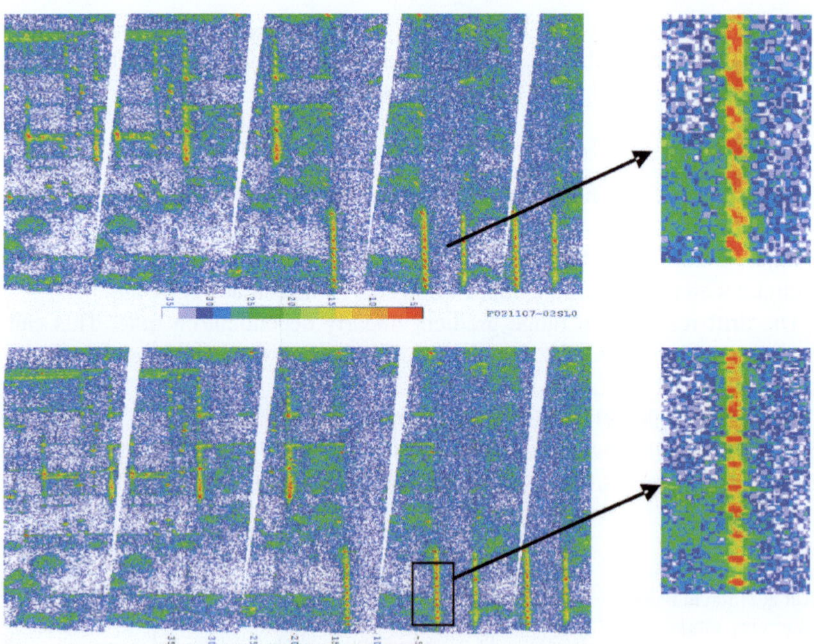

Fig. 11.5 SAR series of range profiles at 35 GHz without and with drift correction

If the SAR sensor is accelerated in a direction perpendicular to the flight path, the point scatterer response is blurred in cross range. While a linear movement gives reason to a constant Doppler shift, resulting in a range walk, a non-linear movement (acceleration) results in a blurring. In the following an estimation is given, for which acceleration a correction is not necessary without notable blurring:

It is quite reasonable, that the excursion due to the acceleration should be less than half of the wavelength, which can be formulated as:

Accelerated path (35 GHz): $s = a/2^*t^2 < \lambda/2 \quad \lambda = c/f, |\lambda = 8.6\,\mathrm{mm}$

$$\text{Max. acceleration: } a < c/(f^*t^2) \tag{11.10}$$

t is the aperture time t_a

which gives: $a < c/(f^* ((R^* c)/(2^* f^* v^* \Delta l))^2)$

or shorter: $a < 4^* f/c^* v^2 * (\Delta l/R)^2 \quad |a < 0.026\,\mathrm{m/s^2}$

The maximum acceleration error increases linearly with the radar frequency but quadratically with the relation resolution/range.

The acceleration for all three axes are given by the Mil-Bus data of the TRANSALL carrier aircraft, which allows the calculation of the acceleration in flight direction. Correction of the data for this acceleration results in a well-focused image.

Tests of the focusing implemented in the MEMPHIS SAR algorithm were conducted to maintain good focusing also over longer ranges. A scene over the Nymphenburg Palace in Munich was chosen. It turned out that the algorithm, as applied initially, is not sufficient for high-resolution processing over the range, relevant for that scene, and a higher sophistication is necessary. The main problem is, that for high-resolution processing a model, which is based upon a constant acceleration is not sufficient. The determination of the effective acceleration by auto-focusing methods only allow to generate optimized single look images, but do not lead to a general improvement of the SAR image. The only way to generate focussed high resolution images is based upon a combination of autofocus (for the determination of the constant offset during one FFT period) and acceleration information of sensors directly incorporated into the radar front-end. The latter deliver the information of the acceleration gradient within a single FFT period. This combined method gives the best focussing and in addition a constant acceleration error of about $0.15\,\mathrm{m/s^2}$, which is related to a depression angle error of about 1° (at 30° depression angle). An image processed with a respective optimized algorithm shows Fig. 11.6 together with the results of three processing steps for a section of that image related to a fence at a parking lot close to the palace.

For the conditions discussed here with the MEMPHIS radar the following statements are true:

1. For slant ranges between sensor and scene below 1 km a SAR processing without the application of correction algorithm delivers images of good quality only under very calm flight conditions.

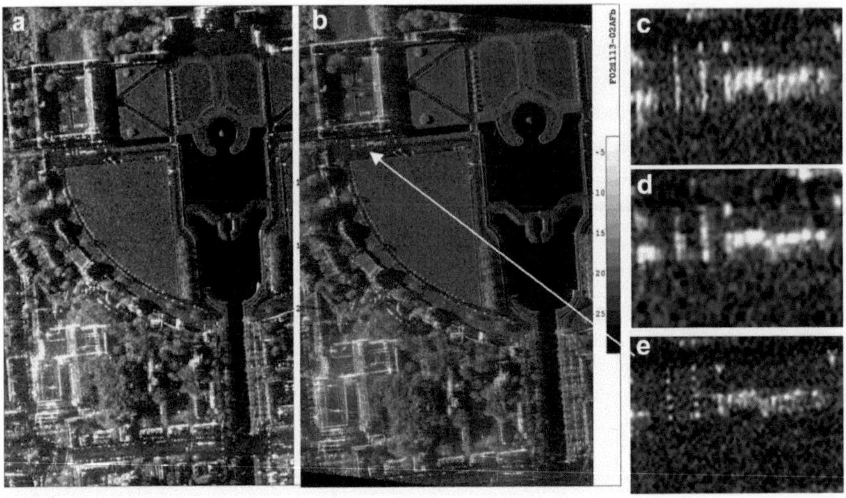

Fig. 11.6 SAR image of Nymphenburg palace at 94 GHz, (**a**) + (**c**) resolution 75 cm, (**b**) with optimized algorithm and resolution 19 cm, (**d**) detail with optimum range processing, (**e**) detail with full range/Doppler correction

2. Simple correction algorithms which solely take into account a constant acceleration deliver images of good quality up to a slant range of 1 km.
3. For slant ranges above 2 km this model is only sufficient for calm flight conditions.
4. For greater height or range a motion compensation process has to be applied which corrects data within one FFT-length.

This is only possible with fast acceleration sensors at the locus of the radar. For these the influence of gravitation has to be taken into account.

A typical MEMPHIS SAR image, with all necessary corrections applied, is shown in Fig. 11.7. It shows an image of the Technical University of Munich.

11.4.4 Millimeter Wave Polarimetry

The MEMPHIS radar is equipped with four receive channels. Two of them are generally dedicated to retain polarimetric information on the measured scene. Whenever polarimetric information is required, a thorough calibration has to be performed. For the data under consideration a technique was employed which uses the data stream itself for calibration and elimination of cross-talk and channel imbalance. This is done in two steps. The first step generates a symmetric data matrix from the not symmetric matrix of measured data. The second step removes the channel imbalance.

Fig. 11.7 High-resolution SAR image of TU Munich at 94 GHz

Fig. 11.8 Pseudo colour representation of polarimetrically weighted SAR image of rural Terrain

Polarimetry plays an important role for the segmentation of different classes of vegetation within a SAR image (Ulaby and Elachi 1990).

A simple way to visualize the capabilities of polarimetry, is to apply a color code to each of the orthogonal polarization components, that is for H–H and H–V channel, and the difference between those components (HV–HH). An example is shown in Fig. 11.8, which some rural terrain.

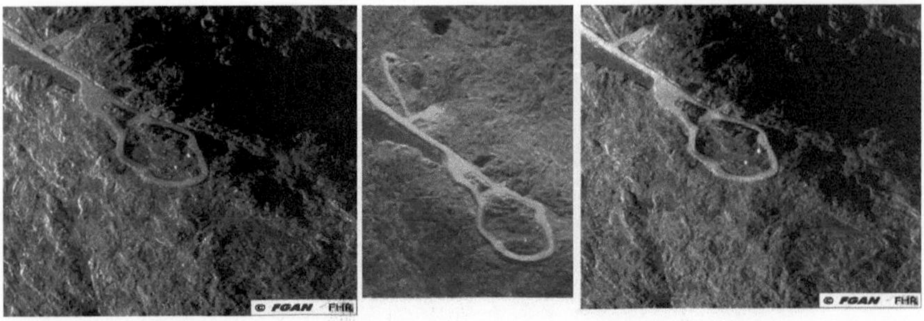

Fig. 11.9 Polarimetric SAR images at 94 GHz for T-R-Polarization L-L (*left*), polarimetric weighting and Polarization L-R

A further case is shown in Fig. 11.9, which gives the characteristics of some rocky terrain in a different polarization state. Specifically it can be seen, that rocks show a higher reflectivity for the polarization left-hand circular/left-hand circular (L/L), but the gravel road has a more dominant signature at left-hand circular/right-hand circular (L/R). The polarimetric differences can be attributed to different micro geometries: For circular polarization, odd returns are sensed by the cross-polarized channel, while the co-polarized channel is sensitive for even numbers of reflections.

For a thorough study of polarization features SAR scenes have to be subdivided into mainly homogenous sub areas. Determination of statistical parameters for these sub-areas and of their specific polarimetric characteristics allow the extraction of knowledge upon the vegetation and even its state.

11.4.5 Multiple Baseline Interferometry with MEMPHIS

Interferometry at millimeter wave frequencies has an important advantage and at the same time exhibits a general shortcoming: The first is a considerably better height estimation accuracy at a fixed interferometric base length, the latter is a lower unambiguity.

For a fixed baseline the height estimation accuracy is linearly dependent on the radar frequency. That means, that at W-band the accuracy for a given interferometric base is by a factor of ten higher than at X-Band. This would be a considerable advantage, as on small aerial vehicles, which can accommodate only small interferometric antenna assemblies the operation at millimeterwaves would be the solution. Unfortunately this advantage is coupled with a disadvantage, namely the unambiguity is also lower by the same factor, which means, that the phase unwrapping is much more time consuming. Figure 11.10 shows the relation between interferometric base and unambiguous range for 10, 35 and 94 GHz.

A solution to this discrepancy between height estimation accuracy and unambiguous range can be found by extending the hardware to a multiple baseline

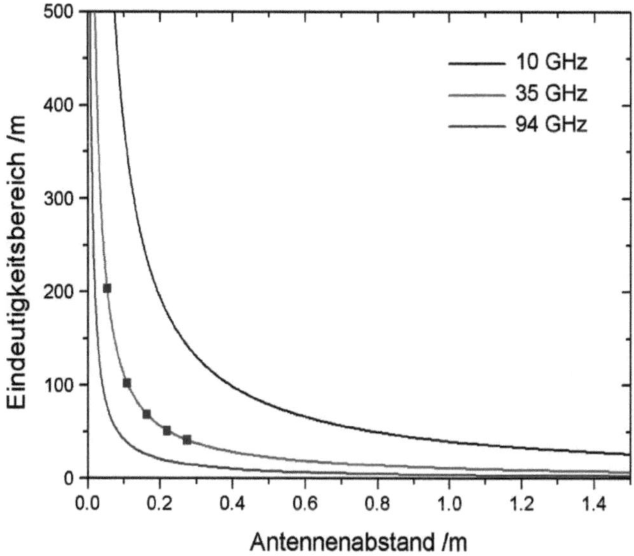

Fig. 11.10 Unambiguous range versus interferometric base (lowest curve 94 GHz, then 35 GHz, then 10 GHz)

antenna as described under Section 4.1. With this approach the advantages of big height estimation accuracy with a wider base length and of a bigger unambiguous range with a smaller base length can be combined. The approach is roughly the following: From the data for the smaller base length a first estimation with lower accuracy but within a wide unambiguous range is given and this is successively improved by using data for wider base lengths. It is obvious, that with increasing interferometric baselength the number of phase periods is increasing.

The phase unwrapping algorithm using multiple baseline data, sorts the interferograms related to different base lengths according to these bases. The interferogram for the smallest base length is suspected to be unambiguous. If this is not the case, it has to be unwrapped with a standard method, like the dipole method. An absolute phase calibration is not necessary, as only phase differences are evaluated. In the next step a scale factor is determined, which is given by the relation between the base length belonging to the reference interferogram and the next, which has to be unwrapped. The reference interferogram is multiplied by this factor and subtracted from the latter, modulo 2π. This procedure leads to the interval chart, which contains the information, how many 2π intervals have to be added to the unwrapped interferogram. A special algorithm takes care upon the amount of phase noise and, if necessary, generates a correction term. If the correction does not deliver a valid value, the original number is taken. For the algorithm it is only tolerable, that single pixels of this kind exist. After all pixels are generated, this interferogram is used as a starting point for the iteration, using the next bigger baselength. This process is consecutively done for all available interferograms.

11.4.6 Test Scenarios

The first test area is a former mine with a conical pit-head stock.

Figure 11.11a–f show the interferograms for the sample area, which are related to the five different interferometric base lengths and additionally a SAR image of that terrain.

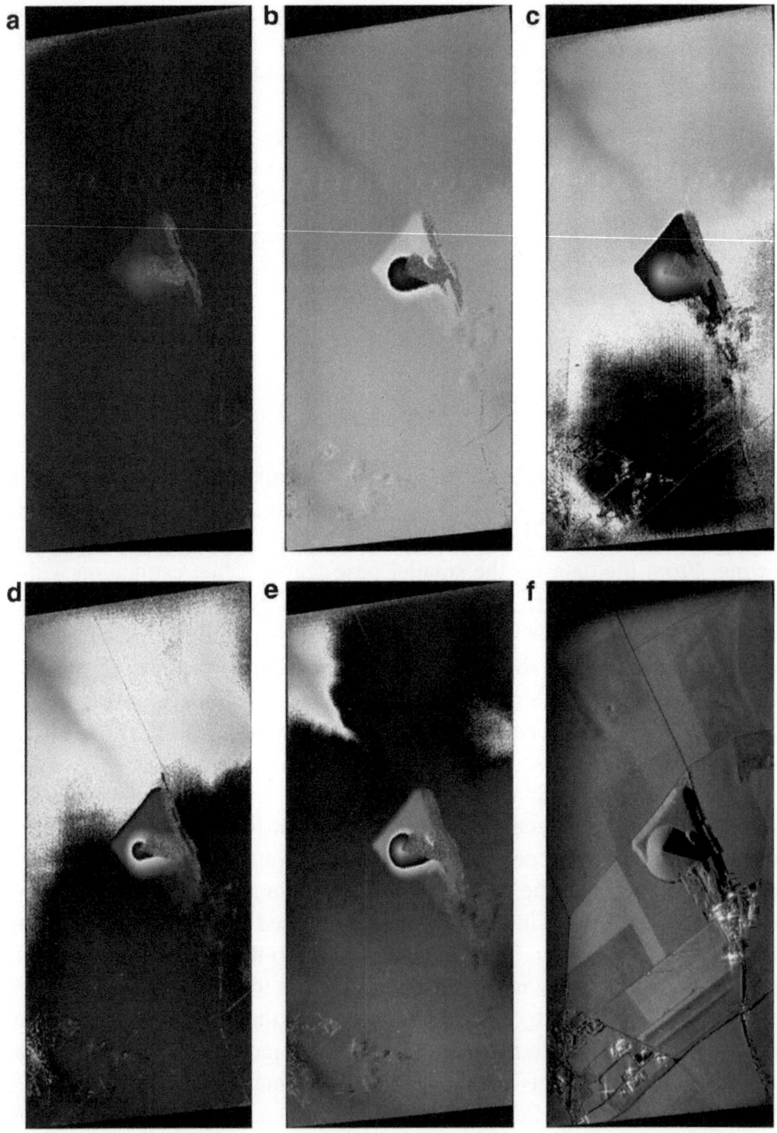

Fig. 11.11 Interferograms for the baselengths 0.055 m (**a**), 0.110 cm (**b**), 0.165 cm (**c**), 0.22 cm (**d**) and 0.275 cm (**e**) and the related SAR Image (**f**)

It has to be noted, that pixels with a reflectivity below −25 dB are cancelled and assigned "black".

To deliver a height, calibrated in meters, an appropriate calculation has to be performed. As additional inputs the flight height, the depression angle and the slant range have to be known. Equation (11.11) has to be solved numerically:

$$\begin{aligned}\frac{\lambda}{2} &\equiv \Delta(\Delta R) = (r_{22} - r_{21}) - (r_{12} - r_{11}) = \\ &= \left[\sqrt{(y - B\sin(\alpha))^2 + (H + B\cos(\alpha) - z)^2} - \sqrt{y^2 + (H - z)^2}\right] \\ &\quad - \left[\sqrt{(y - B\sin(\alpha))^2 + (H + B\cos(\alpha))^2} - \sqrt{y^2 + H^2}\right] \qquad (11.11)\end{aligned}$$

As the range difference $\lambda/2$ is equivalent to a differential phase of π each differential phase value of $\psi_{i,j}$ can be related to a height $h_{i,j}$ and a digital elevation model of the imaged terrain is deduced (DEM). Figure 11.12 shows a respective example for the test area shown in Fig. 11.11.

An interesting application is the interferometry in urban terrain. MEMPHIS was operated over an urban area in Switzerland. The data evaluation was done in co-operation with RSL University Zurich (Magnard et al. 2007). Figure 11.13 shows the respective SAR image at 94 GHz. Figure 11.14 shows the related interferogram, Fig. 11.15 shows details of that scene for a built up area.

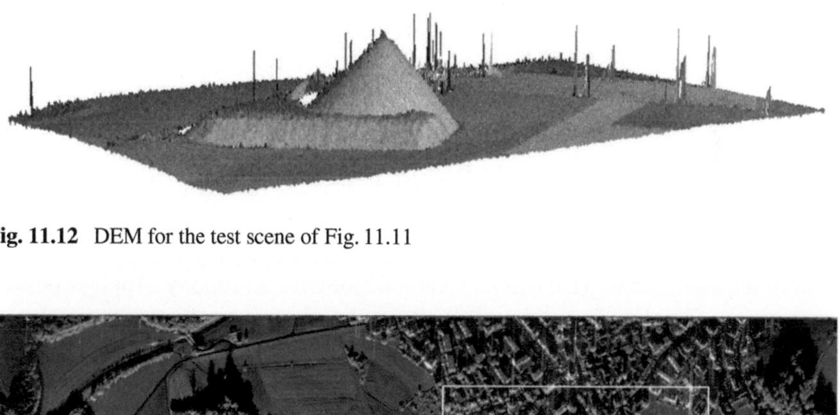

Fig. 11.12 DEM for the test scene of Fig. 11.11

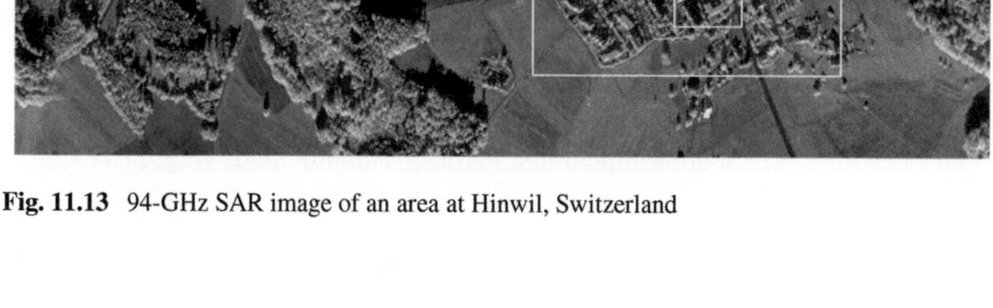

Fig. 11.13 94-GHz SAR image of an area at Hinwil, Switzerland

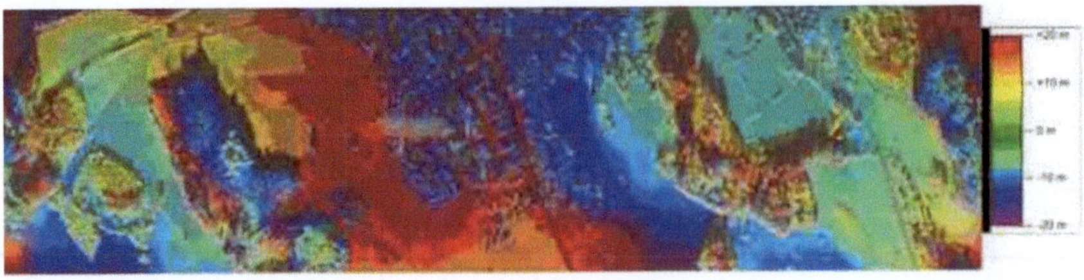

Fig. 11.14 Related interferogram

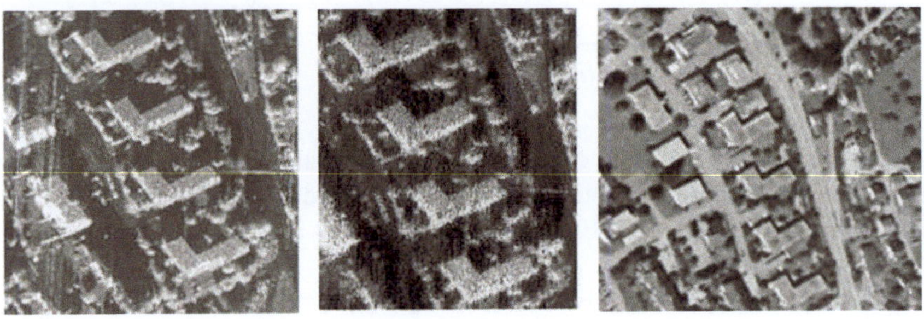

Fig. 11.15 SAR image, DEM and photo of section of Hinwil scene

The example shows very good the height structure of the terrain, calibrated in meters and the geometry of the flat roofed houses in the scene. The shadow regions, which are always critical for urban terrain, are handled quite well. Such data can serve as basis for further investigations on the structure of inhabited areas.

11.4.7 Comparison of InSAR with LIDAR

The standard method to determine digital elevation maps of terrain is the employment of a Laser scanner (LIDAR), as that of TOPOSYS (TopoSys Topographische Systemdaten GmbH). To validate InSAR results some typical areas were investigated using both, InSAR and the TOPOSYS system (Morsdorf et al. 2006). A test area was chosen, which contains urban and rural terrain, forests, rivers, high power lines and other man made structures. Figure 11.16 shows the related SAR image.

For the comparison it has to be noted, that the radar and lidar data have not been taken simultaneously and that a different geometry was used. This leads to some possible referencing errors between the two images.

Due to the depression angle, different from 90° the InSAR images show shadowing effects, which in the interferogram appear as "black", as there are no valid phase values available, as obvious from Fig. 11.17. This is not the case for the TOPOSYS data, which are sampled in a vertical scanning mode (Fig. 11.18). Both

Fig. 11.16 SAR image and map for the test scene "Lichtenau"

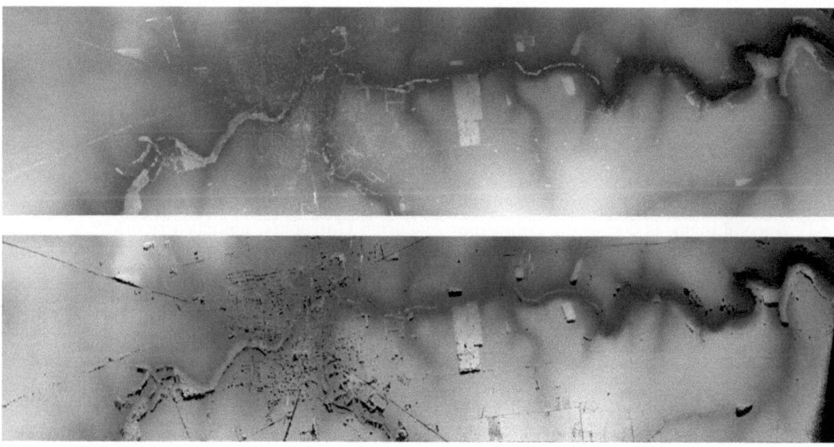

Fig. 11.17 DEM measured with TOPOSYS (*above*) and with InSAR (*below*)

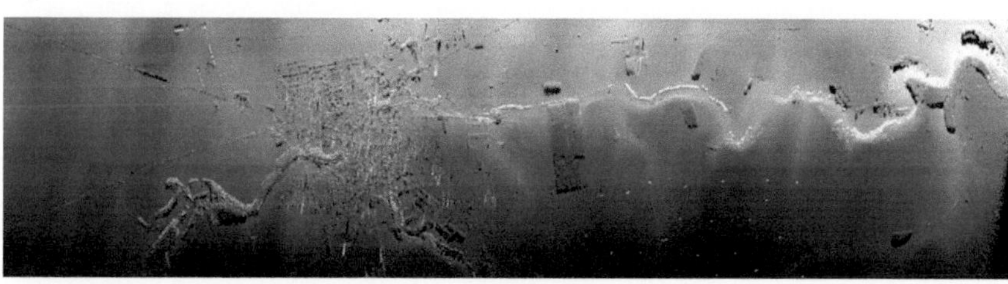

Fig. 11.18 Error map for the data pair TOPOSYS/InSAR

images exhibit ground cells 1.5 × 1.5 m in size with a height estimation accuracy of about 0.15 m.

Qualitatively both elevation maps show a good correspondence. Obvious are the shadow regions, which do not contain height information in the InSAR image. For a quantitative comparison an error map is generated, which is shown in Fig. 11.18. For a numerical evaluation some sample areas were chosen, as a wooded and urban terrain and an open field.

Table 11.4 Height estimation differences for three types of background

	Forest Average/m	Urban Average/m	Open field Average/m
Lidar	322.93	295.05	325.80
InSAR	319.86	270.20	327.87
Δ(Lidar, InSAR)	−1.23	12.75	−2.07

Table 11.4 summarizes the deviations of average height estimations for TOPOSYS and InSAR data for the three different terrain types. It is quite obvious, that both methods to derive a digital elevation map are comparable. The InSAR has the big advantage, that data can be gathered also under bad-weather conditions and, as the ground resolution for radar is independent on range, under considerably longer range.

Acknowledgements The author would like to thank all contributors from FGAN-FHR, Department MHS, namely, Hartmut Schimpf, Thorsten Brehm and Manfred Hägelen. Thank is also due to the former colleague Stephan Boehmsdorff, who is now with the German Procurement Office BWB. Special thank is due to the colleagues of Zurich University, namely Erich Meier, Maurice Ruegg and Christophe Magnard, as well as the technology center of the Swiss Federal Department of Defence (*armasuisse*) and especially Peter Wellig for the wide support and cooperation. Part of the work was done under contract with the German Procurement Office BWB.

References

Almorox-Gonzlez P, Gonzlez-Partida JT, Burgos-Garca M, Dorta-Naranjo BP, de la Morena-Alvarez-Palencia C, Arche-Andradas L (2007) Portable high-resolution LFM-CW radar sensor in millimeter-waveband. In: Proceedings of SENSORCOMM, Valencia, Spain, pp 5–9, October 2007

Berens P (1999) SAR with ultra-high resolution using synthetic bandwidth. In: Proceedings of IGARSS 1999, vol 3. Hamburg, Germany, 28 June–2 July 1999

Boehmsdorff S, Essen H (1998) MEMPHIS an experimental platform for millimeterwave radar, DGON IRS 1998, München, pp 405–411

Boehmsdorff S, Bers K, Brehm T, Essen H, Jäger K (2001) Detection of urban areas in multispectral data. In: IEEE/ISPRS Joint workshop on remote sensing and data fusion over urban areas

Brenner AR, Ender JHG (2002) First experimental results achieved with the new very wideband SAR system Pamir. In: Proceedings of EUSAR 2002, pp 81–86

Brooker GM, Hennessy RC, Lobsey CR, Bishop MV, Widzyk-Capehart E (2007) Seeing through dust and water vapor: millimeter wave radar sensors for mining applications. J Field Robot 24(7):527–557

Curlander JC, McDonough RN (1991) Synthetic aperture radar – systems and signal processing. Wiley, New York

Dreuillet Ph, Cantalloube H, Colin E, Dubois-Fernandezx P, Dupuis X, Fromage P, Garestier F, Heuze D, Oriot H, Peron JL, Peyret J, Bonin G, du Plessis OR, Nouvel JF, Vaizan B (2006) The ONERA RAMSES SAR: latest significant results and future developments. In: Proceedings of 2006 IEEE Conference on Radar, p 7, 24–27 April 2006

Edrich M (2004) Design overview and flight test results of the miniaturised SAR sensor MISAR. In: 1st European Radar Conference, EURAD 2004, pp 205–208

Essen H, Baars EP (1986) Millimeter wave transmission through man-made obscurations in a battlefield environment. AGARD Multifunction Radar for Airborne Applications 1 p (SEE N87–18721 11–32)

Essen H, Schimpf H, Wahlen A (2003) Improvement of the millimeterwave SAR MEMPHIS for very high resolution (in German). GFAN-FHR Technical Report, Werthhoven, May 2003

Hägelen M, Briese G, Essen H, Bertuch T, Knott P, Tessmann A (2008) A millimeterwave landing aid approach for helicopters under brown-out conditions. In: IEEE RadarConference, Rome

Henry JC (1991) The Lincoln laboratory 35 GHz airborne SAR imaging radar system. In: Telesystems Conference, 1991. Proceedings, vol 1, pp 353–358, 26–27 March 1991

Keel BM, Saffold JA, Walbridge MR, Chadwick J (1998) Non-linear stepped chirp waveforms with sub-pulse processing for range sidelobe suppression. In: Proceedings of SPIE, vol 3395. Orlando, pp 87–98

Kendra JR, Sarabandi K, Ulaby FT (1995) Experimental studies of dense media scattering. In: Antennas and propagation society international symposium, 1995. AP-S Digest 4(18–23 June 1995):1712–1715

Koch DB, Tranter WH (1990) Processing considerations for hybrid waveforms utilizing complementary phase coding and linear frequency stepping. In: IEEE Int'l Radar Conference, pp 606–611, May 1990

Kulpa KS, Misiurewicz J (2006) Stretch processing for long integration time passive covert radar. In: International Conference on Radar, Shanghai, 16–19 October 2006

Levanon N (December 2002) Stepped-frequency pulse-train radar signal. IEE Proc Radar Sonar Navig 149:297–309

Magnard C, Meier E, Ruegg M, Brehm T, Essen H (2007) High resolution millimeter wave SAR interferometry. In: Proceedings of GARSS 2007, Barcelona, pp 5061–5064, 23–28 July 2007

Maron DE (1990) Frequency-jumpedburst waveforms with stretch processing. In: IEEE 1990 International Radar Conference, Arlington, VA, pp 274–279, 7–10 May 1990

Marshall JS, Palmer WM (1948) The distribution of raindrops with size. J Meteorol 5:165–166

Morsdorf F, Kötz B, Meier E, Itten KI, Allgöwer B (15 September 2006) Estimation of LAI and fractional cover from small footprint airborne laser scanning data based on gap fraction. Remote Sens Environ 104(1):50–61

Nüßler D, Essen H, von Wahl N, Zimmermann R, Rötzel S, Willms I (2007) Millimeter wave propagation through dust. In: Symposium Digest, SPIE Conference on Remote Sensing, Cardiff

Rangwala M, Wang F, Sarabandi K (2007) Study of millimeter-wave radar for helicopter assisted landing. In: IEEE Proceedings of Geoscience and Remote Sensing, Barcelona

Schimpf H, Essen H, Boehmsdorff S, Brehm T (2002) MEMPHIS – a fully polarimetric experimental radar. In: IGARSS, Toronto, CA, CD; FR08_853, June 2002

Schimpf H, Wahlen A, Essen H (2004) High range resolution by means of synthetic bandwidth generated by frequency stepped chirps. Electron Lett 39(18):1714–1716

Skolnik M (1980) Introduction to RadarSystems. McGraw-Hill, New York, p 581

TopoSys Topographische Systemdaten GmbH,Obere Stegwiesen 26, 88400 Biberach an der Riß (Germany)

Ulaby FT, Elachi CH (1990) Radar polarimetry for geoscience applications. Artech House, Norwood, MA

Weiβ-Wrana K, Jessen W, Kohnle A, Clement D, Höhn DH (January 1995) Atmospheric transmittance measurements of Nd:YAG, iodine and CO_2 laser radiation over 8.6 km, and statistical analysis of extinction coefficients. Infrared Phys Technol 36(1):513–528. In: Proceedings of the sixth international conference on infrared physics

Wikner D (2008) Millimeter-wave propagation measurement through a dust tunnel. Technical Report ARL-TR-4399, Adelphi, 6 August 2008

William JC (1970) Stretch: a time-transformation technique. In: IEEE AES-7, 1970, pp 269–278

Zhou L, Xing M, Sun H (2006) Synthetic bandwidth method integrated with characteristics of SAR. In: International Conference on Radar, Shanghai, pp 1–4, 16–19 October 2006

Index

MIX
Papier aus verantwortungsvollen Quellen
Paper from responsible sources
FSC® C105338

If you have any concerns about our products,
you can contact us on
ProductSafety@springernature.com

In case Publisher is established outside the EU,
the EU authorized representative is:
Springer Nature Customer Service Center GmbH
Europaplatz 3, 69115 Heidelberg, Germany

Printed by Libri Plureos GmbH
in Hamburg, Germany